RANDOM FIELDS

Analysis and Synthesis

Revised and Expanded New Edition

RANDOM FIELDS

Analysis and Synthesis

Revised and Expanded New Edition

Erik Vanmarcke

Princeton University, USA

World Scientific

NEW JERSEY • LONDON • SINGAPORE • BEIJING • SHANGHAI • HONG KONG • TAIPEI • CHENNAI

Published by

World Scientific Publishing Co. Pte. Ltd.
5 Toh Tuck Link, Singapore 596224
USA office: 27 Warren Street, Suite 401-402, Hackensack, NJ 07601
UK office: 57 Shelton Street, Covent Garden, London WC2H 9HE

British Library Cataloguing-in-Publication Data
A catalogue record for this book is available from the British Library.

RANDOM FIELDS
Analysis and Synthesis
(Revised and Expanded New Edition)

ISBN-13 978-981-256-297-5
ISBN-10 981-256-297-4
ISBN-13 978-981-256-353-8 (pbk)
ISBN-10 981-256-353-9 (pbk)

Printed in Singapore.

Preface

Nature is a mutable cloud, which is always and never the same.

Ralph Waldo Emerson, History

Preface to the Second Edition

Since *Random Fields* was first published by The MIT Press in 1983, the reality of complex random variation, and the need to quantify and manage risks in complex systems, has been ever more widely recognized, and fast-growing computing power and database sizes have provided further motivation and empirical support for high-level probabilistic modeling. The main methods and concepts introduced in the first edition, related to the variance function and the scale of fluctuation, and the role of local averages in robust estimation of level-crossing and extreme-value statistics of random fields, have been applied in many areas of science and engineering.

In this revised and expanded edition, the order of presentation of some topics has been changed, as has the manner of exposition, so as to achieve greater clarity and coherence. A much clearer distinction is made between single-scale and multi-scale random fields, facilitated by the use of the scale spectrum. The coverage on statistical estimation of correlation parameters of single-scale random processes has been left out of the 2nd edition. New material is added, in particular on a class of quantum-physics-based probability distributions, described in Section 2.8; these all have a simple analytical form and tractable statistical properties, and promise to be very useful in a variety of applications of random field theory in physics, chemistry and biology.

The book is well suited for use in graduate-level teaching about random media and random fields, and in professional short courses on advanced methods of risk and reliability analysis. The theory's relevance to diverse academic curricula is reflected in my affiliations, at Princeton University, with the Princeton Environmental Institute (PEI), the Bendheim Center for Finance, the Institute for the Science and Technology of Materials, and the Program in Robotics and Intelligent Systems.

The universities where I did the research and teaching leading to this book (and its new edition), MIT, Princeton, and, during sabbatical leaves, Harvard and Stanford, provided inspiring and stimulating environments, and I am indebted to many colleagues and former students. I gratefully acknowledge, in particular, (the late) C. Allin Cornell, Stephen Crandall, Jose Roesset, Daniele Veneziano, Anne Kiremidjian, John Reed, George Gazetas, Mary-Ellen Hynes, Dario Gasparini, Gordon Fenton, Jeen-Shang Lin, Ronald Harichandran, Ernesto Heredia, Henri-Pierre Boissières, Waldemar Hachich, Paul Lai, Ross Corotis, Binod Bhartia, Ricardo Palma, Mark Dobossy, Mircea Grigoriu, Ignacio Rodriguez-Iturbe, Semih Yücemen, Aysen Akkaya, Ove Ditlevsen, Alfredo Ang, Masanobu Shinozuka, George Deodatis, Elizabeth Paté-Cornell, Gregory Baecher, Yin Lu (Julie) Young, Minoru Matsuo, Robert Stark, James Rice, Robert Verhaeghe, and (the late) Emilio Rosenblueth. I am much indebted as well to Hongjun Li, my editor at World Scientific Publishing Company, who patiently and expertly assisted in making this 2nd edition of *Random Fields* a reality.

Princeton, New Jersey

Preface to the First Edition

Random variation is a fact of life that provides substance to a wide range of problems in the sciences, engineering, and economics. In diverse disciplines there is a growing need to quantify aspects of the behavior of complex phenomena that can be modeled as random fields, or distributed disordered systems. This book is dedicated to developing a synthesis of methods to describe and analyze and, where appropriate, predict and control random fields.

Chapters 2 and 3 serve primarily as a review of classical theory of multi-dimensional random processes. Chapter 2 introduces elementary probability concepts and methods in a random field context, while Chapter 3 gives a concise account of second-order analysis of homogeneous random fields, in the space-time domain as well as in the wave number-frequency domain.

Chapter 4 presents a synthesis of results (many new) on level excursions and extremes of Gaussian and related random fields. Spectral moments and associated measures of degree of disorder are introduced and interpreted for different types of stochastic variation.

The main new developments are based on a proposal to treat the correlation structure of one-dimensional random processes in terms of the variance function and the scale of fluctuation. This treatment extends elegantly to multi-dimensional situations and opens the way to considerable expansion of present capabilities to deal in practical and relatively simple ways with problems involving one- and two-dimensional random variation (Chapters 5 and 6, respectively) and general n-dimensional homogeneous fields (Chapter 7). Chapter 8 proposes new methodology in the areas of estimation and prediction, and provides an application-oriented review of new results.

Much of the material is based on my recent research at the Massachusetts Institute of Technology (MIT), and could have been submitted for publication in a series of separate articles to diverse journals. In the process, however, precious bonds between related concepts and approaches would have been broken, and the opportunity offered by the monograph format to stress methodological unity would have been lost.

The book is well suited for use in teaching or self-study of the fundamentals of random fields. While new results are introduced at an appropriate level of mathematical rigor, there is no mathematics requirement beyond basic college level calculus, and every effort has been made to consolidate and simplify theory so as to render it suitable for practical application. The later chapters are substantially self-contained and can be read at different levels. Within the various chapters, each section is written as a unit with respect to the numbering of equations.

The research leading to this book has been carried out, with fluctuating intensity, over the past decade, with a concentration of effort during 1979 and 1980. It follows a course charted by Norbert Wiener and was perhaps touched by his spirit, still alive at MIT. The Institute provided a stimulating environment for the research, and I am indebted to many colleagues and former students. I acknowledge with deep gratitude the support and encouragement of my mentors in the art of applied probability, Robert M. Stark of the University of Delaware, C. Allin Cornell of Stanford University, and Stephen H. Crandall of MIT. I am very grateful to Elizabeth Augustine who patiently and expertly typed the many drafts of this book. Lieven Vanmarcke and Mark Willems assisted with numerical calculations for several tables and figures. Last but foremost, my appreciation goes to my wife and children for their constant support in countless ways.

Cambridge, Massachusetts

Contents

Chapter 1

Introduction

Uncertainty, the sure sense that the ground is shifting at every step, is one of the marks of humanity.

Lewis Thomas, On the Uncertainty of Science

1.1 The Spectrum of Applications

Many physical, biological, economic and social systems have attributes which, viewed on an appropriate scale, exhibit complex patterns of variation in space and time. For example, substances that constitute the earth's environment – air, water, soil, and rock – possess physical properties (such as density, strength, conductivity) which change more or less unpredictably under the influence of metereological variables (such as temperature, pressure, rate of precipitation) that are themselves random processes in time and space. Random variation in the properties of living matter or sociobiological systems, or in valuations in financial markets, is no less great.

When the degree of disorder is sufficiently large, there is usually merit and economy in probabilistic rather than deterministic models. In particular, random field theory seeks to model complex patterns of variation and interdependence in cases where deterministic treatment is inefficient and conventional statistics insufficient. An ideal random field model succeeds in capturing the essential features of a complex random phenomenon in terms of a minimum number of physically meaningful and experimentally accessible parameters.

The methodology of random fields is applicable to phenomena occurring on very different (temporal or spatial) scales. The *time* scale may be the interval between molecular collisions, as in the study of Brownian motion, or may be measured in geological units, for example, to describe the variation of properties and thicknesses of layers of the earth's crust. Similarly, on a *spatial* coordinate axis the scale may be subatomic in the study of superheated plasmas, while an astronomical scale is needed to describe the random fields of temperature, density, or chemical composition of matter in space.

Many economic and social phenomena can also be examined in the light of random field theory. For example, the demand for goods and services and the supply of commodities may be seen as varying randomly (in time and space) throughout a macroeconomic system. Likewise, prior to an election, nationwide voting patterns may in principle be modeled as a kind of spatial random process to be realized on election day. In these cases, one might seek to analyze the behavior or choices of different groupings or aggregations of discrete ("agent-based") units.

To sharpen the focus on applications, consider these varied phenomena:

Depth of snowfall during a winter storm.
Pollutant concentration in a lake.
Shear stress along a fault in the earth's crust.
Height of the surface of the ocean.
Level of toxicity in an organ in the human body.
Amount of recoverable solar energy.
Porosity of a water-bearing stratum.
Areal density of the population of a species.
Insurance claims for losses due to fire.
Unemployment rate in an economy.
Earth's surface temperature.
Agricultural crop yield in a region.

Each phenomenon summons the image of a *distributed disordered system* (d.d. system) whose attributes display complex patterns of variation in space – in two or three dimensions – as well as variation with time. In some cases the attribute may be seen as a random property of a d.d. system, while in others it describes the state of excitation or response of the system. Whether or not formal treatment of uncertainty is warranted in a particular situation depends on such factors as the quality and quantity of information, the intensity of the fluctuations, the system's tolerance to extremes, and the resources at hand.

Random field models of complex stochastic phenomena serve multiple purposes, in particular:

1. in stochastic description they provide a format for efficient characterization of d.d. systems;
2. in system analysis they provide the basis for predicting response (behavior, performance, sustainability) of d.d. systems;
3. in decision situations involving d.d. systems they permit assessment of the impact of alternative strategies (in design, retrofit, data acquisition, or dynamic control in real time).

Differences between types of random fields stem mainly from the nature of the uncertainty and the decisions that the stochastic environment imposes or creates. Uncertainty about the properties of a *random medium* is essentially of a passive type. An attribute at a given location is deterministic, but its value is unknown until accurately measured. Sampling at all locations is usually impractical and uneconomical, and measurement and testing errors tend to dilute the value of the information. Therefore the tasks of prediction, analysis, and decision making must usually proceed on the basis of incomplete information about the medium, and this fact renders its modeling as a random field meaningful. A *space-time process* is characterized by active and inherent (or intrinsic) uncertainty: properties at different points in space change randomly with time. Measurements may be taken at selected points in time and space, or continually (in time) at specific locations. The measurements may be used for forecasting the future values of the process at given locations and times, or the future values of regional or system-wide averages or measures of performance.

When a system consists of separate or discrete units or "agents", such as members of a population or securities comprising a financial market, there is often no obvious "spatial" connotation, but the concept of local averages of their attributes, and how these vary (throughout the d.d. system as a function of time), yields much insight into the fabric and behavior of such complex systems.

The question of scale is of paramount importance in practical applications of random field theory. A phenomenon that appears deterministic on (what might be labeled) the microscale – it may consist of atoms, molecules, helices, or crystals – may at a larger scale exhibit highly variable properties that call for probabilistic description. On an even larger scale (the macroscale) these same structures may be embedded in objects amenable to characterization in terms of average or composite properties. At the subatomic scale, as is well known, fundamental quantum uncertainty governs behavior described by quantum mechanics, obeying Heisenberg's principle of uncertainty. One concludes that there is usually a lower and an upper bound on the range of dimensions (distance or time) within which a random field model has practical value. Outside this range a deterministic description or an entirely different stochastic description may be appropriate.

Scholarly effort has traditionally been directed at phenomena occurring on a scale where determinism dominates behavior. Phenomena occurring on the time or distance scale where chance governs have received relatively scant attention, and not much has been done to connect laws of behavior operative at widely different scales.

One often hears that we live in an age of uncertainty. Increasing awareness of the limits on global resources and the fragility of the environment has accentuated the need for scientists and engineers to understand (and account for) phenomena occurring on a scale where only the laws of probability offer meaningful treatment. Urgent problems in the fields of energy, water and food supply, environmental quality, and natural hazards science fall into this category. The reality of random variation in time and space implies that extreme deviations from expected values will occasionally occur in accordance with laws of probability. In decision situations it offers no option other than careful risk assessment and consideration of the trade-off between benefits and risks. The decisions may involve siting, design, or maintenance of man-made facilities, regulation to enhance environmental quality; actions to stimulate economic activity or prevent system-wide financial collapse; or the design of a network of instruments to monitor or acquire basic data about a distributed disordered system. Random field theory offers a framework within which such problems can be addressed more meaningfully.

1.2 Brief Historical Perspective

Much of the early literature on stochastic processes deals with variation in function of a single parameter, usually time. The most common and tractable models make strong assumptions about the memory of the process: they firmly constrain the degree of probabilistic dependence between past, present, and future values of an uncertain attribute.

The simplest time series model is the purely random process that assumes that successive observations are statistically independent. When integrated, it becomes a random walk, an accumulation of statistically independent increments characterized by single-step Markovian dependence.

When observations are made continuously, the purely random process becomes an "ideal white noise," and its integral is referred to as a Wiener process, after Norbert Wiener [152; 153] who fully developed its mathematical foundation. The white noise process and its integral were introduced much earlier (from all accounts independently), first in economics by Louis Bachelier [7; 8] and then in Brownian motion physics by Albert Einstein [54].

The close relationship between white noise and the Poisson process is highlighted by Stephen Rice's model [113] for "shot noise" in communication systems. Shot noise consists of randomly arriving and randomly sized impulses. The number of impulses in a given time period has a Poisson distribution, and it is statistically independent of the impulse magnitudes which are themselves mutually independent.

Poisson models also apply in multi-dimensional situations. Various geometrical configurations and patterns can be associated with Poisson centers in space – lines, hyperplanes, "Voronoi cells," spheres, and clusters of points. This gives rise to a large family of stochastic models whose chief asset is analytical tractability but whose fundamental liability is the underlying independence assumption (Parzen [103], Feller [57], Torquato [129]). Brownian motion refers to the chaotic movement of microscopic particles suspended in a fluid. In the classical analyses of Brownian movement by Einstein [54] and Smoluchowski [122], a particle experiences a succession of random impulses caused by collisions with other particles. This excitation is modeled as a purely random process which is related to particle velocity by a first-order linear differential equation (the Langevin equation). This renders particle velocity Markovian. The quantity of main interest in the study of Brownian motion is the particle displacement (the integral of velocity) after some period T of exposure to the Brownian chaos. Einstein [54] found that the probability density function of particle displacement obeys a simple diffusion equation with (diffusion) parameter D, and that, after a sufficiently long period T, the probability distribution of the displacement along any given direction becomes Gaussian with mean zero and variance equal to $2DT$.

Ornstein and Uhlenbeck [101] extended Einstein's analysis by modeling the transient part of the solution which acquires importance when T is small. When external forces (gravity or elastic attraction) are added to the equation of Brownian particle motion, the aforementioned "diffusion equation" becomes the Fokker-Planck equation governing the probabilities of transition from state to state of a Markov process (Stratonovitch [125]).

Stochastic models based on the assumption of Markovian dependence are very popular because they offer relative mathematical simplicity and the lure of complete probabilistic characterization. Its essential connection to an underlying purely random process is not always appreciated, however. The Markovian diffusion analysis is based on the assumption that the random forces driving the diffusion process constitute ideal white noise.

The same assumption, that the input is purely random, provides the underpinnings of the state-space approach to stochastic analysis of systems, such as in Zadeh and Desoer [160], Bryson and Ho [25], Brockett [23], and Schweppe [117]. The white noise input and the limited memory inherent in systems described by finite-order differential (or difference) equations guarantee that the time-dependent response of the system will be Markovian. In the state-space approach, transitions between successive system states are described by state transition matrices.

The Markovian assumption is readily understood when a random process depends on *time*. It is a clear and strong statement about probabilistic dependence: given the present value of a random process, its future behavior does not depend on its past. The concept of Markovian dependence does not arise as naturally when a process varies along spatial coordinates which lack the obvious directionality of the time axis. Although some limited use has been made of Markovian models in a multi-dimensional context (Whittle [151]), such models are neither entirely natural nor unambiguous.

In solid-state physics, disorder is often seen as defective order (Ziman [162]). A common approach to problems involving random variation is to consider deviations from the ideal structure, as in a pattern of dislocations in a crystal lattice or an isolated, randomly located inclusion in a homogeneous medium (Cleary *et al.* [35]). In a model of "cellular disorder" one looks at the variation of properties from cell to cell in a topologically ordered lattice. In the simplest so-called Ising model (Ziman [162]) a binary random variable ($+1$ or -1) is associated with each cell; this model is applicable whenever two states dominate behavior and has been used to describe two-phased alloys, impurities in semiconductors, and magnetic properties of materials.

The classical approach to *second-order analysis* of stationary random processes is based on the work of Wiener [152; 153] and Khinchine [77; 78] who established the equivalence of, and relationship between, the autocorrelation function (a.c.f.) and the spectral density function (s.d.f.).

Among the most important features of a random field are the patterns of correlation and persistence reflected in the second-order statistics, and much of the literature on applied stochastic processes offers one candidate or another as a model for the correlation structure (a.c.f. or s.d.f.). A purely random process is characterized by constant spectral density and by a Dirac-delta autocorrelation function.

If a random process is Gaussian, first- and second-order information is sufficient for complete probabilistic characterization. An important related result due to Rice [113], reviewed and extended by Rainal [110; 111], is an expression for the mean rate of crossings of an arbitrary level by a one-dimensional random function. For stationary Gaussian processes the mean threshold crossing rate is simply related to the second moment of the s.d.f., or equivalently to the second derivative of the a.c.f. evaluated at zero separation distance ("lag").

Early and influential applications of second-order stochastic theory to multi-dimensional problems occurred in the field of fluid mechanics (turbulence), where theoretical and experimental work by Geoffrey Taylor [126; 127] and Theodore von Karman (see Collected Works [149]) provided information about the correlation structure of the random field of velocities in turbulent fluid flow.

Random continuum models were also developed in the area of wave propagation physics – as in hydroacoustics and radiophysics (Chernov [32]) and seismology and geophysics (Aki and Richards [5], Bath [10], Zerva [161]) – when wavelengths of interest are much greater than the scale of local discontinuities in the medium. Randomly scattered waves are responsible for such poetic effects as the twinkling of stars. The amount of scattered wave energy and other statistical "ray properties" can be evaluated in terms of the variability and correlation properties of the medium's wave propagation characteristics (such as refractive index and density).

An important practical dimension was added to the theory of random functions by the classical work on *optimum linear estimation* by Wiener [153] and Kolmogorov [79]; see Yaglom [157; 158] for an excellent account. In its simplest form the problem is to make a "best" prediction of the true value of a random function at one point by linearly combining a set of values observed at other points. Assuming the process depends on time, the problem is labeled "forecasting" (or "extrapolation") if future values are to be predicted in terms of past and present observations, while the terms "filtering" and "smoothing" (or "interpolation") refer to prediction of values now and in the past, respectively. The criteria for optimality are that the prediction must be unbiased and the variance of the prediction error minimized. For continuous-time stationary random processes the optimal predictor takes the form of an integral equation called the Wiener-Hopf equation that is usually solved in the frequency domain. The Kalman filtering technique [75; 76] is an efficient alternative time-domain (state-space) approach to the problem of optimal linear estimation.

In response to practical needs to estimate spatial patterns occurring in geology and mining (related to ore body or oil reservoir location), Krige [80] and others such as Matheron [94; 95] and Agterberg [4] developed a class of linear estimation techniques that are fundamentally the same as those developed by Wiener and Kolmogorov but involve spatial rather than temporal variation.

Questions of scale motivate much of the work of Benoit Mandelbrot [88; 89] who observed that temporal and spatial patterns of correlation and persistence occurring in nature often fail to obey common, simple (single-sale) correlation models. At whichever scale a random phenomenon is observed, it tends to contain ripples and "slow trends" which bear witness to the presence of components with lower and higher scales, respectively. Mandelbrot coined the term self-similar to refer to phenomena that display similar patterns of variation regardless of the scale at which they are observed. The spectral density function of such random processes appears (within lower and upper bounds on frequency) to be inversely proportional to frequency, hence the term "$1/f$ noise." In contrast, the spectral density function of white noise is constant (it varies as f^0), while that of its integral, the Wiener process (or Brownian noise), varies in proportion to $1/f^2$.

The brief review in this section has sought to emphasize fundamental contributions and methodological breakthroughs. Much additional information may be found in publications in diverse application areas such as communication engineering (Davenport and Root [47], Lee [82], Middleton [96], Slepian [120]), econometrics (Slutzky [121], Granger [66], Fishman [61]), statistical mechanics (Beran [15], Bolotin [21]), biometrics (Whittle [150], Vogel & Zuckerkandl [148], Geol & Richterdyn [64], random vibration (Crandall & Mark [45], Lin [83], Clarkson [26], Moan & Shinozuka [97], Newland [99], Wirsching *et al.* [154], Lutes & Sarkari [86]), fluid dynamics and turbulence (Batchelor [11], Monin & Yaglom [98], Lumley [85]), earth science and geotechnology (Fenton & Griffiths [59], Hachich & Vanmarcke [68], Andrade *et al.* [6], Bruining *et al.* [24]), environmental science (Rodriguez-Iturbe *et al.* [115], Wunsch [155; 156], Marani [90; 91], She & Nakamoto [119], Chaudhuri & Sekhar [30; 31], Paulson [104]) and empirical analysis of time series (Blackman & Tukey [18], Jenkins & Watts [74], Gelb [63], Brillinger [22], Bendat & Piersol [14]).

1.3 Local Averages and Their Extremes

Central to the development of robust random field models is the concept of the "local average" of a random field. Practical interest often revolves around local *averages* or *aggregates* – temporal or spatial – of random attributes. For example, if the attribute is rainfall depth (as measured by rain gauges), a hydrologist is not interested in "instantaneous" values at a "point" but perhaps in total hourly, daily, or monthly rainfall depth spatially averaged over areas of prescribed size (for example, 1 km^2, or the size of the watershed or a basin prone to flash-flooding).

It is seldom useful or necessary to describe in detail the local point-to-point variation occurring on the "microscale" in time or space. Even if such information were desired, it may be impossible to obtain, as there is a basic trade-off between the accuracy of a measurement and the (time or distance) interval within which the measurement is made. Strain gauges, stress cells, heat sensors, or anemometers (owing to size, inertia, and so on) all measure some kind of local average over space and time. Moreover, through information processing "raw data" are often transformed into average or aggregate quantities such as one-minute averages, daily totals, or yield per hectare.

A pragmatic viewpoint on the question of scale in random field modeling is that there invariably exists a bound on the time or distance scale below which microscale variations of an attribute are:

1. not observed or observable, and/or
2. of no practical interest to the problem at hand, and/or
3. characterized by a mostly deterministic microstructure.

Much the same argument can often be made with respect to very slow variations occurring on a macroscale.

These observations constitute the basis for one of our central theses: random field models need only provide information about random variation on the microscale sufficient to represent behavior under (at least some amount of) local averaging. A more detailed description of the correlation structure on the microscale unnecessarily overloads the model.

A similar principle in solid mechanics holds that erratic patterns of local stresses in a solid body may be replaced by locally averaged stresses for the purpose of overall force-deformation analysis. In other words, details of the microstructure of the material affect behavior on the macroscale only through their effect on local averages. Based on this principle, continuum

Table 1.1 Random field applications in safety assessment of dams.

Hydrology
Distribution of rainfall on the watershed
Inflow into the reservoir
Watershed infiltration characteristics and soil moisture conditions
Geotechnology
Density, porosity, and permeability of embankment and foundation soil
Pore pressure fluctuations
Quantity and velocity of seepage through and under the dam
Random soil compressibility and strength
Earthquake engineering
Foundation resistance to liquefaction during earthquakes
Space-time patterns of earthquake occurrence near the dam (including assessment of possible reservoir induced seismicity)
Intensity of shaking throughout dam foundation and structure during an earthquake
Economic aspects
Downstream development pattern
Uncertain future stream of benefits from the project
Demand for irrigation water downstream
Dam's impact on agricultural yield

mechanics becomes meaningful at the macroscale even for such complex materials as granular soil, crystalline composites, or living tissue.

Practical problems involving complex random phenomena can often be formulated in terms of either *extreme values* of a random field or *excursions* by the field above a high level (threshold). The foregoing discussion about the relationship between point properties and local averages remains relevant when dealing with extremes and level excursions. Highly localized extreme values cause little concern if they do not affect the behavior of the system as a whole. However, such systemwide effects are likely to be felt when the random field is *in the aggregate* high (or low) over a sufficiently large domain in space and/or time. The implication is that two quantities matter: (1) the size of the domain over which the random field is (locally) aggregated and (2) the size of the deviation of the locally aggregated process from its expected value.

Consider, for purposes of illustration, the problem of the safety of dams. Table 1.1 lists the phenomena that relate to diverse aspects of a single technological problem: How to design, construct and maintain safe dams? This problem requires knowledge of hydrology, geology, geotechnical and earthquake engineering, as well as an understanding of economic and environmental aspects of water projects involving dams. The causes and mechanisms of dam failure – overtopping of the structure, rapid internal erosion of foundation or embankment materials, or distress during an earthquake – can be related to extremes of random quantities such as precipitation, reservoir inflow, seepage velocity, and material resistance. In no case, however, is it necessary (or possible) to know the fine details of the point-to-point variation of a multi-dimensional attribute. For instance, shear strength must be sufficiently low (in the aggregate) within a sufficiently large zone before an embankment slide can occur. During earthquake-induced shaking, foundation soil must fail within a sufficiently large volume to cause perceptible distress. Internal erosion, the most dangerous failure mode of an earth dam, is triggered by the occurrence of relatively localized extremes of porosity and flow velocity. Needless to say, the challenge in dam engineering is to conceive (affordable and effective) protective actions, and to monitor performance, so as to minimize the chance that such extremes will occur. The conclusion is that many distributed disordered systems appear to be sensitive to *extremes of local averages.* If the local average exceeds a critical threshold at some (any) location, the entire system may be transformed, perhaps irreversibly and catastrophically.

Chapter 2

Fundamentals of Analysis of Random Fields

Coincidences, in general, are great stumbling-blocks in the way of that class of thinkers who have been educated to know nothing of the theory of probabilities — that theory to which the most glorious objects of human research are endebted for the most glorious of illustration.

Edgar Allan Poe, The Murders in the Rue Morgue

2.1 Types of Random Fields

An experiment is defined as "an act or operation designed to discover some unknown truth or effect." It connotes the setting of a laboratory where an investigator observes and measures experimental outcomes. In elementary probability theory outcomes of experiments are usually described in terms of *random variables*, uncertain quantities to be observed. Thus the outcome $\{X < x\}$ means that "the random variable X is less than the value x." In much the same way it is meaningful to associate with the term "random field" a giant laboratory where an investigator can do numerous experiments. Observing the outcome of all the experiments in the laboratory is equivalent to observing the realization of the random field.

The location of each experimental setup in the laboratory is identified, in the terminology of random fields, by a set of "coordinates" or "parameters." In n-dimensional space, locations are denoted by a vector $\mathbf{t} = (t_1, t_2, \ldots, t_n)$ whose elements t_1, $t_2, \ldots$, t_n are the coordinates or parameters. A random field is also called a random (or stochastic) process, although the term "field" indicates that the parameter space is multi-dimensional. Most of the literature deals with random processes that depend on a single coordinate, usually time. The terms "field" and "process" are used interchangeably throughout the book, and one-dimensional variation is often treated as a special case.

In a simple experiment the investigator may observe the value of the random field at one specific location $\mathbf{t}$, or simultaneously two or three values at different locations. Some other experiment may focus on a limited region R^n in n-dimensional parameter space and seek to determine whether or not any value in that region exceeds a particular threshold.

A random field may be seen as an *indexed* family of random variables $X(\mathbf{t})$. The collective outcome of all the experiments comprising the random field is denoted by $x(\mathbf{t})$, the *realization* of the random field. The outcome of the random variable at location $\mathbf{t}$ defines the *state* of the random field there. Thus a random field is termed "discrete state" or "continuous state" depending upon whether the random variables $X(\mathbf{t})$ are discrete or continuous. Likewise, the adjectives discrete or continuous may be attached to any parameter or coordinate. Depending upon the locations of the observation points (see Fig. 2.1), the random field may be called more specifically:

1. a *random series* if observations are made at discrete (often equally spaced) points on a time axis,
2. a *lattice process* if observations are made at the nodes or sites of a lattice in space,
3. a *continuous random function* if observations are made at all points along one or more spatial coordinate axes (u_1, u_2, u_3) and/or time (t),
4. a *random partition of space* if a discrete random variable (often binary) is observed at every point in space,
5. a *random point process* if points are located in a random pattern in space.

The different descriptions are often closely related. For example, a continuous-parameter, continuous-state random field $X(\mathbf{t})$ may be sampled (observed) at a lattice of locations in the parameter space. At these locations one may choose to assign one of two values, say, $+1$ or -1, depending upon whether $X(\mathbf{t})$ is above or below a threshold. This converts a continuous state field into a binary (two-phased) field. Local averaging of a random field tends to render it continuous in state and parameter.

A regular pattern points in space can be made to fill space by assigning to each point (site, center) its own polyhedral (called the "Wigner-Seitz cell" in crystallography), as illustrated for $n = 2$ in Fig. 2.2. If the pattern

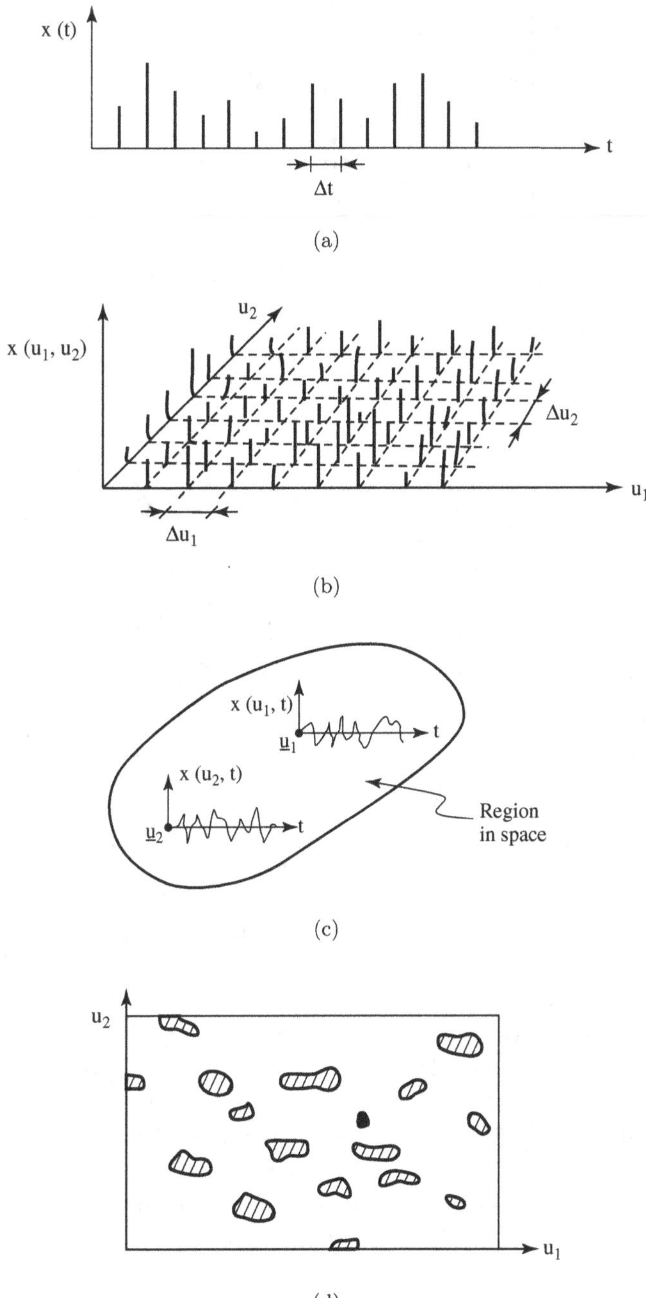

Fig. 2.1 Different types of random fields: (a) random series, (b) lattice process, (c) space-time process, (d) random partition of space, (e) random pattern of points.

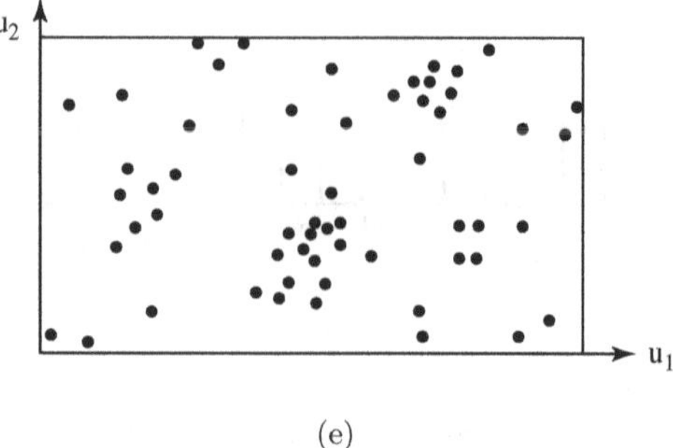

(e)

Fig. 2.1 (*Continued*)

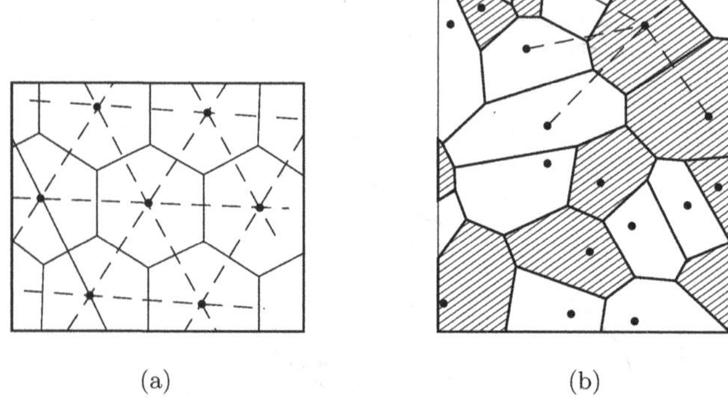

(a) (b)

Fig. 2.2 (a) Wigner-Seitz cells associated with a lattice of points; (b) Voronoi cells, shaded at random, associated with a random pattern of points.

of points is random, this same construction yields so-called "Voronoi cells." By assigning a color to each cell, one obtains a discrete-state, continuous-parameter random field.

The emphasis in this book is on real, scalar random fields. Many physical phenomena can be modeled as space-time processes $X(\mathbf{u}, t)$ that depend on a vector of spatial coordinates $\mathbf{u} = (u_1, u_2, u_3)$ and on time t. Where it is desirable to emphasize the *spatial* character of the coordinates, the symbols u_1, u_2, and u_3 (or vector $\mathbf{u}$) will be used. The time parameter is always denoted by the symbol t. Whenever results are stated for general n-dimensional random fields (or random fields in R^n), the location vector will be denoted by $\mathbf{t} = (t_1, t_2, \ldots, t_n)$.

Time and length parameters are measured on a higher scale of measurement, the so-called interval scale (Churchman and Ratoosh [34]). The increment between 1 m and 2 m (meters or minutes) is exactly the same as the increase from 20 m to 21 m. Increments between units have equal size; and it is possible to define an origin for each coordinate axis – for example, the present time or the observer's location – although the choice of origin is not unique.

The choice of coordinates of a random field is not necessarily limited to length and time. Any scale of measurement might be involved, including the *nominal scale* which classifies observations into mutually exclusive categories of equal rank (Churchman and Ratoosh [34]). These categories may be identified by names, such as type A, type B, or equivalently, by numbers, such as $t_1 = 1, 2, \ldots$. The nominal scale implies little or nothing about relative importance or size: there is no connotation that "3 is greater than 2," or "three times as large as 1." For example, the nominal coordinate may identify the components of a vector of attributes at every point $(\mathbf{u}, t)$ in a space-time continuum, such as the components of the velocity vector in a turbulent fluid. At each location and instant, three scalar quantities corresponding to three principal directions may be observed. In other cases the scalar quantities may be different material properties at a "point" in a random medium; these properties may be seen as "coupled" random fields; or one might introduce a new integer-valued parameter, say, t_n, that labels the different scalar quantities (such as $t_n = 1, 2$ and 3).

As an example of a discrete-unit or "agent-based" coordinate system, consider a simple random field model of a *two-candidate election*. The election outcome is viewed as the realization of a discrete state random field. Each registered voter has the choice (1) not to vote, (2) to vote for candidate A, or (3) to vote for candidate B. This choice can be represented by an indexed discrete random variable X which can take the values $x = -1$ (vote for A), $x = 0$ (no vote), or $x = +1$ (vote for B). The vector of parameters of the random process might identify voters by geographic (residential) location (t_1, t_2), income (t_3), age (t_4), and so on. By summing the random field over t_1 and t_2, one obtains random variable Y. Candidate B wins if Y is positive; candidate A if Y is negative. Constraining the summation to specific values of t_3 or t_4 permits analysis, before or after the election, of voting patterns.

2.2 Basic Probabilistic Description

Brief Review of Elementary Probability Theory

The basic tools to analyze random fields are provided by elementary probability theory of which this section gives a brief overview. The quantity $X(\mathbf{t})$ may be interpreted either as a single random variable at location $\mathbf{t}$ in the parameter space or as the family of all such random variables in the parameter space. Each random field offers the analyst the opportunity to design all kinds of experiments, some simple, involving $X(\mathbf{t})$ at one or two locations, others complex and based on combinations of many or all of the random variables comprising the field.

Possible outcomes of an experiment are called *events*, and the collection of all possible outcomes is called the *sample space* of that experiment. Probability theory guides the assignment of measures of likelihood to points or regions in the sample space. The ground rules for probability assignment are the *axioms* of probability theory, stated below for easy reference:

1. For any event A, the probability of A must lie between zero and one:

$$0 \leq P[A] \leq 1. \tag{2.2.1}$$

2. The probability associated with the entire sample space S, a certain event, is unity:

$$P[S] = 1. \tag{2.2.2}$$

3. If the events $A_1, A_2, \ldots, A_M$ are mutually exclusive, the probability of their *union* is a sum of probabilities:

$$P[A_1 \cup A_2 \cup \cdots \cup A_M] = \sum_{k=1}^{M} P[A_k]. \tag{2.2.3}$$

The *joint event* that both A and B occur is denoted by $\{A \cap B\}$. The event that A or B occurs is the *union of events* $\{A \cup B\}$, whose probability can be expressed as

$$P[A \cup B] = P[A] + P[B] - P[A \cap B], \tag{2.2.4}$$

which follows directly from the Venn diagram representation, shown in Fig. 2.3, of events A and B . If the events A and B are mutually exclusive – if they cannot occur simultaneously – then $P[A \cap B] = 0$ and $P[A \cup B] = P[A] + P[B]$. If all possible outcomes of an experiment are represented by

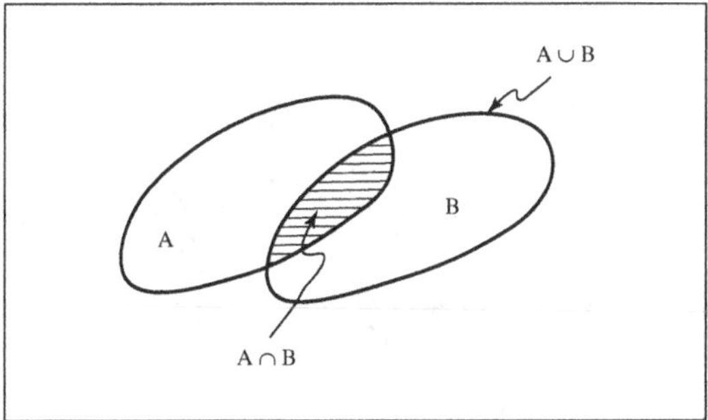

Fig. 2.3 Union and intersection of two events.

a set of mutually exclusive events A_1, A_2,..., A_M, then it follows from axioms 2 and 3 that the associated probabilities sum to one:

$$P[A_1] + P[A_2] + \cdots + P[A_M] = 1. \tag{2.2.5}$$

The events A_i, $i = 1, \ldots, M$, are then referred to as *simple events*, which are by definition "mutually exclusive and collectively exhaustive." For $M = 2$, Eq. (2.2.5) implies

$$P[A^c] = 1 - P[A], \tag{2.2.6}$$

where the event A^c is the *complement* of A.

The degree of probabilistic dependence between two events A and B is expressed by the *conditional probability* of A *given* B:

$$P[A|B] = \frac{P[A \cap B]}{P[B]}. \tag{2.2.7}$$

Of course, A and B are interchangeable, and one may also write

$$P[A \cap B] = P[A|B]P[B] = P[B|A]P[A]. \tag{2.2.8}$$

Two events A and B are *independent* if $P[A|B] = P[A]$ or $P[B|A] = P[B]$, or alternately, if $P[A \cap B] = P[A]P[B]$.

The Venn diagram in Fig. 2.4 shows a set of simple events A_i, $i = 1, 2, \ldots, M$, intersecting another event B. The relationship

$$P[B] = \sum_{i=1}^{M} P[A_i \cap B] = \sum_{i=1}^{M} P[B|A_i]P[A_i] \tag{2.2.9}$$

is referred as the *total probability theorem.* Combining Eqs. (2.2.7) and

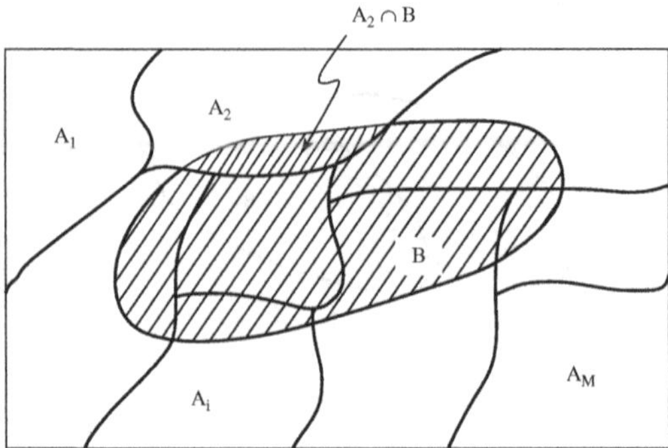

Fig. 2.4 Relationship between event B and a collection of mutually exclusive and collectively exhaustive events $A_1, A_2, \ldots, A_M$.

(2.2.9) yields the conditional probability of A_i given B,

$$P[A_i|B] = \frac{P[B|A_i]P[A_i]}{\sum_{j=1}^{M} P[B|A_j]P[A_j]}. \tag{2.2.10}$$

This is a statement of *Bayes' theorem* which provides a mechanism for updating the "prior" probabilities $P[A_i]$ based on the information that event B occurred. $P[A_i|B]$ is the updated or "posterior" probability of A_i (given B), and $P[B|A_i]$ is the likelihood of observing B provided A_i is true.

Where a choice of philosophy about probability – frequentist vs. subjective – must be made, developments in this book are consistent with the Bayesian viewpoint according to which probability assignment is broadly interpreted as "a numerical encoding of one's state of knowledge" (see, for example, Savage [116] and Tribus [130]). Bayes' theorem is the analysis tool that enables combining different kinds of knowledge – about relative frequency as well as subjective judgment – and updating one's state of knowledge based on newly acquired information that is relevant to the probability assessment.

Random Variables

The value of a random field at location $\mathbf{t}$ in the parameter space is a random variable $X = X(\mathbf{t})$, and $x = x(\mathbf{t})$ denotes an observed or realized value. The probability of the event $A = \{X \leq x\}$ is by definition the *cumulative distribution function* (c.d.f.) of X:

$$P[A] = P[X \leq x] = F_X(x) \equiv F(x). \tag{2.2.11}$$

$F(x)$ increases monotonically from 0 to 1 as x increases from $-\infty$ to $+\infty$. The probability of the complement of A,

$$P[A^c] = P[X > x] = F^c(x) = 1 - F(x), \tag{2.2.12}$$

defines the *complementary* c.d.f. For *continuous-state* random fields, $X = X(\mathbf{t})$ is a continuous random variable; taking the derivative of $F(x)$ yields the *probability density function* (p.d.f.) of X,

$$f(x) = \frac{dF(x)}{dx}, \tag{2.2.13}$$

a nonnegative function of x that satisfies

$$F(x) = \int_{-\infty}^{x} f(x_1)dx_1 \tag{2.2.14}$$

and

$$\int_{-\infty}^{+\infty} f(x)dx = 1. \tag{2.2.15}$$

For *discrete-state* random fields, the random variable $X = X(\mathbf{t})$ takes on one of a set of discrete values x_1, x_2, ... with finite probability, and it is convenient to introduce the *probability mass function*

$$p(x_i) = P[X = x_i]. \tag{2.2.16}$$

One avoids the dual notation (discrete vs. continuous) by permitting the p.d.f. to include Dirac delta functions. The probability mass $p(x_i)$ is then interpreted as the area "under the delta function" at point x_i.

Now consider an experiment in which the random field $X(\mathbf{t})$ is to be observed at two different locations $\mathbf{t}_1$ and $\mathbf{t}_2$. The two random variables are $X_1 = X(\mathbf{t}_1)$ and $X_2 = X(\mathbf{t}_2)$. Any outcome of this experiment, such as $X_1 = x_1$ and $X_2 = x_2$, can be represented by a point in a two-dimensional sample space. The probability of the joint event $A = \{(X_1 \leq x_1) \cap (X_2 \leq x_2)\}$ is by definition the *joint* cumulative distribution function of $X_1 = X(\mathbf{t}_1)$ and $X_2 = X(\mathbf{t}_2)$:

$$F_{x_1,x_2}(x_1, x_2) \equiv F(x_1, x_2) = P[\{X_1 \leq x_1\} \cap \{X_2 \leq x_2\}]. \tag{2.2.17}$$

For continuous random variables differentiation of $F(x_1, x_2)$ with respect to x_1 and x_2 yields the *joint* p.d.f.

$$f(x_1, x_2) = \frac{\partial^2 F(x_1, x_2)}{\partial x_1 \partial x_2}. \tag{2.2.18}$$

The *marginal* p.d.f. of x_1 is obtained by integrating the joint p.d.f. over all possible values of x_2:

$$f(x_1) = \int_{-\infty}^{+\infty} f(x_1, x_2)\, dx_2. \tag{2.2.19}$$

These results remain valid for discrete-state random fields (or discrete random variables) provided the p.d.f. is understood to include Dirac delta functions. Increases in $F(x_1, x_2)$ then occur in jumps at the discrete values of X_1 and X_2.

By imposing conditions on the random variable X_1 or X_2, or both, a new conditional sample space can be constructed. In particular, the condition $\{X_1 = x_1\}$ leads to the conditional c.d.f. of X_2 given $X_1 = x_1$:

$$F(x_2|x_1) = P[X_2 \leq x_2 | X_1 = x_1]. \tag{2.2.20}$$

The corresponding conditional p.d.f. is obtained by differentiation:

$$f(x_2|x_1) = \frac{d}{dx_2} F(x_2|x_1). \tag{2.2.21}$$

It satisfies

$$\int_{-\infty}^{+\infty} f(x_2|x_1)\, dx_2 = 1. \tag{2.2.22}$$

The conditional, joint, and marginal probability density functions are related as follows

$$f(x_1, x_2) = f(x_2|x_1) f(x_1) = f(x_1|x_2) f(x_2). \tag{2.2.23}$$

Bayes' theorem [see Eq. (2.2.10)] may be rewritten as follows:

$$f(x_2|x_1) = \frac{f(x_1, x_2)}{f(x_1)} = \frac{f(x_1|x_2) f(x_2)}{\int_{-\infty}^{\infty} f(x_1|x_2) f(x_2)\, dx_2}. \tag{2.2.24}$$

In Eqs. (2.2.20) through (2.2.24), the conditional p.d.f. is made to depend on a specific value of the random variable X_1, namely $X_1 = x_1$. If X_1 remains random, however, the conditional p.d.f. becomes a *function of the random variable* X_1. This alternate interpretation is highly valuable in applications of the theory of conditional expectation, discussed in Sec. 2.6.

A parallel set of relations can be stated in terms of the probability mass functions of a discrete-state random field. Provided the probability density functions are permitted to include Dirac delta functions, it is unnecessary to distinguish between discrete-state and continuous-state random fields.

Two random variables $X_1 = X(\mathbf{t}_1)$ and $X_2 = X(\mathbf{t}_2)$ are called *independent* if and only if

$$f(x_1, x_2) = f(x_1)f(x_2), \tag{2.2.25}$$

for all values of x_1 and x_2. An alternate and equivalent statement of the independence condition follows from the definition of conditional probability: two random variables X_1 and X_2 are independent if and only if

$$f(x_1|x_2) = f(x_1), \tag{2.2.26}$$

where the subscripts 1 and 2 may of course be interchanged.

An entirely analogous description is possible if the random field is observed at more than two locations during a single "compound" experiment. A random vector $\mathbf{X}$, that is, an array of random variables $X_1 = X(\mathbf{t}_1)$, $X_2 = X(\mathbf{t}_2), \ldots, X_M = X(\mathbf{t}_M)$, may be characterized by the joint c.d.f.

$$\begin{aligned} F_X(\mathbf{x}) &\equiv F(\mathbf{x}) \equiv F(x_1, \ldots, x_M) \\ &= P[\{X_1 \leq x_1\} \cap \cdots \cap \{X_M \leq x_M\}], \end{aligned} \tag{2.2.27}$$

from which one can derive assorted lower-order cumulative distribution functions, probability density (or mass) functions, and conditional probability distributions for a continuous-state random field. The joint p.d.f. of the random vector $\mathbf{X}$ is

$$f_X(\mathbf{x}) \equiv f(\mathbf{x}) \equiv f(x_1, \ldots, x_M) = \frac{\partial^M F(x_1, \ldots, x_M)}{\partial x_1 \cdots \partial x_M}. \tag{2.2.28}$$

The random variables $X_1, \ldots, X_M$ are *independent* if and only if

$$f(x_1, \ldots, x_M) = f(x_1) \cdots f(x_M). \tag{2.2.29}$$

More generally, one may view the random vector $\mathbf{X}$ as being partitioned into two vectors $\mathbf{X}_1$ and $\mathbf{X}_2$. (The component vectors may represent an array of observations either from a single random field or from two different random fields.) The *joint* p.d.f. of $\mathbf{X}_1$ and $\mathbf{X}_2$ is $f(\mathbf{x}_1, \mathbf{x}_2)$; the *marginal* p.d.f. is $f(\mathbf{x}_i)$, $i = 1, 2$; and $f(\mathbf{x}_1|\mathbf{x}_2)$ is the *conditional* p.d.f. of $\mathbf{X}_1$ given $\mathbf{X}_2 = \mathbf{x}_2$. The generalization of Bayes' theorem is

$$f(\mathbf{x}_1|\mathbf{x}_2) = \frac{f(\mathbf{x}_1, \mathbf{x}_2)}{f(\mathbf{x}_2)} = \frac{f(\mathbf{x}_2|\mathbf{x}_1)f(\mathbf{x}_1)}{\int_{\text{All}\, x_i} f(\mathbf{x}_2|\mathbf{x}_1)f(\mathbf{x}_1)dx_1}. \tag{2.2.30}$$

In these equations the vector of "observations" $\mathbf{x}_2$ may also be replaced by a random vector $\mathbf{X}_2$, in which case the conditional p.d.f. on the left side of Eq. (2.2.30) becomes a *function of the random vector* $\mathbf{X}_2$.

Conceptual Extension to Random Fields

The parameter space of a random field usually contains an infinite and uncountable number of points. Therefore the "full distribution approach", which would require $M \to \infty$, is only of theoretical and conceptual value. There are never sufficient data to validate assumed full multivariate probability distributions. Several key properties of random fields – homogeneity, isotropy, and ergodicity – permit the analyst to make more efficient use of limited data available to estimate probability distributions and their statistical parameters. The *strict* definitions of these properties, in terms of the multivariate probability density function $f(\mathbf{x})$, are introduced next. In practice, random fields that are assumed to possess these properties to any degree often do so in a limited or weak sense, not in the strict sense.

A random field is called *homogeneous* if all the joint probability distribution functions remain the same when the set of locations $\mathbf{t}_1, \mathbf{t}_2, \ldots, \mathbf{t}_M$ is translated (but not rotated) in the parameter space. This implies that all the probabilities depend only on the *relative*, not the absolute, locations of the points $\mathbf{t}_1, \ldots, \mathbf{t}_M$. For one-dimensional random processes the term "stationary" is commonly used, instead of "homogeneous." The field is *isotropic* if the joint probability density functions remain the same when the constellation of points $\mathbf{t}_1, \mathbf{t}_2, \ldots, \mathbf{t}_M$ is rotated in the parameter space. A random field is *ergodic* if all information about its joint probability distributions (and all their statistical parameters) can be obtained from a single realization of the random field.

These properties are mainly of conceptual value. In practical applications there are usually upper and lower bounds on the time and distance intervals over which they apply. Much more is said about these properties in Chap. 3 on second-order analysis of random fields.

Transformation of Random Variables

Elementary Transformations

In stochastic analysis of random fields, interest frequently focuses *not* on the random variables $X_k = X(\mathbf{t}_k)$ themselves but on a *function* of one or more of the random variables comprising the field. For example, the one-to-one transformation

$$Y = \ln X, \tag{2.2.31}$$

may transform the random field $X(\mathbf{t})$ at just one specific location $\mathbf{t}$ or at *all* locations in the parameter space. If the random variables X_1, X_2, and X_3 are the orthogonal components of a random vector at a point, interest may focus on the field of vector lengths:

$$Y = [X_1^2 + X_2^2 + X_3^2]^{1/2}. \tag{2.2.32}$$

Functions of random variables are themselves random variables. For example, given the relations $Y_1 = g_1(X_1, X_2, X_3)$ and $Y_2 = g_2(X_1, X_2, X_3)$ *and* the joint probability distribution of X_1, X_2, and X_3, it is in principle possible to derive the joint distribution of Y_1 and Y_2. Although there is no closed-form analytical solution for the general case, it is possible, by careful accounting of the probability density (or mass) in the relevant sample spaces, to solve derived distribution problems numerically.

The simplest case is when a monotonic one-to-one relationship $Y = g(X)$ exists between the random variables being transformed, as in Eq. (2.2.31). This insures the existence of the inverse relations [such as $X = e^Y$, relative to Eq. (2.2.31)]. The probability density function $f_M(y)$ is then:[1]

$$f_Y(y) = f_X(x)\left|\frac{dx}{dy}\right| \equiv f_X(x)\left|\frac{dg^{-1}(y)}{dy}\right|, \tag{2.2.33}$$

where $x = g^{-1}(y)$ is the inverse transformation. The derivative is positive or negative depending on whether $g(x)$ is monotonically increasing or decreasing (see Fig. 2.5). For example, in case $Y = \ln X$, the c.d.f. and p.d.f. of Y are, respectively,

$$F_Y(y) = F_X(e^y), \quad y \geq 0, \text{ and} \tag{2.2.34}$$

$$f_Y(y) = e^y f_X(e^y), \quad y \geq 0. \tag{2.2.35}$$

If the relationship is linear, $Y = aX + b$, then

$$f_Y(y) = \frac{1}{|a|} f_X\left(\frac{y-b}{a}\right). \tag{2.2.36}$$

[1]The subscripts X and Y are added here to avoid confusion about which probability distribution is involved.

If the same monotonically increasing one-to-one transformation $Y = g(X)$ is applied to the entire random field, the multivariate cumulative distribution function is obtained as follows:

$$F_Y(y_1, \ldots, y_M) = P[\{g(X_1) \geq y_1\}, \ldots, \{g(X_M) \leq y_M\}]$$

$$= P[\{X_1 \leq g^{-1}(y_1)\} \cdots \{X_M \leq g^{-1}(y_M)\}] = F_X(x_1, \ldots, x_M), \quad (2.2.37)$$

where it is understood that $x_k = g^{-1}(y_k)$, $k = 1, \ldots, M$. The joint p.d.f. is obtained by iterative partial differentiation of Eq. (2.2.37), yielding

$$f_Y(y_1, \ldots, y_M) = f_X(x_1, \ldots, x_M) \prod_{k=1}^{M} \left| \frac{dx_k}{dy_k} \right|, \quad (2.2.38)$$

where again, $x_k = g^{-1}(y_k)$, $k = 1, \ldots, M$. For example, if the transformation $Y = \log X$ is applied, it suffices to introduce $x_k = \exp(y_k)$; $k = 1, \ldots, M$, into Eq. (2.2.38).

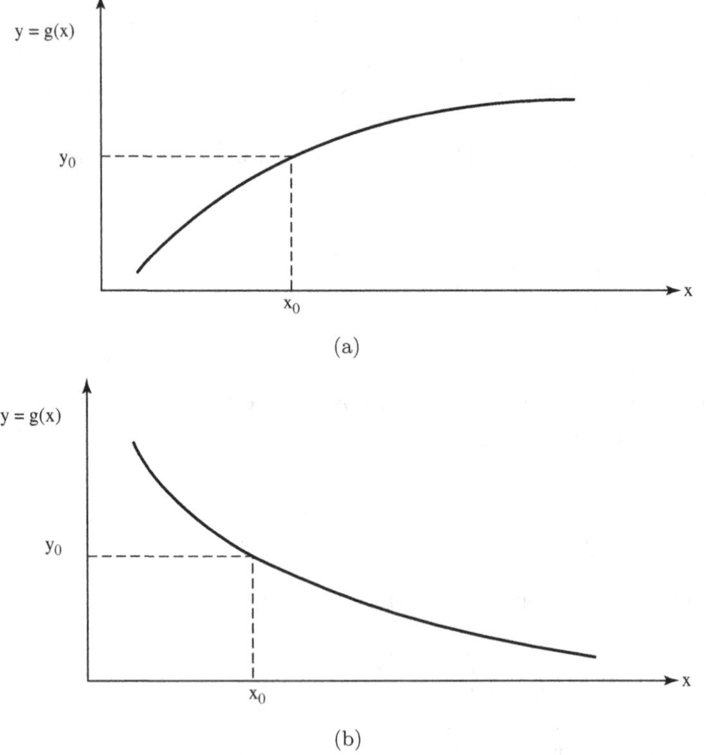

Fig. 2.5 Monotonic one-to-one transformation: (a) y is increasing with x; (b) y is decreasing with x.

Binary Random Fields

An important role is played by *two-valued* or *binary* random fields. A family of such fields can be derived from any continuous- or discrete-state random field $X(\mathbf{t})$ by considering its position relative to a fixed threshold x, as illustrated in Fig. 2.6. If $X(\mathbf{t}) > x$, we define $Z(\mathbf{t}) = 1$; if $X(t) \leq x$, then $Z(\mathbf{t}) = 0$. Changing the value of the threshold produces a family of derived binary fields. Each member of the family is identified by the threshold-level exceedance probability $p = p_Z(1) = F_X^c(x)$. The probability mass function of $Z = Z(\mathbf{t})$ is

$$p_Z(z) = \begin{cases} 1 - F_X(x), & z = 1, \\ F_X(x), & z = 0. \end{cases} \tag{2.2.39}$$

The joint p.m.f. of $Z_1 = Z(\mathbf{t}_1)$ and $Z_2 = Z(\mathbf{t}_2)$ is (see Fig. 2.7):

$$p_{Z_1,Z_2}(z_1, z_2) = \begin{cases} F_{X_1,X_2}(x, x), & z_1 = z_2 = 0, \\ F_{X_1}(x) - F_{X_1,X_2}(x, x), & z_1 = 0,\ z_2 = 1, \\ F_{X_2}(x) - F_{X_1,X_2}(x, x), & z_1 = 1,\ z_2 = 0, \\ 1 - F_{X_1}(x) - F_{X_2}(x) \\ \quad + F_{X_1,X_2}(x, x), & z_1 = z_2 = 1. \end{cases} \tag{2.2.40}$$

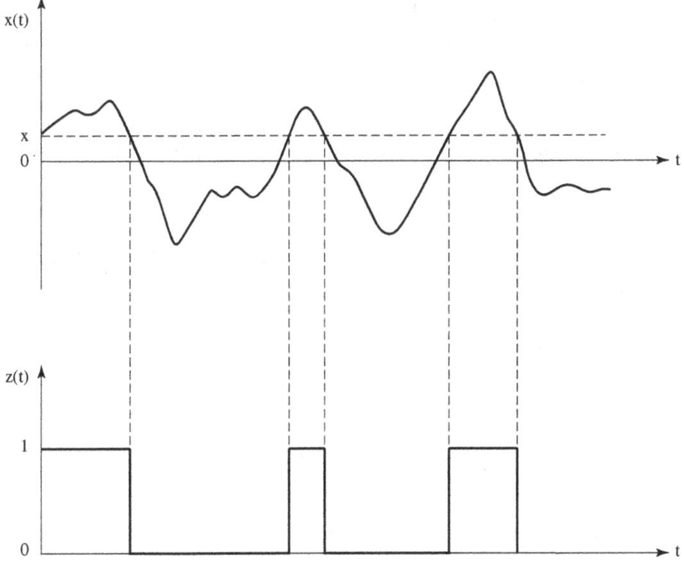

Fig. 2.6 Sample functions of the random processes $x(t)$ and $z(t)$.

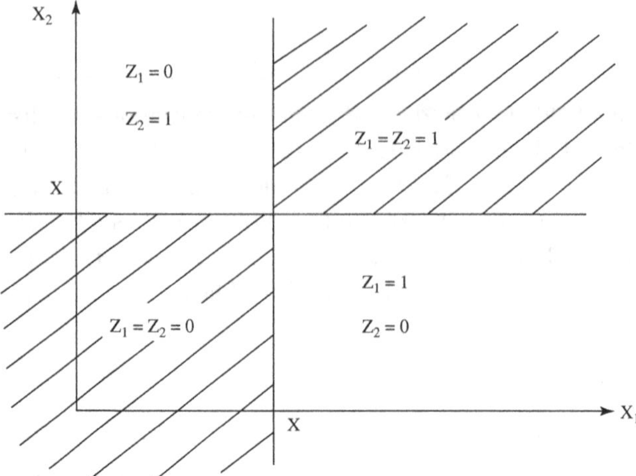

Fig. 2.7 Relationship between the two-state sample spaces (x_1, x_2) and (z_1, z_2).

It is easy to verify that the derived binary random variables Z_1 and Z_2 will be independent if X_1 and X_2 are independent.

Sums of Independent Random Variables

Thus far we have examined only one-to-one transformations, in which $Y(\mathbf{t})$ is related directly to $X(\mathbf{t})$. Full analytical treatment of multivariate transformations is seldom possible, owing to lack of mathematical tools and/or information about joint probability distributions. The problem is often rendered tractable by *assuming* that the random field is "purely random" (i.e. by ignoring probabilistic dependence). Consider specifically the *sum*

$$Y = X_1 + X_2 + \cdots + X_M, \tag{2.2.41}$$

where $X_k = X(\mathbf{t}_k)$, $k = 1, \ldots, M$. Starting with $M = 2$, referring to Fig. 2.8, we may write

$$F_Y(y) = P[X_1 + X_2 \le y] = \iint\limits_{x_1+x_2 \le y} f_{X_1,X_2}(x_1, x_2) dx_1 dx_2$$

$$= \int_{-\infty}^{+\infty} F_{X_1|X_2}(y - x|x) f_{X_2}(x) dx. \tag{2.2.42}$$

It is evident that, in order to proceed, the joint (or conditional) probability distribution must be available. Only if X_1 and X_2 are *independent* can the integrand in Eq. (2.2.42) be expressed in terms of marginal distributions:

$$F_Y(y) = \int_{-\infty}^{+\infty} F_{X_1}(y - x) f_{X_2}(x)\, dx. \tag{2.2.43}$$

Differentiating with respect to y then yields the relation

$$f_Y(y) = \int_{-\infty}^{+\infty} f_{X_1}(y - x) f_{X_2}(x)\, dx, \tag{2.2.44}$$

which is a *convolution operation.* Repeated application of Eq. (2.2.44) (adding random variables one at a time) permits the derivation of the p.d.f. of a sum of independent random variables.

Another important derived random variable that depends on many values of a random field is the *maximum value*

$$V = \text{Max}\{X_1, X_2, \ldots, X_M\}, \tag{2.2.45}$$

where $X_k = X(\mathbf{t}_k)$, $k = 1, \ldots, M$. The c.d.f. of V is

$$\begin{aligned} F_V(v) &= P[V \le v] = P[\{X_1 \le v\} \cap \cdots \cap \{X_M \le v\}] \\ &= F_X(v, \ldots, v). \end{aligned} \tag{2.2.46}$$

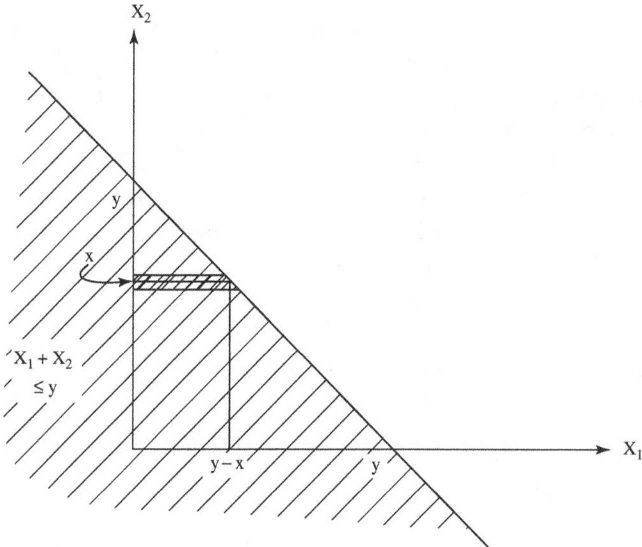

Fig. 2.8 The shaded area indicates the event $\{x_1 + x_2 \le y\}$ in the two-state sample space (x_1, x_2).

In general, the *joint* c.d.f. $F_X(\mathbf{x})$ must be available. If the random variables $X_1, \ldots, X_M$ are *independent*, Eq. (2.2.46) becomes

$$F_V(v) = \prod_{k=1}^{M} P[X_k \leq v] = \prod_{k=1}^{M} F_X(v). \tag{2.2.47}$$

If the random variables $X_1, X_2, \ldots, X_M$ are also *identically distributed*, the c.d.f. is

$$F_V(v) = [F_X(v)]^M. \tag{2.2.48}$$

When $M \to \infty$, for most (but not all) distributions $F_X(x)$, the form of $F_V(v)$ will tend toward one of the following three *largest value distributions* (see Gumbel [67], Feller [57], de Haan [49], Leadbetter *et al.* [81], Reinick [112], Embrechts *et al.* [55], and Coles [36]):

$$\begin{aligned}
&(EX_{I,L}) && F(v) = \exp\{-\exp(-v)\}, \quad -\infty \leq v \leq +\infty; \\
&(EX_{II,L}) && F(v) = \begin{cases} \exp\{-v^{-k}\}, & v \geq 0,\ k > 0, \\ 0, & v < 0; \end{cases} \\
&(EX_{III,L}) && F(v) = \begin{cases} 1, & v \geq 0, \\ \exp\{-(-v)^k\}, & v < 0,\ k > 0. \end{cases}
\end{aligned} \tag{2.2.49}$$

These are often referred to as the Gumbel (Type I), Frechét (Type II), and Reversed Weibull (Type III) extreme values distributions. The third law is usually applied to minimum values (to the maximum of $-X_1$, $-X_2$, and so on). If a particular probability law leads to the Type I largest value distribution, it is said to "belong to the domain of attraction" of $EX_{I,L}$. The normal, exponential, and Gamma distributions belong to the domain of attraction of $EX_{I,L}$. Gnedenko [65] gives necessary and sufficient conditions for $F_X(x)$ to belong to the domain of attraction of one of the three extreme value distributions. In particular, an extreme value distribution belongs to its *own* domain of attraction.

The three types of extreme value distributions may be seen as special cases of the Generalized Extreme Value (GEV) distribution having the form:

$$F(v) = \exp\{-(1 + kv)^{-1/k}\}, \tag{2.2.50}$$

where the cases $k \to 0$, $k > 0$ and $k < 0$ correspond, respectively, to the Gumbel (Type I), Frechét (Type II), and Reversed Weibull (Type III) distributions. The respective intervals within which Eq. (2.2.50) is valid are: $-\infty \leq v \leq +\infty$ for Gumbel; $v \geq -1/k$ for Frechét ; and $v \leq -1/k$ for Reversed Weibull. Finally, we can substitute in each case $v = (x - a)/b$, where a is a location parameter and b is a nonnegative scale (or dispersion) parameter.

Numerical Approaches to Transformation Problems

In a general multivariate transformation problem, the random vector $\mathbf{X}$ must be mapped into the random vector $\mathbf{Y}$ to which it is linked by a set of "input-output relations". One of the following methods of numerical transformation may be used. In the *enumeration* method all the input and output random variables are discretized. The probability mass associated with each point $(x_1, \ldots, x_N)$ of the input sample space is assigned to the corresponding point $(y_1, \ldots, y_M)$ of the output sample space. The problem reduces to accounting for, and transferring, all the probability mass from the original to the new sample-space coordinates.

In *Monte Carlo Simulation*, the digital computer repeatedly simulates sets of joint observations of the input random variables and transforms them into joint observations of the output random variables. When a sufficient synthetic data base has been assembled, one estimates the frequency of occurrence of joint events like $\{Y_1 \leq y_1\} \cap \cdots \cap \{Y_M \leq y_M\}$, defining the joint cumulative distribution function $F(y_1, \ldots, y_M)$.

These two methods, although highly efficient owing to vast gains in computing power and speed, nevertheless require the joint probability distribution of $X_1, \ldots, X_N$ to be available. This is seldom the case, however, and the situation is often poorly remedied by assuming (for convenience and without clearly understanding the implications) that the input random variables are statistically independent. A compromise is to seek only the first- and second-order moments, not the joint probability distribution, of the derived random variables.

2.3 Expectation and Conditional Expectation

In this section we briefly state, for easy reference, some well-known definitions and relationships involving *expectations* of functions of one, two, or more random variables. Each subsection concludes with an interpretation of the results in the context of random fields.

Expectation Involving One Random Variable

If X is a random variable with probability density function $f(x)$, the *expectation* of the function $g(X)$ is defined by

$$E[g(X)] = \int_{-\infty}^{+\infty} g(x)f(x)\,dx, \tag{2.3.1}$$

provided the integral exists.[2] If the random variable X is described by a probability mass function $p(x_i)$, then

$$E[g(X)] = \sum_{\text{All } x_i} g(x_i)p(x_i). \tag{2.3.2}$$

Henceforth no distinction will be made between the discrete and continuous cases, as the p.d.f. is permitted to include Dirac delta functions at discrete values. The expectation operation, being linear, may generally be interchanged with other linear operations such as summation, integration, and differentiation.

Taking $g(X) = X^n$, yields the *moments* of the random variable X. The first moment ($n = 1$) is the *mean* or the *expectation* of X:

$$m \equiv m_X = E[X]. \tag{2.3.3}$$

The second moment $E[X^2]$ is the *mean-square* of X. The *central moments* of X are obtained by taking $g(X) = (X - m)^n$. The first central moment is zero, and the second central moment is by definition the *variance* of X:

$$\sigma^2 \equiv \sigma_X^2 = \text{Var}[X] = E[(X - m)^2] = E[X^2] - m^2. \tag{2.3.4}$$

The square root of the variance, σ, is the *standard deviation* of X. If $m = 0$, the standard deviation equals the root-mean-square (r.m.s.) value. If $m \neq 0$, it is common to quantify the degree of uncertainty of X in terms of the *coefficient of variation*:

$$V \equiv V_X = \frac{\sigma}{m}. \tag{2.3.5}$$

The coefficients of skewness (γ_1) and kurtosis or flatness (γ_2) are defined in terms of the third and fourth central moments, respectively:

$$\gamma_1 = \frac{E[(X - m)^3]}{\sigma^3}, \quad \gamma_2 = \frac{E[(X - m)^4]}{\sigma^4}. \tag{2.3.6}$$

[2]For the integral to exist, the product of the functions $g(x)$ and $f(x)$ must be "absolutely integrable," that is, $E[|g(X)|]$ must be finite.

Two other common measures of central tendency of a random variable are: (1) the *median* whose value of X corresponds to $F(x) = 0.5$, and (2) the *mode* which is the value of x for which the p.d.f. is maximum.

In the context of random fields it is appropriate to replace X by $X(\mathbf{t})$ in Eqs. (2.3.3) through (2.3.6). In particular, the mean of the random field at location $\mathbf{t}$ is

$$E[X(\mathbf{t})] = m_X(\mathbf{t}) \equiv m(\mathbf{t}), \tag{2.3.7}$$

and the variance of $X(\mathbf{t})$ is

$$\mathrm{Var}[X(\mathbf{t})] = E[(X(\mathbf{t}) - m(\mathbf{t}))^2] = \sigma_X^2(\mathbf{t}) \equiv \sigma^2(\mathbf{t}). \tag{2.3.8}$$

The mean and variance are, in general, functions of location $\mathbf{t}$. They become *constants* if the random field is *homogeneous*, namely:

$$\begin{aligned} E[X(\mathbf{t})] &= m(\mathbf{t}) = m, \\ \mathrm{Var}[X(\mathbf{t})] &= \sigma^2(\mathbf{t}) = \sigma^2. \end{aligned} \tag{2.3.9}$$

To obtain the expectation of a derived random variable, say, $Y = g(X)$, it is unnecessary to derive $f_Y(y)$ first, since

$$E[Y] = E[g(X)]. \tag{2.3.10}$$

For example, for a *binary* random field $Z(\mathbf{t})$ defined in relation to the threshold level x [see Eq. (2.2.39)], the mean and variance are, respectively:

$$m_Z(\mathbf{t}) = P[Z(\mathbf{t}) = 1] = P[X(\mathbf{t}) > x] \equiv p(\mathbf{t}) \tag{2.3.11}$$

and

$$\sigma_Z^2(\mathbf{t}) = p(\mathbf{t})[1 - p(\mathbf{t})], \tag{2.3.12}$$

where $p(\mathbf{t})$ is the complementary c.d.f. of $X(\mathbf{t})$ evaluated at x. Of course, $p(\mathbf{t}) \equiv p$ if $X(\mathbf{t})$ is homogeneous.

Expectation Involving Two Random Variables

The expectation of a function $g(X_1, X_2)$ of two random variables X_1 and X_2 is defined by

$$E[g(X_1, X_2)] = \int_{-\infty}^{+\infty} \int_{-\infty}^{+\infty} g(x_1, x_2) f(x_1, x_2) dx_1 dx_2. \tag{2.3.13}$$

For convenience, we denote the means and variances by m_i and σ_i (short for m_{X_i} and σ_{X_i}) throughout this section. The *covariance* of X_1 and X_2 is defined as the expectation of the product of the deviations from their respective means,

$$\begin{aligned} B_{12} &= \mathrm{Cov}[X_1, X_2] = E[(X_1 - m_1)(X_2 - m_2)] \\ &= E[X_1 X_2] - m_1 m_2. \end{aligned} \tag{2.3.14}$$

Dividing B_{12} by $\sigma_1\sigma_2$ yields the dimensionless *coefficient of correlation* between X_1 and X_2:

$$\rho_{12} \equiv \rho_{X_1,X_2} = \frac{\mathrm{Cov}[X_1, X_2]}{\sigma_1\sigma_2} = \frac{B_{12}}{\sigma_1\sigma_2}. \tag{2.3.15}$$

It follows from the definition that $B_{12} = B_{21}$ and $\rho_{12} = \rho_{21}$. Also ρ_{12} must be bounded by -1 and $+1$, or $|\rho_{12}| \leq 1$ (and hence $B_{12} \leq \sigma_1\sigma_2$). Two jointly distributed random variables are said to be *perfectly correlated* (positively or negatively) if $\rho_{12} = \pm 1$ and *uncorrelated* if $\rho_{12} = 0$. Finally, *independence* between two random variables implies lack of correlation, but not *vice versa.*

The coefficient of correlation measures the degree of *linear* dependence. To see this, consider two linearly related random variables X and $Y = aX + b$, where a and b are constants and $a \neq 0$. The variance of Y is

$$\sigma_Y^2 = a^2\sigma_X^2 \tag{2.3.16}$$

and the covariance of X and Y is

$$\begin{aligned} \mathrm{Cov}[X, Y] &= E[X(aX + b)] - m_X(am_X + b) \\ &= a(E[X^2] - m_X^2) = a\sigma_X^2. \end{aligned} \tag{2.3.17}$$

Inserting Eqs. (2.3.16) and (2.3.17) into the definition of the correlation coefficient [Eq. (2.3.15)] confirms:

$$\rho_{X,Y} = \begin{cases} 1, & \text{for } a > 0, \\ -1, & \text{for } a < 0. \end{cases} \tag{2.3.18}$$

Hence, if a *linear* functional relationship exists between two random variables, they will be perfectly (positively or negatively) correlated.

The Covariance Matrix of a Random Vector

The covariances of M random variables X_1, X_2,..., X_M (comprising the random vector $\mathbf{X}$) can be arranged into an M by M symmetric covariance matrix $\mathbf{B} \equiv \mathbf{B}_X$:

$$\mathbf{B_X} = \begin{bmatrix} B_{11} & B_{12} & \cdots & B_{1M} \\ B_{21} & B_{22} & \cdots & B_{2M} \\ \vdots & \vdots & \ddots & \vdots \\ B_{M1} & B_{M2} & \cdots & B_{MM} \end{bmatrix}. \tag{2.3.19}$$

The principal property of the matrix $\mathbf{B}_X$ is that it is *positive definite.* Given any set of coefficients $a_1,a_2,\ldots,a_M$ (not all zero), one can construct the linear combination

$$\mathbf{a}^t\mathbf{X} = a_1X_1 + a_2X_2 + \cdots + a_MX_M. \tag{2.3.20}$$

The variance of this linear combination must be positive:

$$\mathbf{a}^t\mathbf{B_X}\mathbf{a} = \sum_{k=1}^{M}\sum_{l=1}^{M} a_k a_l B_{kl} > 0. \tag{2.3.21}$$

Any symmetric matrix $\mathbf{B_X}$ that satisfies Eq. (2.3.21) is called positive definite. Such a matrix has a set of M positive eigenvalues λ_k, $k = 1,\ldots,$ M. Specifically, there exists a linear transformation $\mathbf{Z} = \mathbf{CX}$ such that the $M \times M$ covariance matrix $\mathbf{B_Z}$ becomes a *diagonal* matrix:

$$\mathbf{B}_Z = \mathbf{C}\mathbf{B}_X\mathbf{C}^t = \begin{bmatrix} \lambda_1 & & & 0 \\ & \lambda_2 & & \\ & & \ddots & \\ 0 & & & \lambda_M \end{bmatrix}, \tag{2.3.22}$$

where λ_k, $k = 1,\ldots, M$, are the eigenvalues of the matrix $\mathbf{B}_X$. The new random variables $Z_1, \ldots, Z_M$ are uncorrelated, and their variances are the eigenvalues of $\mathbf{B}_X$.

Covariance, Correlation and Cross-covariance Functions

The covariance of the values of a random field at two different locations, $X_1 = X(\mathbf{t}_1)$ and $X_2 = X(\mathbf{t}_2)$, is by definition the *covariance function* (of

$\mathbf{t}_1$ and $\mathbf{t}_2$):

$$B(\mathbf{t}_1, \mathbf{t}_2) \equiv B_X(\mathbf{t}_1, \mathbf{t}_2) = \mathrm{Cov}[X(\mathbf{t}_1), X(\mathbf{t}_2)]$$
$$= E[X(\mathbf{t}_1)X(\mathbf{t}_2)] - m(\mathbf{t}_1)m(\mathbf{t}_2). \tag{2.3.23}$$

Likewise, the *correlation function*

$$\rho(\mathbf{t}_1, \mathbf{t}_2) \equiv \rho_X(\mathbf{t}_1, \mathbf{t}_2) = \frac{B(\mathbf{t}_1, \mathbf{t}_2)}{\sigma(\mathbf{t}_1)\sigma(\mathbf{t}_2)} \tag{2.3.24}$$

is by definition the coefficient of correlation between $X(\mathbf{t}_1)$ and $X(\mathbf{t}_2)$. The covariance of two values associated with different (coupled) random fields defines the *cross-covariance function*

$$B_{X_1 X_2}(\mathbf{t}_1, \mathbf{t}_2) = \mathrm{Cov}[X_1(\mathbf{t}_1), X_2(\mathbf{t}_2)]. \tag{2.3.25}$$

Evidently, $B_{X_1 X_1}(\mathbf{t}_1, \mathbf{t}_2) \equiv B_{X_1}(\mathbf{t}_1, \mathbf{t}_2)$. The principal properties (symmetry, bounds) of these functions follow directly from those of B_{12} and ρ_{12}. Another important property, that the covariance function is "positive definite," is a direct consequence of the fact that the covariances of values of a random field associated with a set of points $\mathbf{t}_1$, $\mathbf{t}_2, \ldots, \mathbf{t}_M$ can be arranged in the form of a covariance matrix.

Consider the linear combination of M values of the random field:

$$a_1 X(\mathbf{t}_1) + a_2 X(\mathbf{t}_2) + \cdots + a_M X(\mathbf{t}_M), \tag{2.3.26}$$

where the coefficients $a_1, \ldots, a_M$ are non-zero and the locations $\mathbf{t}_1, \ldots, \mathbf{t}_M$ are arbitrary. The variance of the linear combination must be nonnegative, which implies [as in Eq. (2.3.21)]:

$$\sum_{k=1}^{M} \sum_{l=1}^{M} a_k a_l B_X(\mathbf{t}_k, \mathbf{t}_l) > 0. \tag{2.3.27}$$

Any function that satisfies this condition is called positive definite. More generally, the array of coefficients a_k, $k = 1, \ldots, M$, may be replaced by some function $a(\mathbf{t})$, for which

$$\int_t \int_{t'} a(\mathbf{t})a(\mathbf{t}')B_X(\mathbf{t}, \mathbf{t}')\, d\mathbf{t}\, d\mathbf{t}' > 0. \tag{2.3.28}$$

Also, there exists a function $c(\mathbf{t}, \mathbf{t}')$ [which parallels the matrix $\mathbf{C}$ in Eq. (2.3.22)] such that $X(\mathbf{t})$ can be expressed as

$$X(\mathbf{t}) = \int_{t'} c(\mathbf{t}, \mathbf{t}')Z(\mathbf{t}')\, d\mathbf{t}', \tag{2.3.29}$$

where $Z(\mathbf{t}')$ is an *uncorrelated* random field. The variances of the components of the process $Z(\mathbf{t}')$, equivalent to the eigenvalues in Eq. (2.3.22), are directly related to the spectral representation of random fields, introduced in Chap. 3.

Expectations Involving Many Random Variables

The expectation of a function of several values of a random field, $X_k = X(\mathbf{t}_k)$, $k = 1, \ldots, M$, is

$$E[g(X_1, \ldots, X_M)] = \int_{-\infty}^{+\infty} \cdots \int_{-\infty}^{+\infty} g(X_1, \ldots, X_M) f_X(x_1, \ldots, x_M)\, dx_1 \ldots dx_M, \quad (2.3.30)$$

provided the multiple integral remains finite when the function g in the integrand is replaced by $|g|$.

Linear combinations of random variables are of particular interest. Consider, for example, two derived random variables Y_1 and Y_2, both linear combinations of different sets of values of the random field:

$$Y_1 = a_0 + \sum_{k=1}^{M_1} a_k X_k \quad \text{and} \quad Y_2 = b_0 + \sum_{l=1}^{M_2} b_l X_l, \quad (2.3.31)$$

where $X_k = X(\mathbf{t}_k)$ and $X_l = X(\mathbf{t}_l)$. The covariance of Y_1 and Y_2 is

$$\text{Cov}[Y_1, Y_2] = E\left[\left\{\sum_{k=1}^{M_1} a_k(X_k - m_k)\right\}\left\{\sum_{l=1}^{M_2} b_l(X_l - m_l)\right\}\right]$$

$$= \sum_{k=1}^{M_1}\sum_{l=1}^{M_2} a_k b_l \,\text{Cov}[X_k, X_l] = \sum_{k=1}^{M_1}\sum_{l=1}^{M_2} a_k b_l \sigma_k \sigma_l \rho_{kl}, \quad (2.3.32)$$

where ρ_{kl} denotes the coefficient of correlation between X_k and X_l. If the random variables are mutually uncorrelated, then $\rho_{kl} = 0$ for all $k \neq l$, and all the cross-terms in Eq. (2.3.32) vanish. In particular, the variance of the sum Y of a set of uncorrelated random variables is the sum of the variances.

These results for linear combinations also provide the basis for useful approximations of the lower-order moments of nonlinear functions of random variables, $Y = g(X_1, \ldots, X_M)$. The Taylor series expansion about the mean values of $X_1, \ldots, X_M$ provides the linear approximation

$$Y = g(X_1, \ldots, X_M) = g(m_1, \ldots, m_M) + \sum_{k=1}^{M} a_k (X_k - m_k) + \cdots, \quad (2.3.33)$$

where each coefficient a_k is the partial derivative of the function g with respect to X_k evaluated at the mean values $X_k = m_k$, $k = 1, \ldots, M$.

Conditional Expectation

Consider two random variables X_1 and X_2 with conditional probability density function $f(x_1|X_2)$. The conditional expectation of $g(X_1)$ given X_2 is defined by

$$E[g(X_1)|X_2] = \int_{-\infty}^{+\infty} g(x_1) f(x_1|X_2)\, dx_1. \quad (2.3.34)$$

In particular, the conditional mean of X_1 given X_2 is

$$E[X_1|X_2] = \int_{-\infty}^{+\infty} x_1 f(x_1|X_2) dx_1, \quad (2.3.35)$$

and the conditional variance of X_1 given X_2 is

$$\text{Var}[X_1|X_2] = \int_{-\infty}^{+\infty} (x_1 - E[X_1|X_2])^2 f(x_1|X_2) dx_1. \quad (2.3.36)$$

It is important to emphasize that these conditional statistics *are themselves random variables*, as they depend on X_2. If the random variable X_2 takes a specific value, then the conditional statistics become constants. When viewed as a random variable, the conditional expectation $E[g(X_1)|X_2]$ possesses many interesting properties. In particular

$$E[E[g(X_1)|X_2]] = E[g(X_1)], \quad (2.3.37)$$

where the first expectation is with respect to X_2. Also,

$$E[g_1(X_1) g_2(X_2)|X_2] = g_2(X_2) E[g_1(X_1)|X_2]. \quad (2.3.38)$$

If X_1 and X_2 are independent, the conditional and marginal expectations are the same:

$$E[g(X_1)|X_2] = E[g(X_1)]. \quad (2.3.39)$$

Eqs. (2.3.34) through (2.3.39) remain valid when restated in terms of *two random vectors* $\mathbf{X}_1 = (X_{1,1}, \ldots, X_{1,M})$ and $\mathbf{X}_2 = (X_{2,1}, \ldots, X_{2,N})$ instead of the two random variables X_1 and X_2.

Important applications of the concept of conditional expectation to *optimal linear prediction* of random fields are described in Sec. 2.6.

Geometric Mean and Related Statistics

Various random quantities expressed as *products* of random variables can be described efficiently in terms of *geometric* means and related statistics. The geometric mean $\overline{X}$ of any non-negative random variable X is

$$\overline{X} \equiv \exp\{E[\ln X]\} = \exp\left\{\int_0^{\infty} \ln x\, f_X(x)\, dx\right\}, \tag{2.3.40}$$

where $f_X(x)$ is the p.d.f. of X. It follows that the geometric mean of a product of random variables, $X = X_1 \ldots X_n$, equals the product of their geometric means, $\overline{X} = \overline{X}_1 \ldots \overline{X}_n$, and hence that the geometric mean of a power of X equals $\overline{X}$ raised to that power.

The degree of variability of X can be quantified in term of its "characteristic value", defined as

$$\widetilde{X} \equiv \overline{X} \exp\left\{\frac{1}{2}\mathrm{Var}[\ln X]\right\}, \tag{2.3.41}$$

which depends on the "logarithmic variance", or the variance of the logarithm, of X:

$$\mathrm{Var}[\ln X] = E[(\ln X)^2] - E^2[\ln X] = 2\ln(\widetilde{X}/\overline{X}), \tag{2.3.42}$$

where the term on the right side is just the inverse of Eq. (2.3.41). Both the logarithmic variance and the "stochasticity factor" $(\widetilde{X}/\overline{X})$, the quotient of the characteristic value $\widetilde{X}$ and the geometric mean $\overline{X}$, measure the spread of the p.d.f. $f_X(x)$ of any non-negative random variable X.

2.4 Characteristic Functions

Characteristic Functions

The expectation of the function $\exp\{iuX\}$ (where $i = \sqrt{-1}$) is by definition the characteristic function of the random variable X:

$$E[e^{iuX}] = \int_{-\infty}^{+\infty} f(x)e^{iux}\, \mathrm{dx} \equiv \varphi_X(u) \equiv \varphi(u). \tag{2.4.1}$$

$\varphi(u)$ is in fact the Fourier transform of the probability density function $f(x)$, a transform that is sure to exist since the integral under $f(x)$ equals one. The inverse Fourier transform is

$$f(x) = \frac{1}{2\pi} \int_{-\infty}^{+\infty} e^{-iux} \varphi(u) \, du. \tag{2.4.2}$$

The definition of the characteristic function of a discrete random variable X is

$$\varphi(u) = \sum_{\text{All } x} p(x) e^{iux} = E[e^{iuX}]. \tag{2.4.3}$$

Note that $\varphi(0) = 1$. Expressing the exponential function as a series,

$$e^{iuX} = 1 + iuX + \frac{1}{2}(iuX)^2 + \cdots, \tag{2.4.4}$$

and taking the expectation on both sides of Eq. (2.4.4) yields:

$$\varphi(u) = E[e^{iuX}] = 1 + iuE[X] + \frac{1}{2}(iu)^2 E[X^2] + \cdots. \tag{2.4.5}$$

This series expansion allows the moments of X to be expressed in terms of the derivatives of the characteristic function evaluated at $u = 0$. In particular, the mean and the mean-square of X are, respectively:

$$m = E[X] = \frac{1}{i} \left[\frac{d\varphi(u)}{du} \right]_{u=0}, \tag{2.4.6}$$

$$E[X^2] = - \left[\frac{d^2 \varphi(u)}{du^2} \right]_{u=0}, \tag{2.4.7}$$

provided these moments exist.

The joint characteristic function of two random variables X_1 and X_2 is defined as the expectation of the function $\exp\{i(u_1 x_1 + u_2 x_2)\}$:

$$\begin{aligned} \varphi(u_1, u_2) &= \int_{-\infty}^{+\infty} \int_{+\infty}^{+\infty} f(x_1, x_2) e^{i(u_1 x_1 + u_2 x_2)} \, dx_1 \, dx_2 \\ &= E[e^{i(u_1 x_1 + u_2 x_2)}], \end{aligned} \tag{2.4.8}$$

which is identical to the two-dimensional Fourier transform of the joint p.d.f. $f(x_1, x_2)$. It is easy to verify the following relations:

$$\begin{cases} \varphi(u_1) = \varphi(u_1, 0), \\ \varphi(u_2) = \varphi(0, u_2). \end{cases} \tag{2.4.9}$$

Also $\varphi(0,0) = 1$, and provided the joint moment exists,

$$E[X_1X_2] = -\left[\frac{\partial^2\varphi(u_1,u_2)}{\partial u_1\partial u_2}\right]_{u_1=u_2=0}. \tag{2.4.10}$$

More generally, the characteristic function of M random variables $X_1,\ldots,$ X_M is

$$\varphi_X(\mathbf{u}) \equiv \varphi(\mathbf{u}) \equiv \varphi(u_1,\ldots,u_M) = E[e^{i(u_1X_1+\cdots+u_MX_M)}], \tag{2.4.11}$$

and the joint moment of order K is

$$E[X_1^{k_1}X_2^{k_2}\cdots X_M^{k_M}] = \left(\frac{1}{i}\right)^K \left[\frac{\partial^K\varphi(u_1,u_2,\ldots,u_M)}{\partial u_1^{k_1}\partial u_2^{k_2}\cdots\partial u_M^{k_M}}\right]_{u_1=\cdots=u_M=0}, \tag{2.4.12}$$

where

$$K = k_1 + k_2 + \cdots + k_M.$$

Characteristic functions provide a simple and elegant approach to finding the moments and the probability distribution of *sums of independent random variables*. Consider the sum of M *independent* and *identically distributed* random variables with (common) characteristic function $\varphi(u)$:

$$Y = X_1 + X_2 + \cdots + X_M = \sum_{k=1}^{M} X_k. \tag{2.4.13}$$

Since the expectation of the product of functions of independent random variables equals the product of the expectations of those functions, we can express the characteristic function of Y as follows:

$$\varphi_Y(u) = E\left[\exp\left\{iu\sum_{k=1}^{M}X_k\right\}\right] = E\left[\prod_{k=1}^{M}\exp\{iuX_k\}\right] = [\varphi(u)]^M. \tag{2.4.14}$$

Cumulant Functions

The multiplicative form of Eq. (2.4.14) motivates the introduction of the *logarithm* of the characteristic function, often called the *cumulant function* (c.f.) or the log-characteristic function:

$$K_X(u) \equiv K(u) = \ln\varphi(u). \tag{2.4.15}$$

Since $\varphi(0) = 1$, $K(0) = 0$. The series expansion of $K(u)$ about $u = 0$ has the form

$$K(u) = \sum_{n=1}^{\infty} \frac{(iu)^n}{n!} K_n, \tag{2.4.16}$$

where

$$K_n = \frac{1}{i^n} \left[\frac{d^n K(u)}{du^n} \right]_{u=0}, \quad n = 1, 2, \ldots. \tag{2.4.17}$$

Direct comparison of the series for $\varphi(u)$ and $K(u)$ yields simple relations between the "cumulants" K_n and the moments of X. The first three cumulants are:

$$\begin{aligned} K_1 &= m, \\ K_2 &= \sigma^2, \\ K_3 &= E[(X - m)^3] = \gamma_1 \sigma^3. \end{aligned} \tag{2.4.18}$$

The cumulants of a sum of *independent* random variables are all equal to the sum of the respective cumulants (hence the name "cumulant function").

The joint cumulant function of two random variables X_1 and X_2 is similarly defined:

$$K(u_1, u_2) = \ln \varphi(u_1, u_2). \tag{2.4.19}$$

Its derivatives are related to the joint central moments, in particular:

$$\begin{aligned} \mathrm{Cov}[X_1, X_2] &= E[(X_1 - m_1)(X_2 - m_2)] \\ &= -\left[\frac{\partial^2 K(u_1, u_2)}{\partial u_1 \partial u_2} \right]_{u_1 = u_2 = 0}. \end{aligned} \tag{2.4.20}$$

The cumulant functions of some common probability distributions are listed in Table 2.1.

2.5 Gaussian and Related Probability Distributions

The Gaussian (Normal) Distribution

The probability density function of a Gaussian (or normal) random variables is

$$f(x) = \frac{1}{\sqrt{2\pi}\sigma} \exp\left\{ -\frac{1}{2} \left(\frac{x - m}{\sigma} \right)^2 \right\}, \quad -\infty \le x \le +\infty. \tag{2.5.1}$$

Table 2.1 Characteristics of some common probability models.

Probability distribution	Probability density function or probability mass function	Cumulant function $K(u) = \ln \varphi(u)$	Mean m	Variance σ^2
Gaussian or normal	$f(x) = \frac{1}{\sqrt{2\pi}\sigma} \exp\left\{-\frac{1}{2}\left(\frac{x-m}{\sigma}\right)^2\right\}$	$ium - \frac{1}{2}u^2\sigma^2$	m	σ^2
Exponential	$f(x) = \lambda e^{-\lambda x}, \quad x \geq 0$	$-\ln\left(1 - \frac{iu}{\lambda}\right)$	$\frac{1}{\lambda}$	$\frac{1}{\lambda^2}$
Gamma	$f(x) = e^{-\lambda x}\frac{(\lambda x)^{n-1}}{(n-1)!}, \quad x \geq 0$	$-n\ln\left(1 - \frac{iu}{\lambda}\right)$	$\frac{n}{\lambda}$	$\frac{n}{\lambda^2}$
Chi-square	$f(x) = \frac{(\frac{x}{2})^{(n/2)-1}e^{-x/2}}{2\Gamma(\frac{n}{2})}, \quad x \geq 0$	$-\frac{n}{2}\ln(1-2iu)$	n	$2n$
Binomial	$p(x) = \binom{n}{x} p^x(1-p)^{n-x}$, $x = 0, 1, \ldots, n$	$n\ln[pe^{iu} + (1-p)]$	np	$np(1-p)$
Poisson	$p(x) = e^{-\lambda t}\frac{(\lambda t)^x}{x!}$, $x = 0, 1, 2, \ldots$	$\lambda(e^{iu} - 1)$	λ	λ

Note:

$$\varphi(u) = \int_{-\infty}^{+\infty} f(x)e^{iux}\,dx, \quad (i = \sqrt{-1}) \qquad m = \frac{1}{i}\left[\frac{dK(u)}{du}\right]_{u=0} \qquad \sigma^2 = -\left[\frac{d^2K(u)}{du^2}\right]_{u=0}.$$

The characteristic function of X is

$$\varphi(u) = E[e^{iuX}] = \exp\left\{ium - \frac{1}{2}u^2\sigma^2\right\}. \tag{2.5.2}$$

The mean and variance are, respectively:

$$m_X = E[X] = m, \tag{2.5.3}$$

$$\sigma_X^2 = E[(X - m)^2] = \sigma^2. \tag{2.5.4}$$

All the odd central moments of X are zero:

$$E[(X - m)^{2k+1}] = 0, \quad k = 0, 1, \ldots, \tag{2.5.5}$$

while the even central moments depend on σ only:

$$E[(X - m)^{2k}] = 1 \cdot 3 \cdot \cdots \cdot (2k - 1)\sigma^{2k}, \quad k = 1, 2, \ldots. \tag{2.5.6}$$

Hence $\gamma_{1,x} = 0$ and $\gamma_{2,x} = 3$. Any p.d.f. for which Eqs. (2.5.5) and (2.5.6) hold is Gaussian.

A Gaussian random variable U is said to be Standard Gaussian (normal) if $m_U = 0$ and $\sigma_U = 1$. The standard normal cumulative distribution function $F_U(u)$ is widely tabulated, thus permitting evaluation of the c.d.f. of any Gaussian random variable by means of the relationship:

$$P[X \leq x] = F_X(x) = F_U\left(\frac{x - m}{\sigma}\right) = P\left[U \leq \frac{x - m}{\sigma}\right]. \tag{2.5.7}$$

The fact that $f_U(u)$ is symmetric about $u = 0$ insures that

$$F_U(-u) = 1 - F_U(u), \tag{2.5.8}$$

and, for $u \geq 0$,

$$P[-u \leq U \leq u] = 1 - 2F_U(-u) = 2F_U(u) - 1. \tag{2.5.9}$$

Interest often focuses on relatively large values of $u = (x - m)/\sigma$. There exists an approximate series (Dwight [52]) useful for evaluating $F_U(u)$ at relatively large values of u, namely:

$$f_U(u) = 1 - \frac{1}{\sqrt{2\pi}u} e^{-u^2/2}\,\Phi(u),$$

where

$$\Phi(u) \simeq 1 - \frac{1}{u^2} + \frac{3}{u^4} - \frac{1 \cdot 3 \cdot 5}{u^6} + \cdots. \tag{2.5.10}$$

Table 2.2 Standard normal cumulative distribution function $F_U(u)$ and the associated function $\Phi(u)$.

u	$F_U(u)$	$\Phi(u)$
0.0	0.5000	0.0000
0.1	0.5398	0.1159
0.2	0.5793	0.2152
0.3	0.6179	0.3006
0.4	0.6554	0.3744
0.5	0.6915	0.4382
0.6	0.7257	0.4940
0.7	0.7580	0.5426
0.8	0.7881	0.5853
0.9	0.8159	0.6228
1.0	0.8413	0.6560
1.1	0.8643	0.6853
1.2	0.8849	0.7114
1.3	0.90320	0.7345
1.4	0.91924	0.7553
1.5	0.93319	0.7739
1.6	0.94520	0.7906
1.7	0.95543	0.8057
1.8	0.96407	0.8193
1.9	0.97128	0.8318
2.0	0.97725	0.8429
2.1	0.98214	0.8529
2.2	0.98610	0.8622
2.3	0.98928	0.8706
2.4	0.99180	0.8789
2.5	0.99379	0.8859
3.0	0.99865	0.9140
3.5	0.999767	0.9347
4.0	0.9999683	0.9476
4.5	0.9999966	0.9574
5.0	0.99999971	0.9751

The error in this series expansion is less than the last term used. The functions $F_U(u)$ and $\Phi(u)$ are presented in Table 2.2 for values of u ranging between 0 and 5.

The Multivariate Gaussian Distribution

A vector $\mathbf{X}$ of M Gaussian (or normal) random variables is characterized by the multivariate normal p.d.f.

$$f_X(\mathbf{x}) = (2\pi)^{-M/2}|\mathbf{B}|^{-1/2}\exp\left\{-\frac{1}{2}(\mathbf{x}-\mathbf{m})^t\mathbf{B}^{-1}(\mathbf{x}-\mathbf{m})\right\}, \qquad (2.5.11)$$

where $\mathbf{m}$ is the vector of mean values and $\mathbf{B}$ is the $M \times M$ covariance matrix

$$\mathbf{B} = [B_{kl}] \equiv [\mathrm{Cov}[X_k, X_l]] \equiv [\rho_{kl}\sigma_k\sigma_l]. \tag{2.5.12}$$

If the argument in the exponential expression is set equal to a constant, c^2, one obtains the equation

$$(\mathbf{x} - \mathbf{m})^t \mathbf{B}^{-1} (\mathbf{x} - \mathbf{m}) = c^2, \tag{2.5.13}$$

which describes ellipsoidal contours of equal probability density. In the one-dimensional case the equation reduces to $(x - m)^2/\sigma^2 = c^2$; a given value of c corresponds to a pair to c-sigma bounds, that is: $x = m \pm c\sigma$. In general, the center of each ellipsoid is the vector $\mathbf{m}$ of mean values; the lengths along the major axes are determined by the eigenvalues of $\mathbf{B}$; and the orientation (directions of axes) of the ellipsoids is determined by the eigenvectors of $\mathbf{B}$. The multivariate Gaussian characteristic function is

$$\varphi(\mathbf{u}) = \exp\left\{ i\mathbf{u}^t\mathbf{m} - \frac{1}{2}\mathbf{u}^t\mathbf{B}\mathbf{u} \right\}. \tag{2.5.14}$$

(If the uncertainty in $\mathbf{X}$ is described by the characteristic function, $\mathbf{B}^{-1}$ need not necessarily exist; the covariance matrix is permitted to be semi-definite.)

Consider specifically the bivariate case ($M = 2$). The vector of random variables and its mean are, respectively,

$$\mathbf{X} = \begin{bmatrix} X_1 \\ X_2 \end{bmatrix}, \quad \mathbf{m} = \begin{bmatrix} m_1 \\ m_2 \end{bmatrix}, \tag{2.5.15}$$

and the covariance matrix is

$$\mathbf{B} = \begin{bmatrix} B_{11} & B_{12} \\ B_{21} & B_{22} \end{bmatrix} = \begin{bmatrix} \sigma_1^2 & \rho\sigma_1\sigma_2 \\ \rho\sigma_1\sigma_2 & \sigma_2^2 \end{bmatrix}. \tag{2.5.16}$$

The determinant of $\mathbf{B}$ equals

$$|\mathbf{B}| = B_{11}B_{22} - B_{12}^2 = \sigma_1^2\sigma_2^2(1 - \rho^2). \tag{2.5.17}$$

The inverse of $\mathbf{B}$ is

$$\mathbf{B}^{-1} = |\mathbf{B}|^{-1} \begin{bmatrix} B_{22} & -B_{12} \\ -B_{12} & B_{11} \end{bmatrix}$$

$$= \frac{1}{\sigma_1^2 \sigma_2^2 (1-\rho^2)} \begin{bmatrix} \sigma_2^2 & -\rho\,\sigma_1\sigma_2 \\ -\rho\,\sigma_1\sigma_2 & \sigma_1^2 \end{bmatrix}. \tag{2.5.18}$$

Combining Eqs. (2.5.11) and (2.5.18) leads to the bivariate normal distribution (for $-\infty \le x_1 \le \infty$ and $-\infty \le x_2 \le \infty$):

$$f(x_1, x_2) = \frac{(1-\rho^2)^{-1/2}}{2\pi\sigma_1\sigma_2} \exp\left\{ \frac{-1}{2(1-\rho^2)} \left[\left(\frac{x_1 - m_1}{\sigma_1}\right)^2 \right.\right.$$
$$\left.\left. - 2\rho \left(\frac{x_1 - m_1}{\sigma_1}\right)\left(\frac{x_2 - m_2}{\sigma_2}\right) + \left(\frac{x_2 - m_2}{\sigma_2}\right)^2 \right] \right\}. \tag{2.5.19}$$

From Eqs. (2.5.14) and (2.5.16) the bivariate characteristic function is

$$\varphi(u_1, u_2) = \exp\left\{ i(u_1 m_1 + u_2 m_2) \right.$$
$$\left. - \frac{1}{2}(u_1^2\sigma_1^2 + u_2^2\sigma_2^2 + 2u_1u_2\rho\sigma_1\sigma_2) \right\}. \tag{2.5.20}$$

Integrating $f(x_1, x_2)$ over all values of x_2 yields the univariate normal distribution $N(m_1, \sigma_1^2)$, while integration over x_1 generates $N(m_2, \sigma_2^2)$. Likewise $\varphi(u_1, 0) \equiv \varphi(u_1)$, and so on [see Eq. (2.4.9)]. If two Gaussian random variables X_1 and X_2 are uncorrelated ($\rho = 0$), they are also independent, that is, $f(x_1, x_2) = f(x_1)f(x_2)$ and $\varphi(u_1, u_2) = \varphi(u_1)\varphi(u_2)$.

Higher-order joint moments can be obtained from the multivariate characteristic function [see Eq. (2.4.12)]. All the central moments of odd order are zero. The central moments of even order can be evaluated using the following procedure proposed by Isserlis [73]. Assume we have $2M$ random variables X_1, X_2, ..., X_{2M}, jointly normally distributed and all having zero mean (for convenience and without loss of generality). The moment of order $2M$ can be expressed as a sum of products of covariances:

$$E[X_1 X_2 \cdots X_{2M}] = \sum E[X_{k_1} X_{k_2}] E[X_{k_3} X_{k_4}] \cdots E[X_{k_{2M-1}} X_{k_{2M}}], \tag{2.5.21}$$

where the summation is over all possible arrangements of the indices 1, 2, ..., $2M$ into exactly M pairs. The number of such arrangements is

$1 \cdot 3 \cdot 5 \cdots (2M-1)$. Eq. (2.5.21) leads to the following result for the fourth-order joint central moment:

$$E[(X_1 - m_1)(X_2 - m_2)(X_3 - m_3)(X_4 - m_4)]$$
$$= B_{12}B_{34} + B_{13}B_{24} + B_{14}B_{23}, \tag{2.5.22}$$

where $B_{kl} = \text{Cov}[X_k, X_l]$. If some of the random variables are identical, this result can be further specialized. For example, we have

$$E[(X_1 - m_1)^4] = 3B_{11}^2 = 3\sigma_1^4, \tag{2.5.23}$$

and

$$E[(X_1 - m_1)^2(X_2 - m_2)^2] = B_{11}B_{22} + 2B_{12}^2. \tag{2.5.24}$$

Conditional Gaussian Distribution

Assume the vector $\mathbf{X}$ of Gaussian variables to be partitioned into two vectors $\mathbf{X}_1$ and $\mathbf{X}_2$ with dimensions M_1 and M_2, respectively. The vector of means $\mathbf{m}$ and the covariance matrix $\mathbf{B}$ may be partitioned accordingly:

$$\mathbf{X} = \begin{bmatrix} \mathbf{X}_1 \\ \hline \mathbf{X}_2 \end{bmatrix}, \qquad \mathbf{m} = \begin{bmatrix} \mathbf{m}_1 \\ \hline \mathbf{m}_2 \end{bmatrix}, \qquad \mathbf{B} = \left[\begin{array}{c|c} \mathbf{B}_{11} & \mathbf{B}_{12} \\ \hline \mathbf{B}_{12}^t & \mathbf{B}_{22} \end{array}\right]. \tag{2.5.25}$$

The conditional probability density function of $\mathbf{X}_1$ given $\mathbf{X}_2$ can be found by substituting the joint p.d.f. $f(\mathbf{x}_1, \mathbf{x}_2) \equiv f(\mathbf{x})$ and the "marginal" p.d.f. $f(\mathbf{x}_2)$ into Bayes' theorem [Eq. (2.2.30)]. The result is the conditional (or posterior) Gaussian p.d.f. of $\mathbf{X}_1$ given $\mathbf{X}_2$:

$$f(x_1|\mathbf{X}_2) = (2\pi)^{-M_1/2}|\mathbf{B}_{11|2}|^{-1/2}$$
$$\times \exp\left\{-\frac{1}{2}(\mathbf{x}_1 - \mathbf{m}_{1|2})^t \mathbf{B}_{11|2}^{-1}(\mathbf{x}_1 - \mathbf{m}_{1|2})\right\}, \tag{2.5.26}$$

where $\mathbf{m}_{1|2} = E[\mathbf{X}_1|\mathbf{X}_2]$ is the conditional (or posterior) mean of $\mathbf{X}_1$ given $\mathbf{X}_2$:

$$\mathbf{m}_{1|2} = \mathbf{m}_1 + \mathbf{B}_{12}\mathbf{B}_{22}^{-1}[\mathbf{X}_2 - \mathbf{m}_2], \tag{2.5.27}$$

and $\mathbf{B}_{11|2}$ is the conditional (posterior) covariance matrix:

$$\mathbf{B}_{11|2} = \mathbf{B}_{11} - \mathbf{B}_{12}\mathbf{B}_{22}^{-1}\mathbf{B}_{12}^t. \tag{2.5.28}$$

If $\mathbf{X}_2$ is viewed as a random vector (prior to observation), then the conditional expectation $\mathbf{m}_{1|2} = E[\mathbf{X}_1|\mathbf{X}_2]$ becomes itself a Gaussian random vector, linearly related to $\mathbf{X}_2$ as expressed by Eq. (2.5.27). The mean of $\mathbf{m}_{1|2}$ then equals the prior mean $\mathbf{m}_1$, and the covariance matrix is $\mathbf{B}_{12}\mathbf{B}_{22}^{-1}\mathbf{B}_{22}(\mathbf{B}_{12}\mathbf{B}_{22}^{-1})^t = \mathbf{B}_{12}\mathbf{B}_{22}^{-1}\mathbf{B}_{12}^t$. If the random variables comprising $\mathbf{X}_2$ are actually observed, the random vector $\mathbf{X}_2$ may be replaced by the vector of observed values $\mathbf{x}_2$ in Eqs. (2.5.26) and (2.5.27). This permits replacement (updating) of the prior p.d.f. $f(\mathbf{x}_1)$ by the posterior p.d.f. $f(\mathbf{x}_1|\mathbf{x}_2)$. A key feature of the updating process is that the posterior covariance matrix $\mathbf{B}_{11|2}$ does *not* depend on the observations and can therefore be evaluated beforehand (enabling assessment of the effectiveness of a data acquisition program in reducing uncertainty).

The Lognormal Distribution

If $X = \ln Y$, the natural logarithm of Y, has a normally distribution $N(m, \sigma^2)$, then the random variable $Y = e^X$ will be lognormally distributed, with probability density function

$$f_Y(y) = \frac{1}{\sqrt{2\pi}y\sigma_X} \exp\left\{-\frac{1}{2}\left(\frac{\ln y - \ln \overline{m}}{\sigma_X}\right)^2\right\}, \quad y \geq 0, \tag{2.5.29}$$

where $\overline{m}$ denotes the *median* (or fiftieth percentile) of Y, and $\sigma_X \equiv \sigma$ is the standard deviation of $\ln Y$. The mean value of Y is [see Eq. (2.5.2)]:

$$E[Y] = E[e^X] = \varphi_X\left(\frac{1}{i}\right) = \exp\left\{m + \frac{\sigma^2}{2}\right\} = \overline{m}\exp\left\{\frac{\sigma^2}{2}\right\}. \tag{2.5.30}$$

The higher-order moments of Y can also be found from the characteristic function of X:

$$E[Y^k] = E[e^{kX}] = \varphi_X\left(\frac{k}{i}\right) = \exp\left\{km + \frac{1}{2}k^2\sigma^2\right\}, \tag{2.5.31}$$

where m and σ^2 represent the mean and variance of $\ln Y$, respectively. The variance of Y equals

$$\sigma_Y^2 = [\exp\{\sigma^2\} - 1]\exp\{2m + \sigma^2\}. \tag{2.5.32}$$

The square of the coefficient of variation of Y may be expressed in terms of $\sigma_X = \sigma$, the standard deviation of $\ln Y$, as follows:

$$V_Y^2 = \left(\frac{\sigma_Y}{m_Y}\right)^2 = \frac{E[Y^2] - E^2[Y]}{E^2[Y]} = \exp\{\sigma^2\} - 1. \tag{2.5.33}$$

The inverse relation expresses $\sigma^2 = \sigma^2_{\ln Y}$ in terms of V_Y:

$$\sigma^2 = \sigma^2_{\ln Y} = \ln(1 + V_Y^2). \tag{2.5.34}$$

Note that $\sigma_{\ln Y} \simeq V_Y$ when the dispersion of Y is small. The coefficient of skewness [Eq. (2.3.6)] equals $\gamma_{1,Y} = 3V_Y + V_Y^3$. Observe also that $\overline{Y}$, the geometric mean Y, equals the median $\overline{m}$; the characteristic value $\widetilde{Y}$ equals the arithmetic mean $E[Y]$; and the stochasticity factor $\widetilde{Y}/Y$ equals $1 + V_Y^2$. For related definitions, see Eqs. (2.3.40) - (2.3.42).

The cumulative distribution function of Y can be evaluated in terms of the standard normal c.d.f.:

$$F_Y(y) = P[Y \leq y] = P[X \leq \ln y] = F_U\left(\frac{\ln y - m}{\sigma}\right). \tag{2.5.35}$$

If two Gaussian random variables X_1 and X_2 (distributed as $N(m_1, \sigma_1^2)$ and $N(m_2, \sigma_2^2)$, respectively, and with coefficient of correlation ρ_{12}) are both transformed into lognormal random variables, $Y_1 = e^{X_1}$ and $Y_2 = e^{X_2}$, one can obtain the joint moment of second order:

$$\begin{aligned} E[Y_1 Y_2] &= E[e^{X_1+X_2}] = \varphi_X\left(\frac{1}{i}, \frac{1}{i}\right) \\ &= \exp\left\{m_1 + m_2 + \frac{1}{2}(\sigma_1^2 + \sigma_2^2 + 2\rho_{12}\sigma_1\sigma_2)\right\}, \end{aligned} \tag{2.5.36}$$

and, from Eqs. (2.5.32) and (2.5.36), the coefficient of correlation between Y_1 and Y_2 may be expressed as:

$$\rho_{Y_1,Y_2} = \frac{\mathrm{Cov}[Y_1, Y_2]}{\sigma_{Y_1}\sigma_{Y_2}} = \frac{\exp\{\rho_{12}\sigma_1\sigma_2\} - 1}{[\exp\{\sigma_1^2\} - 1]^{1/2}[\exp\{\sigma_2^2\} - 1]^{1/2}}. \tag{2.5.37}$$

If $X(\mathbf{t})$ is a homogeneous Gaussian random field with mean m, variance σ^2, and correlation function $\rho(\boldsymbol{\tau})$, the derived "lognormal" random field $Y(\mathbf{t}) = \exp\{X(\mathbf{t})\}$ will have mean and variance as given by Eqs. (2.5.30) and (2.5.32), respectively, and [based on Eq. (2.5.37)] its correlation function will have the following form:

$$\rho_Y(\boldsymbol{\tau}) = \frac{\exp\{\rho(\boldsymbol{\tau})\sigma^2\} - 1}{\exp\{\sigma^2\} - 1}, \tag{2.5.38}$$

enabling second-order analysis of homogeneous *lognormal* random fields.

Certain power law transformations of Gaussian random fields can be analyzed by means of Eq. (2.5.21). For example, it can be used to obtain the covariance function of the derived random field $Y(\mathbf{t}) = X^2(\mathbf{t})$, where

$X(\mathbf{t})$ is a Gaussian field with known covariance function. A similar method of analysis is employed to obtain the statistics of the *envelope* (and the partial derivatives of the envelope) of homogeneous Gaussian random fields in Sec. 4.3. The marginal probability distribution of this envelope is the Rayleigh distribution, given below.

Distributions of Random Distances

If two independent random variables X_1 and X_2 are normally distributed with zero mean and common variance σ^2, the derived random variable

$$R = [X_1^2 + X_2^2]^{1/2} \tag{2.5.39}$$

will follow a Rayleigh distribution with p.d.f.

$$f(r) = \frac{r}{\sigma^2} \exp\left\{-\frac{1}{2}\left(\frac{r}{\sigma}\right)^2\right\}, \quad r \geq 0. \tag{2.5.40}$$

The mean of R is $E[R] = \sqrt{\pi/2}\sigma$, and its mean square is $E[R^2] = 2\sigma^2$. In the analysis of random trajectories in two-dimensional space, the Rayleigh distribution characterizes the vector distance from the point of origin when the movements X_1 and X_2 along two orthogonal coordinates are Gaussian with mean zero and common variance σ^2.

The travel distance from the point of origin in three-dimensional space is expressed by

$$R = [X_1^2 + X_2^2 + X_3^2]^{1/2}. \tag{2.5.41}$$

If the three components X_i, $i = 1, 2, 3$, are independent and identically distributed Gaussian random variables, $N(0, \sigma^2)$, then the probability density function of the distance R is

$$f(r) = \frac{1}{(2\pi)^{3/2}\sigma^3} \exp\left\{-\frac{1}{2}\left(\frac{r}{\sigma}\right)^2\right\}, \quad r \geq 0. \tag{2.5.42}$$

The mean square of R may be obtained directly from Eq. (2.5.41); it equals $E[R^2] = 3\sigma^2$. A simple extension allows the Gaussian directional components to have different means and variances (see Chandrasekhar [29]).

The Central Limit Theorem

It is well known that the probability distribution of the sum $Y = \sum X_k$ of M independent, identically distributed random variables tends to become

Gaussian when the number M increases. In the limit when $M \to \infty$, the p.d.f. of the sum becomes Gaussian; this implies that its cumulant function converges to

$$K_Y(u) = ium_Y - \frac{1}{2}u^2\sigma_Y^2, \tag{2.5.43}$$

where m_Y and σ_Y^2 are, respectively, the mean and the variance of the sum $Y = \sum X_k$. The classical proof of (this weak statement of) the Central Limit Theorem relies on the fact that a unique relationship exists between the p.d.f. $f_Y(y)$ and the cumulant function $K_Y(u)$, and that the cumulants of Y of third and higher order become negligibly small (in relation to the first two cumulants, $K_1 = m_Y$ and $K_2 = \sigma_Y^2$) when $M \to \infty$. (See, for example, Rice [113]).

If the independent random variables $X_1, \ldots, X_M$ are *not identically distributed,* convergence of the p.d.f. of their sum $Y = \sum X_k$ toward the Gaussian distribution (when $M \to \infty$) is subject to an additional requirement: the fractional contribution of any one of the random variables to the total variance σ_Y^2 must vanish when $M \to \infty$.

$$\frac{\sigma_k^2}{\sigma_Y^2} = \frac{\sigma_k^2}{\sum_{k=1}^{M} \sigma_k^2} \to 0, \quad \text{for any } k, \text{ for } M \to \infty. \tag{2.5.44}$$

This requirement is implied by the "Lindeberg condition" which is necessary and sufficient for the sum of independent random variables to converge to the Gaussian distribution (Pugachev [109]).

A very similar analysis enables one to show that the bivariate p.d.f. $f_{I_1 I_2}(i_1, i_2)$ of two random variables $I_1 = I(\mathbf{t}_1)$ and $I_2 = I(\mathbf{t}_2)$ obtained by *local integration* of an *uncorrelated* random field $X(\mathbf{t})$ will tend toward the bivariate Gaussian p.d.f. (regardless of the p.d.f. of $X(\mathbf{t})$). Likewise, higher-order probability density functions, $f_{I_1 I_2 I_3}(i_1, i_2, i_3)$, and so on, will tend toward the multivariate Gaussian p.d.f. By the same argument, the derived random field $I(\mathbf{t})$ obtained by *local integration* of an uncorrelated random field $X(\mathbf{t})$ will tend toward a Gaussian field as the size of the aggregation domain increases; and the same may be said about random fields obtained by local averaging.

The Central Limit Theorem does not require complete statistical independence between the random variables being summed. The key requirements are that there must be an aggregation of *many weakly correlated* random effects and that no single effect (or small subset of effects) accounts for a dominant fraction of the total variance. These same requirements govern convergence toward normality for a random field $I(\mathbf{t})$ derived by locally aggregating a weakly correlated field $X(\mathbf{t})$. The size of the domain D over which $X(\mathbf{t})$ is aggregated must be sufficiently large compared to the correlation distance(s) of $X(\mathbf{t})$. Also, if the field $X(\mathbf{t})$ is homogeneous, contributions to the variance of $I(\mathbf{t})$ are guaranteed to be evenly distributed and condition (2.5.44) will be satisfied in the limit when $D \to \infty$.

2.6 Optimal Linear Prediction and Updating

Classical Estimation Problems

A classical *estimation problem* arises when a random field with known correlation structure is observed at one or more specific locations with the aim of learning more about the details of the field in the neighborhood of the observation points. The new information may then be put to use, where appropriate, in updating, prediction, and decision analysis about data acquisition or data processing.

This kind of estimation problem is known under different names depending on what kind of information is given or sought. In the writings of Kolmogorov [79], Yaglom [158] and Wiener [153] on optimal linear estimation the following types of problems are posed.

The Problem of Interpolation

Observations of a stationary random process $X(t)$ are made at equal time intervals Δt and the data so obtained are used to predict, or rather, reconstruct, the complete "time history" (see Fig. 2.9). Linear prediction theory is concerned with providing the coefficients a_k in the equation:

$$E[X(t)] = \sum_{k=-\infty}^{\infty} a_k x(k\Delta t), \tag{2.6.1}$$

where $x(k\Delta t)$ is an observation of the random process at time $k\Delta t$. The objective is to minimize the mean square interpolation error, or equivalently, minimize the "posterior" or updated variance of $X(t)$.

This classical problem statement may be extended in various ways. The number of observation points may be finite and need not be regularly spaced. Also the random process may depend on more than one coordinate. For example, in the spatial two-dimensional situation illustrated in Fig. 2.10, the problem is to obtain a "best" estimate of the random prop-

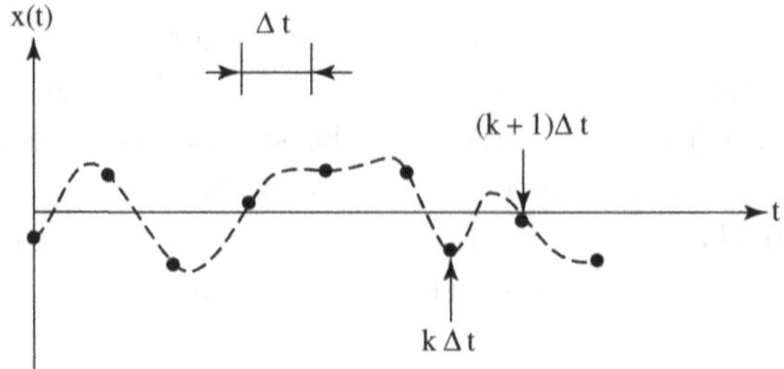

Fig. 2.9 Interpolation of a random function (or a time history).

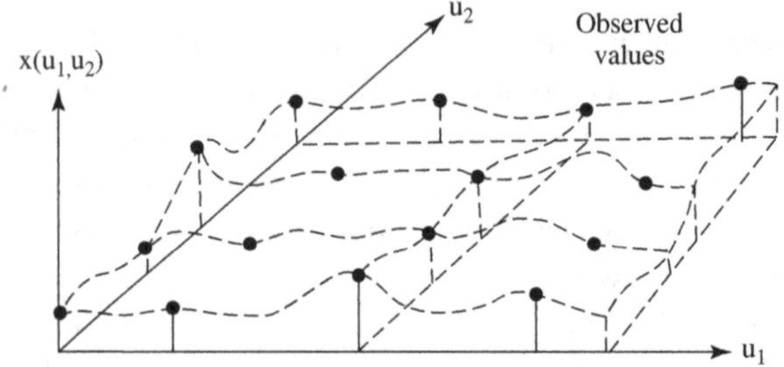

Fig. 2.10 Interpolation of a two-dimensional random field.

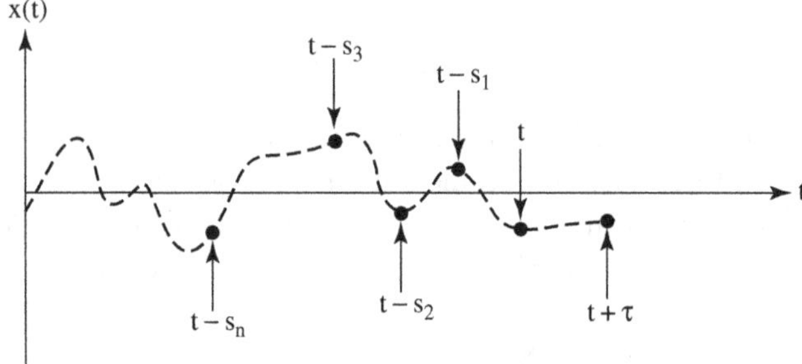

Fig. 2.11 Extrapolation of a random function (or a time history).

erty X at each of a set of locations $\mathbf{u}_j$ ($j = 1, \ldots, M$) by linearly combining the observations of X at a number of other locations $\mathbf{u}_k$ ($k = 1, \ldots, N$). The purpose is to find the coefficients a_{kj} in the prediction equations

$$E[X(\mathbf{u}_j)] = a_{0j} + \sum_{k=1}^{N} a_{kj} x(\mathbf{u}_k), \quad j = 1, \ldots, M, \tag{2.6.2}$$

and estimate the "posterior" variances and covariances of $X(\mathbf{u}_j)$, $j = 1, \ldots, M$. This problem is referred to as "kriging" in the literature on mathematical geology (Krige [80], Matheron [95]). More generally, the quantities to be observed or estimated need not be "point values" but may be, for example, local averages or "concentrations."

The Problem of Extrapolation

Given a number of *past* error-free observations, $x(t - s_1)$, $x(t - s_2)$, $\ldots$, $x(t - s_N)$, as illustrated in Fig. 2.11, make a "best" prediction of future values $x(t + \tau), \tau \geq 0$, using the following linear prediction equation:

$$E[X(t+\tau)] = \sum_{k=1}^{N} a_k x(t - s_k), \tag{2.6.3}$$

and evaluate the posterior variance of $X(t + \tau)$.

This problem statement can be generalized when the random process under study depends not only on time but also on one or more spatial coordinates: a space-time process is observed at a limited number of locations for a limited period of time. Based on this information, the analyst seeks to predict certain quantities that depend on the future behavior of the process (such as spatial averages or concentrations, space-time averages, or values at specific locations and times).

The Problem of Filtering

The random process $X(t)$ has been measured indirectly (for example, by a noisy measuring device), and one seeks to predict $X(t)$ from the measurements $z(t)$. The objective of the "linear filtering problem" is to provide the "best" coefficients a_k for equations such as

$$E[X(t)] = \sum_{k=-\infty}^{\infty} a_k\, z(k\Delta t), \tag{2.6.4}$$

and evaluate the variance of the prediction error. Eq. (2.6.4) may be viewed as an extension of Eq. (2.6.1). Similar extensions are possible for other estimation problems; for example, the observations $x(\mathbf{u}_k)$ in Eq. (2.6.2) may be replaced by the *measurements* $z(\mathbf{u}_k)$ in case $z(u_k) \neq x(u_k)$, enabling accounting for errors in measurement.

Bivariate Linear Estimation

Consider two random variables Y_1 and Y_2. The statistics of Y_1 are to be updated based on the observation $Y_2 = y_2$. The *prior* means and variances of Y_1 and Y_2 are, respectively, m_1, m_2, σ_1^2, and σ_2^2. For example, Y_2 may be the value of a random field $X(\mathbf{t})$ at an observation point $\mathbf{t}_2$, and Y_1 may be the local average of $X(\mathbf{t})$ within a region surrounding location $\mathbf{t}_1$.

Interest focuses on the form of the "prediction equation", expressing the dependence of the *conditional expectation* of Y_1 on the observation Y_2. We may write

$$\hat{Y}_1 = E[Y_1|Y_2 = y_2] = g(y_2). \tag{2.6.5}$$

Before the observation is made, the outcome of the prediction is unknown, and the prediction equation may be viewed as a function of the random variable Y_2:

$$\hat{Y}_1 = E[Y_1|Y_2] = g(Y_2). \tag{2.6.6}$$

In *linear* estimation theory the functional form of the prediction equation is linear, as follows:

$$\hat{Y}_1 = E[Y_1|Y_2] = aY_2 + b, \tag{2.6.7}$$

and the *prediction error* is the difference between the actual value and the prediction:

$$Y_1 - \hat{Y}_t = Y_1 - aY_2 - b. \tag{2.6.8}$$

Its mean should be zero (indicating lack of bias) and its variance should be as small as possible. These conditions provide two equations from which the two unknowns a and b can be obtained. The first equation is

$$E[Y_1 - \hat{Y}_1] = m_1 - am_2 - b = 0, \tag{2.6.9}$$

yielding

$$b = m_1 - am_2. \tag{2.6.10}$$

The "posterior variance" or the mean-square prediction error may be calculated as follows:

$$\begin{aligned}\mathrm{Var}[Y_1|Y_2] &= E[(Y_1 - \hat{Y}_1)^2] = \mathrm{Var}[aY_2 - Y_1]\\ &= a^2\sigma_2^2 + \sigma_1^2 - 2aB_{12},\end{aligned} \tag{2.6.11}$$

where $B_{12} = \text{Cov}[Y_1, Y_2]$. The value of a minimizing the posterior variance is found by setting its derivative with respect to a equal to zero:

$$\frac{d}{da}\text{Var}[Y_1|Y_2] = 2a\sigma_2^2 - 2B_{12} = 0. \tag{2.6.12}$$

The resulting "optimal" value of a, denoted by a_0, is

$$a_0 = \frac{B_{12}}{\sigma_2^2} = \frac{\rho_{12}\sigma_1}{\sigma_2}, \tag{2.6.13}$$

where $\rho_{12} = B_{12}/\sigma_2\sigma_1$ is the coefficient of correlation between Y_2 and Y_1. Combining Eqs. (2.6.7), (2.6.10) and (2.6.13) yields the optimal linear prediction equation

$$\begin{aligned}\hat{Y}_1 &= E[Y_1|Y_2] = m_1 + \frac{B_{12}}{\sigma_2^2}(Y_2 - m_2)\\ &= m_1 + \rho_{12}\sigma_1\left(\frac{Y_2 - m_2}{\sigma_2}\right).\end{aligned} \tag{2.6.14}$$

The associated minimum posterior variance of Y_1 is

$$\text{Var}[Y_1|Y_2] = \sigma_1^2 - a_0^2\sigma_2^2 = \sigma_1^2 - \frac{B_{12}^2}{\sigma_2^2} = \sigma_1^2(1 - \rho^2{}_{12}). \tag{2.6.15}$$

If Y_1 and Y_2 are the values of a homogeneous random field $Y(\mathbf{t})$ with mean m, variance σ^2, and correlation function $\rho(\boldsymbol{\tau})$, Eq. (2.6.14) becomes:

$$\hat{Y}_1 = E[Y_1|Y_2] = (Y_2 - m)\rho(\boldsymbol{\tau}) + m, \tag{2.6.16}$$

where $\boldsymbol{\tau}$ is the vector distance between the locations of Y_1 and Y_2. The associated posterior variance [Eq. (2.6.15)] is

$$\text{Var}[Y_1|Y_2] = \sigma^2[1 - \rho^2(\boldsymbol{\tau})]. \tag{2.6.17}$$

If $\rho(\boldsymbol{\tau}) = 0$, the prior and posterior statistics of Y_1 are the same. If $\rho(\boldsymbol{\tau}) \to 1$, the posterior variance vanishes, and the "prediction" of Y_1 (the conditional mean of Y_1 given $Y_2 = y_2$) converges toward the observed value y_2, as expected.

Multivariate Linear Estimation

Much the same procedure can be used to determine the coefficients of the prediction equation, as well as the expression for the posterior variance, in a more general case when the vector $\mathbf{Y}_1$ comprising M_1 random variables is to be *estimated* based on the observation of the vector $\mathbf{Y}_2$ comprising M_2 random variables. In a random field context, the vectors $\mathbf{Y}_1$ and $\mathbf{Y}_2$ may represent "point" values or local averages of a random field $X(\mathbf{t})$ at different locations. Let $\mathbf{m}_1$ and $\mathbf{m}_2$ denote the mean values of $\mathbf{Y}_1$ and $\mathbf{Y}_2$, respectively. The covariance matrix of all random variables contained in $\mathbf{Y}_1$ and $\mathbf{Y}_2$ may be partitioned as follows:

$$\mathbf{B_Y} = \left[\begin{array}{c|c} \mathbf{B}_{11} & \mathbf{B}_{12} \\ \hline \mathbf{B}_{12}^t & \mathbf{B}_{22} \end{array}\right], \tag{2.6.18}$$

where $\mathbf{B}_{11} \equiv \mathbf{B}_{\mathbf{Y}_1}, \mathbf{B}_{22} \equiv \mathbf{B}_{\mathbf{Y}_2}$, and $\mathbf{B}_{12} \equiv \mathbf{B}_{\mathbf{Y}_1,\mathbf{Y}_2}$. The set of linear prediction equations may be expressed in matrix form:

$$\hat{\mathbf{Y}}_1 = E[\mathbf{Y}_1|\mathbf{Y}_2 = \mathbf{y}_2] = \mathbf{b} + \mathbf{A}\mathbf{y}_2. \tag{2.6.19}$$

A priori, before observations are made, the outcomes of the predictions are random variables, namely linear combinations of the random vector $\mathbf{Y}_2$:

$$\hat{\mathbf{Y}}_1 = E[\mathbf{Y}_1|\mathbf{Y}_2] = \mathbf{b} + \mathbf{A}\mathbf{Y}_2. \tag{2.6.20}$$

The predictions will be bias-free if

$$E[\mathbf{Y}_1 - \hat{\mathbf{Y}}_1] = \mathbf{m}_1 - \mathbf{b} - \mathbf{A}\mathbf{m}_2 = \mathbf{0}. \tag{2.6.21}$$

The posterior covariance matrix of $\mathbf{Y}_1$ given $\mathbf{Y}_2$ is

$$\begin{aligned} \mathbf{B}_{11|2} \equiv \mathbf{B}_{\mathbf{Y}_1|\mathbf{Y}_2} &= E[(\mathbf{Y}_1 - \hat{\mathbf{Y}}_1)(\mathbf{Y}_1 - \hat{\mathbf{Y}}_1)^t] \\ &= \mathbf{A}\mathbf{B}_{22}\mathbf{A}^t + \mathbf{B}_{11} - \mathbf{A}\mathbf{B}_{12}^t - \mathbf{B}_{12}\mathbf{A}^t. \end{aligned} \tag{2.6.22}$$

The value of $\mathbf{A}$ that minimizes the posterior variances is again found by differentiation. The matrix of optimal coefficients is $\mathbf{A} \equiv \mathbf{A}_0$, where

$$\mathbf{A}_0 = \mathbf{B}_{12}\mathbf{B}_{22}^{-1}. \tag{2.6.23}$$

Combining Eqs. (2.6.20), (2.6.21) and (2.6.23) yields the optimal prediction equation:

$$\hat{\mathbf{Y}}_1 = E[\mathbf{Y}_1|\mathbf{Y}_2 = \mathbf{y}_2] = \mathbf{m}_1 + \mathbf{B}_{12}\mathbf{B}_{22}^{-1}[\mathbf{y}_2 - \mathbf{m}_2]. \tag{2.6.24}$$

The associated posterior covariance matrix of $\mathbf{Y}_1$ given $\mathbf{Y}_2$, is obtained by inserting Eq. (2.6.23) into (2.6.22):

$$\mathbf{B}_{11|2} = \mathbf{B}_{11} - \mathbf{A}_0\mathbf{B}_{22}\mathbf{A}_0^t = \mathbf{B}_{11} - \mathbf{B}_{12}\mathbf{B}_{22}^{-1}\mathbf{B}_{12}^t. \tag{2.6.25}$$

Note that Eqs. (2.6.19) through (2.6.25) are very similar in form to the equations for the two-variable case [Eqs. (2.6.7) through (2.6.15)]. The minimum posterior variance is always less than or equal to the prior variance. Also, the posterior covariance matrix does not depend on the observations, so that the degree of variance reduction to be gained from a measurement program can be assessed in advance. The methodology thus provides a powerful tool for evaluating alternative measurement strategies (e.g., where to best locate instruments) in the context of random fields.

If some system response measure R is functionally related to the random vectors $\mathbf{Y}_1$ and $\mathbf{Y}_2$, the prior and posterior variances of R can also be evaluated by means of the analysis procedures discussed in Sec. 2.3. The solution method will be *exact* if the relationship between R and $\mathbf{Y}$ happens to be linear, and *approximate* (for instance, based on Taylor series expansion about the mean values) when the relationship is nonlinear.

Prediction and updating of random fields are constant themes throughout the book. In the later chapters (5 through 7) the theory just presented is considerably enriched and rendered more practical by the development of more efficient procedures to describe the correlation structure of random fields and evaluate the covariances between the random variables comprising the vectors $\mathbf{Y}_1$ and $\mathbf{Y}_2$. The problem of how best to choose instrument types, locations or sampling intervals to reduce uncertainty about a random field can also be formally addressed within this framework.

2.7 Purely Random Fields and Markov Processes

This section is devoted to a review of some common random process models based, directly or indirectly, on the assumption of statistical *independence*. The purposes of this review are: to illustrate how to construct simple random field models of real-life phenomena; to show how different models of the same random phenomenon interrelate; to derive a number of common probability distributions in the context of random fields; and last but not least, to demonstrate the role of, and the need for, the methodological developments advanced in subsequent chapters of this book.

Purely Random Fields

Purely Random Series and Lattice Processes

The simplest "purely random process" is a random series $X(t)$ whose values are *independent* and *identically distributed* random variables. Assume that the observation points are equally spaced on the time axis. This random process is completely characterized by the "first-order" probability density function $f(x)$. (No distinction will be made between continuous-state and discrete-state series; we assume the p.d.f. may contain Dirac delta functions representing discrete probability masses.) The joint probability density function of the random variables $X_k \equiv X(k\Delta t)$, $k = 0, 1, \ldots$, may be expressed as the product of marginal probability density functions such as

$$f(x_0, x_1, \ldots, x_M) = f(x_0)f(x_1)\cdots f(x_M). \tag{2.7.1}$$

Since the values of $X(t)$ at different times are independent, the covariance of $X(t)$ consists of a "delta function":

$$B(t,t') = B(\tau) = \sigma^2\delta(\tau) = \begin{cases} \sigma^2, & \tau = 0, \\ 0, & \tau \neq 0, \end{cases} \tag{2.7.2}$$

in which σ^2 is the variance of $X(t)$ and $\tau = t - t'$ is the time lag. The correlation function is

$$\rho(\tau) = \frac{B(\tau)}{\sigma^2} = \delta(\tau) = \begin{cases} 1, & \tau = 0, \\ 0, & \tau \neq 0. \end{cases} \tag{2.7.3}$$

Homogeneous purely random *lattice processes* have similar properties. For a random process $X(u_1, u_2)$ on the (u_1, u_2) plane, the covariance function is

$$B(\nu_1, \nu_2) = \sigma^2\rho(\nu_1)\rho(\nu_2) = \sigma^2\delta(\nu_1)\delta(\nu_2), \tag{2.7.4}$$

where $\nu_1 = (u_1 - u_2')$ and $\nu_2 = (u_2 - u_2')$ denote the separation distances along the respective coordinate axes.

Extension to processes with more than two parameters is straightforward. For example, if the values at each nodal point of the lattice in Fig. 2.1(b) vary randomly and independently with time and location, we then have a purely random space-time lattice process $X(u_1, u_2, t)$. Such a random process could serve as a simple probabilistic model for the (discrete-parameter) observations of a (continuous-parameter) space-time process, for example, crop yield or rainfall depth sampled at discrete geographic locations on a yearly basis.

Purely random fields owe their simplicity to the fact that the *location* of the observations in the parameter space need not be considered in the analysis. Information about the field at one location does not influence probability assessments about nearby values. The independence assumption is particularly attractive (albeit by no means always justified) when random fields depend upon parameters defined on a *nominal scale* (i.e., when the parameter implies no intrinsic ordering or value). In fact the techniques of classical statistics are largely based on the assumption that observations are independent and identically distributed, and consequently, in a random field context, that it is unnecessary to retain information about where in the parameter space the observations were made. Most of the work on statistics of extreme values, in particular, relates to sets of identical, independent random variables (see Sec. 2.2).

Two-Valued Independent Lattice Processes

By introducing a fixed threshold level $X = x$, a random series $X(t)$ gives rise to a two-valued (binary) independent random series

$$Z(t) = \begin{cases} 0, & \text{if } X(t) \leq x, \\ 1, & \text{if } X(t) > x. \end{cases} \tag{2.7.5}$$

If t is fixed, $Z(t)$ is a Bernoulli random variable with mean $m_Z = p = P[X(t) > x] = 1 - F_X(x)$, and variance $\sigma_Z^2 = p(1-p)$. The correlation function of $Z(t)$ is a Dirac delta function, $p_Z(\tau) = \delta(\tau)$.

Likewise, the purely random lattice process $X(u_1, u_2)$ generates a family of two-valued lattice processes $Z(u_1, u_2)$ whose characteristics depend on $p = P[X(u_1, u_2) > x]$. In particular, the mean of $Z(t_1, t_2)$ is $m_Z = p$, and its covariance is

$$B_Z(\nu_1, \nu_2) = \begin{cases} p(1-p), & \nu_1 = \nu_2 = 0, \\ 0, & \nu_1 \text{ or } \nu_2 \neq 0. \end{cases} \tag{2.7.6}$$

Extension to lattice fields of higher dimension is straightforward.

White Noise in One or More Dimensions

The purely random series $X(t)$ (observed at regular intervals 0, Δt, $2\Delta t$, ...) can be derived from ideal white noise, a *continuous parameter* random process, by local averaging within successive non-overlapping time intervals Δt. By letting the interval Δt become very small, the ideal white noise process is approached. The limiting operation is delicate, however, since it involves going from a discrete- to a continuous-parameter process.

Consider a second random series $X'(t)$ observed n times as often as $X(t)$, the corresponding averaging interval being $\Delta t' = \Delta t/n$. The series $X(t)$ may then also be seen as the *average* of n consecutive values of the (more rapidly sampled) series $X'(t)$. The means and variances of the two series are related as follows:

$$m = m', \qquad \text{and} \qquad \sigma^2 = \frac{1}{n}(\sigma')^2. \tag{2.7.7}$$

Inserting $n = \Delta t/\Delta t'$, the right side of Eq. (2.7.7) yields

$$\sigma^2 \Delta t = (\sigma')^2 \Delta t' \equiv w. \tag{2.7.8}$$

Eq. (2.7.8) implies that there is something arbitrary or indefinite about the variance of any purely random series or white noise model. One can obtain essentially any value for the point variance σ^2 depending upon the choice of the "digitization" interval Δt, yet this choice is arbitrary. Equally significant, the product $\sigma^2 \Delta t$ remains "invariant" under the operation of local averaging. This invariant, denoted here by w, is a measure of the "white noise intensity". To realize ideal white noise, the interval Δt must go to zero, and consequently the associated variance σ^2 must go to infinity. This renders ideal white noise, whose correlation function is given by Eq. (2.7.3), physically unrealizable.

In a two-dimensional parameter space, the same kind of limiting operation can be applied to the purely random lattice process $X(u_1, u_2)$ mentioned earlier. The equivalent of Eq. (2.7.8) is

$$\sigma^2 \Delta u_1 \Delta u_2 = (\sigma')^2 \Delta u_1' \Delta u_2' \equiv w, \tag{2.7.9}$$

where the prime superscript refers to a second purely random 2-D lattice process $X'(u_1, u_2)$, and w is by definition the white noise intensity. The 2-D ideal white noise field (obtained by letting Δu_1 and Δu_2 vanish) has the correlation function

$$\rho(\nu_1, \nu_2) = \delta(\nu_1)\delta(\nu_2) = \begin{cases} 1, & \nu_1 = \nu_2 = 0, \\ 0, & \nu_1 \text{ or } \nu_2 \neq 0, \end{cases} \tag{2.7.10}$$

where $\delta(\nu_i)$, for $i = 1$ or 2, is a Dirac delta function. In particular, if time is added as a third parameter, we obtain a three-dimensional "rain-on-the-roof" white noise process $X(u_1, u_2, t)$, considered by Crandall [43].

Random Fields Involving Sums of Independent Increments

When consecutive observed values of an independent random process accumulate, the resulting sum or integral becomes a process with independent increments. The types of processes introduced earlier in this section are now examined with respect to their behavior under (local) aggregation.

Sums of Purely Random Series and Lattice Processes

Consider a stationary purely random series $X(t)$, where $t = 0, \Delta t, 2\Delta t, \ldots$, with mean m and variance σ^2. It is convenient to make $X(t)$ an integer-parameter series by taking $\Delta t = 1$. By summing the values of the series $X(t)$, a nonstationary process $I(t)$ is obtained:

$$I(t) = \sum_{v=0}^{t} X(v). \tag{2.7.11}$$

The relationship between the means of $I(t)$ and $X(t)$ is

$$m_I = mt, \quad t = k\,\Delta t,\ k = 0, 1, 2, \ldots. \tag{2.7.12}$$

The lack of correlation between consecutive values of $X(t)$ ensures that an equally simple relationship links the respective variances:

$$\sigma_I^2 = \sigma^2 t, \quad t = k\,\Delta t,\ k = 0, 1, 2, \ldots. \tag{2.7.13}$$

The covariance of $I(t)$ may be evaluated by means of Eq. (2.3.32) (in which all cross-terms vanish, and $a_k = b_l = 1$). We obtain $B_I(t, t') = \sigma^2 \min(t, t')$, and the correlation function of $I(t)$ is

$$\rho_I(t, t') = \frac{B_I(t, t')}{\sigma_{I(t)}\sigma_{I(t')}} = \frac{\min(t, t')}{\sqrt{t\,t'}} = \begin{cases} \sqrt{\dfrac{t}{t'}}, & \text{if } t \le t', \\[2ex] \sqrt{\dfrac{t'}{t}}, & \text{if } t \ge t'. \end{cases} \tag{2.7.14}$$

The process $I(t)$ has *independent increments.* Assuming $t' > t$, the fact that the increment $I(t') - I(t)$ is independent of $I(t)$ follows directly from the definition [Eq. (2.7.11)]. The increment $I(t') - I(t)$ has the same probability distribution as $I(t'-t) - I(0) = I(t'-t)$; its mean is $(t'-t)m$ and its variance $(t'-t)\sigma^2$. Evidently, only the size of the time interval matters, not its location on the time axis.

Closely related to the "integral process" $I(t)$, defined by Eq. (2.7.12), is the "process of averages"

$$Y(t) = \frac{1}{t}I(t) = \frac{1}{t}\sum_{v=0}^{t} X(v). \tag{2.7.15}$$

It has the same correlation function as $I(t)$ [see Eq. (2.7.14)], but its mean and variance differ:

$$m_Y = m, \qquad \sigma_Y^2 = \frac{\sigma^2}{t}. \tag{2.7.16}$$

From the probability density function $f_X(x)$, the p.d.f. of the sum I or the average Y may be obtained by repeated application of the convolution theorem [Eq. (2.2.44)]. From the Central Limit Theorem we know that $f_I(i)$ and $f_Y(y)$ will tend toward the Gaussian p.d.f. [Eq. (2.5.1)] as the number of contributing independent random variables grows large. Similar results are easily obtained for sum and average processes derived from homogeneous independent lattice processes.

Discrete-Parameter Counting Process

Recall that crossings above a fixed threshold x_0 by a stationary integer-parameter random series $X(t)$ give rise to a binary random series $Z(t)$ which is *one* if $X(t) > x_0$ and *zero* if $X(t) \leq x_0$. The total number of crossings during the interval $(0, t)$ is a *counting process* $I(t)$ defined by

$$I(t) = \sum_{k=0}^{t} Z(k), \quad t = 0, 1, 2, \ldots. \tag{2.7.17}$$

$Z(t)$ is a Bernoulli random variable [as defined by Eq. (2.7.5)] whose parameter is the probability $p = P[X(t) > x_0]$. From elementary probability theory we know that $I(t)$ has a binomial probability mass function:

$$p_I(i) = P[I(t) = i] = \binom{t}{i} p^i (1-p)^{t-i}, \quad i = 0, 1, \ldots, t, \quad t = 0, 1, 2, \ldots. \tag{2.7.18}$$

where $\binom{t}{i} = t!/i!(t-i)!$ is the binomial coefficient. The binomial law is a function of i, and its parameters are p and t (where $0 \le p \le 1$ and $t = 0$, 1, 2, ...). The mean of $I(t)$ is

$$E[I(t)] = pt, \quad t = 0, 1, 2, \ldots, \tag{2.7.19}$$

and the variance of $I(t)$ is t times the variance of the Bernoulli random variable:

$$\text{Var}[I(t)] = p(1-p)t, \quad t = 0, 1, 2, \ldots. \tag{2.7.20}$$

These results remain valid, *mutatis mutandis*, if the independent binary process is multi-dimensional. Only the *number* of Bernoulli random variables included in the sum matters, not their origin or location in the parameter space. Therefore, if a region in the parameter space contains exactly t observation points, the number of x_0-crossings in that region will obey the binomial distribution given by Eq. (2.7.18).

In the one-dimensional case, two related probability distributions deserve mention. The number of zeros encountered before the kth crossing of the level x_0 is a discrete random variable, R_k, which follows a *negative binomial* distribution:

$$P[R_k = r] = \binom{k+r-1}{r} p^k (1-p)^r, \quad r = 0, 1, 2, \ldots. \tag{2.7.21}$$

The case $k = 1$ is of particular interest: R_1 is the number of zeros encountered before the first x_0-crossing. We have

$$P[R_1 = r] = p(1-p)^r, \quad r = 0, 1, 2, \ldots. \tag{2.7.22}$$

The quantity $R_1 + 1$, the time to the first x_0-crossing, obeys the so-called *geometric* probability law. In a multi-dimensional situation, if the observer follows a path connecting nodes on a multi-dimensional grid (never returning to any point previously visited), then the number of zero-encounters until the kth exceedance will follow the negative binomial distribution.

Processes with Independent Increments Derived from Ideal White Noise

The partial integral of a one-dimensional ideal white noise process $X(t)$ is a process with independent increments

$$I(t) = \int_0^t X(u)du, \tag{2.7.23}$$

often called the Wiener process or Wiener-Einstein process. The French economist Bachelier [7; 8] first proposed its use as a mathematical model for the fluctuations of equity or option prices, and Einstein [54] applied it to the study of Brownian motion.

If m is the mean of $X(t)$, the mean of the Wiener process $I(t)$ is

$$m_I = mt. \tag{2.7.24}$$

The variance of $I(t)$ is also proportional to t:

$$\sigma_I^2 = wt, \tag{2.7.25}$$

where w measures the intensity of the white noise. The analogy between Eqs. (2.7.13) and (2.7.25) motivates some to call w (improperly) the *variance* of the white noise. The correlation function of $I(t)$ has the same form as Eq. (2.7.14):

$$\rho_I(t,t') = \begin{cases} \sqrt{\dfrac{t}{t'}}, & \text{if } t \le t', \\[2ex] \sqrt{\dfrac{t'}{t}}, & \text{if } t \ge t'. \end{cases} \tag{2.7.26}$$

By virtue of the Central Limit Theorem, the probability density function of the Wiener process will tend to be Gaussian regardless of the probability distribution of the impulses $X(t)$.

Much the same analysis applies to integrals of a multi-dimensional white noise process. The key concept is that integrals over disjoint (nonoverlapping) regions in the parameter space are statistically independent. This guarantees that the variance of the integral over any region will be proportional to the size of that region, the proportionality constant being the white noise intensity.

As an example of a multi-dimensional white noise process, consider the rain-on-the-roof process $X(u_1, u_2, t)$ mentioned earlier. If X represents the rate of rainfall (per unit of time and per unit of area), practical interest is likely to focus on various derived integral processes, such as

$$I(t) = \iint_{\text{Area } a} X(u_1, u_2, t)\, du_1 du_2. \tag{2.7.27}$$

If $X(u_1, u_2, t)$ is a homogeneous white noise with mean m and intensity w, the mean of the derived process $Y(t)$ becomes $m_I = ma$ and its variance equals $\sigma_I^2 = wa$.

It is meaningful in some applications to integrate $X(u_1, u_2, t)$ along some trajectory in the parameter space (u_1, u_2, t). Owing to the independence between adjacent values of the white noise field, only the total duration of exposure (or, if the speed is constant, the length of the trajectory) has an influence on the statistics of the integral process. A classical example (discussed further in the next section) is encountered in the study of Brownian motion where the variance of the length of the path traveled by a particle in suspension is proportional to the duration of exposure (Einstein [54]). The proportionality constant, a measure of the intensity of the random process of particle velocities, was determined empirically by Perrin [105] from electron-microscope observations of the travel-path-length variance.

Poisson and Related Random Fields

Poisson Process and Related Distributions

Recall the model introduced earlier in this section, that considers the crossings of a fixed threshold level x_0 by a homogeneous purely random field $X(\mathbf{t})$. As the level x_0 is raised to higher and higher values, crossings become increasingly rare, and the exceedance probability $P[X(\mathbf{t}) > x_0]$ becomes very small ($p \ll 1$). In the one-dimensional case, in the limit, the probability p becomes the mean rate of crossings per unit time (assuming $\Delta t = 1$), and the crossings occur according to a Poisson process with mean rate $\lambda \simeq p$. It is well known that the binomial distribution of $N(t)$, the number of crossings in t time steps, converges toward the Poisson distribution. The Poisson probability mass function is

$$p_N(n) = P[N(t) = n] = e^{-\lambda t}\frac{(\lambda t)^n}{n!}, \qquad n = 0, 1, 2, \ldots. \tag{2.7.28}$$

$N(t)$ is a *counting process* with mean

$$E[N(t)] = \lambda t. \tag{2.7.29}$$

Since $p \ll 1$, $(1 - p) \to 1$, and $\lambda \simeq p$, the binomial variance [Eq. (2.7.20)] converges to

$$\text{Var}[N(t)] = \lambda t. \tag{2.7.30}$$

Let T denote the random time until the level x_0 is crossed for the first time, the so-called "first passage time." The event $\{T \text{ exceeds } t\}$ is equivalent to the event {the time interval $(0, t)$ contains no crossings}, so that the corresponding probabilities are equal, yielding:

$$P[T > t] = 1 - F_T(t) = P[N(t) = 0] = e^{-\lambda t}. \tag{2.7.31}$$

It follows that the first passage time T is exponentially distributed with probability density function

$$f_{_T}(t) = \lambda e^{-\lambda t}, \quad t \geq 0. \tag{2.7.32}$$

The waiting time until the nth crossing occurs, W_n, may be expressed as the sum of n independent and identically distributed random variables T_i, $i = 1, \ldots, n$, with common p.d.f. $f_T(t)$:

$$W_n = T_1 + T_2 + \cdots + T_n, \quad \text{for } n \geq 1. \tag{2.7.33}$$

The p.d.f. of W_n can be obtained from Eq. (2.7.32) by repeated application of the convolution theorem [see Eq. (2.2.44)]. However, a simpler and more elegant derivation is as follows.

Note that the events $\{W_n > w\}$ and $\{N(w) \leq n-1\}$ are equivalent: the fact that the waiting time until the nth crossing is greater than w implies that no more than $(n-1)$ crossings have occurred in the interval $(0, w)$. The complementary cumulative distribution function of W_n is therefore

$$P[W_n > w] = P[N(w) \leq n-1] = \sum_{i=0}^{n-1} e^{-\lambda w} \frac{(\lambda w)^i}{i!}, \quad w \geq 0. \tag{2.7.34}$$

Taking the derivative with respect to w and reversing the sign yields the *Gamma* p.d.f. for W_n:

$$f_{W_n}(w) = e^{-\lambda w} \frac{(\lambda w)^{n-1}}{(n-1)!}, \quad w \geq 0. \tag{2.7.35}$$

The mean and the variance of W_n are, respectively, $E[W_n] = n/\lambda$, and $\text{Var}[W_n] = n/\lambda^2$. Note that $W_1 = T_1$; the Gamma distribution reduces to the exponential distribution when $n = 1$.

Poisson Process in Two Dimensions

In the two-dimensional case, if the value of the threshold level x_0 is high enough so that $p \ll 1$, the homogeneous two-valued random field $Z(u_1, u_2)$ (defined in Sec. 2.7) becomes a homogeneous random pattern of points. Each point corresponds to the location of a threshold exceedance. Taking $\Delta u_1 = \Delta u_2 = 1$, we can say that the probability p converges toward λ, the mean number of points per unit area. The number of points in an area of specified size a is again Poisson distributed, with mean λa.

It is also possible to determine the probability distribution of the distance D to the *nearest* crossing from any point (see, for example, Feller [57]). The distance D will exceed a given value r if no crossings occur inside a circle of radius r around the point. We denote by $N(a)$ the number of crossings inside $a = \pi r^2$. The complementary cumulative distribution function of D can now be expressed as:

$$\begin{aligned} F_D^c(r) &= P[D > r] = P[N(a) = 0] \\ &= \exp\{-\lambda a\} = \exp\{-\pi\lambda r^2\}. \end{aligned} \tag{2.7.36}$$

By differentiating and reversing the sign, the p.d.f. of D is obtained:

$$f_D(r) = 2\pi\lambda r \exp\{-\pi\lambda r^2\}, \quad r \geq 0. \tag{2.7.37}$$

Similarly, we call D_k the distance to the kth closest crossing. The fact that the distance D_k is greater than r imply the occurrence of no more than $(k-1)$ excursions in the circle with radius r. Hence the events $\{D_k > r\}$ and $\{N(\pi r^2) \leq k-1\}$ are equivalent, and the complementary cumulative distribution function of D_k, for $r \geq 0$, is given by:

$$P[D_k > r] = P[N(\pi r^2) \leq k-1] = \sum_{i=0}^{k-1} \frac{(\lambda\pi r^2)^i}{i!} \exp\{-\lambda\pi r^2\}. \tag{2.7.38}$$

The associated probability density function of D_k is

$$f_{D_k}(r) = 2\pi\lambda r\, e^{-\pi\lambda r^2} \frac{(\pi\lambda r^2)^{k-1}}{(k-1)!}, \quad r \geq 0 \tag{2.7.39}$$

which reduces to Eq. (2.7.37) for $k = 1$. Parallel results for the Poisson process in 3-D space are given by, among others, Chandrasekhar [29]. For a two- or three-dimensional field it is *not* possible to express D_k as a sum of independent, identically distributed random variables [as in Eq. (2.7.33)].

The main asset of these Poisson-based models, namely mathematical simplicity stemming from assumed independence, is also their principal weakness: actual patterns of correlation, interdependence, and persistence may not be properly accounted for.

Markov Processes and Diffusion Theory

In the classical analysis of Brownian motion of colloidal matter suspended in a viscous fluid, individual particles experience a succession of random impulses caused by collisions with other particles. This "purely random" excitation is related by a first-order linear differential equation (the Langevin equation) to particle velocity. The quantity of main interest in the theory is particle displacement (the integral of velocity) after some period t. Owing to the fact that the excitation is purely random and to the causal nature of the physical phenomena involved, the velocity and the displacement of particles are both Markovian, that is, characterized by limited memory: their future behavior depends only on their present state, not on past states. This is the Markov property.

Stochastic analysis becomes quite tractable when a random process possesses the Markov property (see Bartlett [9], Parzen [103], Feller [57], Drake [51], Spitzer [124]). We demonstrate this here first for a discrete-time Markov process. The discrete-time solution is then extended in the context of the analysis of Brownian motion, whose stochastic behavior parallels that of physical processes such as diffusion, heat conduction, or transport through porous media.

Markov Chains

A discrete-state, discrete-parameter Markov process, $X(t)$, also called a Markov chain, is characterized by a set of discrete states $i = 1, 2, \ldots$ (the number of states need not be finite) and discrete observation times $t = 0$, 1, 2, etc. The probability of being in state i at time t is

$$q_i(t) = P[X(t) = i], \quad i = 1, 2, \ldots. \tag{2.7.40}$$

The probability of transition from state i at time t to state j at time $t + \tau$ is denoted by

$$p_{ij}^{(t)}(\tau) = P[X(t+\tau) = j | X(t) = i]. \tag{2.7.41}$$

The Markov property implies

$$\begin{aligned} P[X(t+\tau) = j | (X(0) = i_0) \cap (X(1) = i_1) \cap \cdots \cap (X(t) = i)] \\ = P[X(t+\tau) = j | X(t) = i] = p_{ij}^{(t)}(\tau). \end{aligned} \tag{2.7.42}$$

The state probabilities at time $(t + \tau)$ may be expressed in terms of the state probabilities at time t and the transition probabilities for the interval

$(t, t+\tau)$, as follows:

$$q_j(t+\tau) = \sum_i q_i(t) p_{ij}^{(t)}(\tau), \quad j = 1, 2, \ldots. \tag{2.7.43}$$

For a homogeneous (or stationary) Markov chain, the transition probabilities in Eq. (2.7.41) do not depend on t,

$$p_{ij}^{(t)}(\tau) = p_{ij}(\tau), \tag{2.7.44}$$

and the one-step ($\tau = 1$) transition probabilities may be denoted simply by p_{ij}. The transition probabilities may also be found recursively; in the case of a homogeneous chain, we may write

$$p_{ij}(\tau) = \sum_k p_{ik}(\tau - 1) p_{kj}. \tag{2.7.45}$$

Equilibrium or steady-state probabilities $q_i^* = \lim_{t\to\infty} q_i(t)$ may exist for homogeneous Markov chains. If so, they can be found from Eq. (2.7.43) by taking $t \to \infty$ and $\tau = 1$, yielding:

$$q_j^* = \sum_i q_i^* p_{ij}, \quad j = 1, 2, \ldots. \tag{2.7.46}$$

For an N-state Markov chain, Eq. (2.7.46) yields N linear equations, of which one is always redundant. Accounting for the condition $\sum_i q_i^* = 1$, one can then solve for the steady-state probabilities, provided they exist.

Another result readily obtainable for Markov processes is the probability distribution of the time to first arrival at a specified state j given that the process is in state i at time $t = 0$. Taking $i = j$, in particular, yields the distribution of recurrence or renewal times.

Random Walk and Brownian Motion

The simplest model for Brownian motion is a Markov chain called a "random walk". In the one-dimensional case, the Markov chain $X(t)$ describes a particle's position at time t. The particle moves along a straight line in steps of unit size at discrete times $t = 0, 1, 2, \ldots$. (The time step is actually an interval Δt and the state increment is an interval Δx). Successive movements are statistically independent, and there is an equal chance of movement to the right or to the left. The probability that a particle is in state i at time t obeys a difference equation that follows directly from

Eq. (2.7.43):

$$q_i(t) = \frac{1}{2}q_{i+1}(t-1) + \frac{1}{2}q_{i-1}(t-1). \tag{2.7.47}$$

It can be solved iteratively if the initial condition is specified. If the particle starts in state j, the initial state probabilities are

$$q_j(0) = 1, \quad q_k(0) = 0, \quad k \neq j. \tag{2.7.48}$$

By subtracting the term $q_i(t-1)$ from both sides of Eq. (2.7.47), one obtains a difference equation of first order in the time coordinate and of second order in the space coordinate:

$$q_i(t) - q_i(t-1) = \frac{1}{2}[q_{i-1}(t-1) - 2q_i(t-1) + q_{i+1}(t-1)]. \tag{2.7.49}$$

Consider now that the time step is actually Δt and the increment of the space coordinate is Δx. In the limit when both time and space coordinates become continuous, Eq. (2.7.49) takes the form of the classical diffusion equation with diffusion coefficient $D = \lim[(\Delta x)^2/(2\Delta t)]$:

$$\frac{\partial f(x,t)}{\partial t} = D\frac{\partial^2 f(x,t)}{\partial x^2}, \tag{2.7.50}$$

where $f(x,t)\,dx$ is the probability that $X(t)$ lies in the interval $(x, x + dx)$. In the macroscopic theory of diffusion, $f(x,t)$ is interpreted as the concentration of the diffusing substance at x and at time t. By solving Eq. (2.7.50) subject to the initial condition $X(0) = 0$, one finds that, after a sufficiently long time t, the particle displacement distribution becomes Gaussian with mean zero and variance:

$$E[X^2(t)] = 2Dt. \tag{2.7.51}$$

By considering random flights instead of (one-dimensional) random walks, the analogy with diffusion theory can be extended to three dimensions. The analysis yields the probability distribution for the displacement of a particle starting at the origin. Its form is that of a random travel distance, given by Eq. (2.5.42). Einstein [54] found that after a period of time of the order of β^{-1}, the mean square displacement along any given direction is given by

$$E[X^2(t)] = 2Dt \quad \text{where} \quad D = \frac{k_B T^\circ}{m\beta}, \tag{2.7.52}$$

where k_B is Boltzmann's constant, T° is the temperature, m the particle mass, and β the dynamic friction coefficient of the medium in which the Brownian motion occurs.

Ornstein and Uhlenbeck [101] extended Einstein's analysis by modeling the transient part of the solution, which becomes important when t is relatively small. When external forces (gravitational, elastic or magnetic attraction) are added to the equation of particle motion, the Fokker-Planck equation [an extension of Eq. (2.7.50)] governs the probabilities of transition from state to state. However, the process remains Markovian (Kolmogorov [79], Smoluchowski [122]), and the displacement variance retains the simple form of Eqs. (2.7.51) and (2.7.52).

The key result of the analysis of Brownian motion is that the variance of particle displacement is proportional to duration of exposure [Eq. (2.7.51)] at all but very small exposure times. Since the displacement is the integral of the random process of particle velocities, we may also write [adopting the notation of Eq. (2.7.25)]:

$$\sigma_x^2 = w_V t, \tag{2.7.53}$$

where the "white noise intensity" $w_V = 2D$ is a characteristic of the random process $V(t)$ of particle velocities. It is of interest to note that Brownian motion particle velocity has a known marginal probability distribution referred to as Maxwellian (see Chandrasekhar [29]); actually, it is a Gaussian distribution with mean zero and variance

$$\sigma_V^2 = \frac{k_B T^\circ}{m}. \tag{2.7.54}$$

In the Brownian motion model the stationary random process of particle velocities is actually Markovian with correlation function (see Ornstein and Uhlenbeck [101]):

$$\rho_V(\tau) = e^{-\beta|\tau|}, |\tau| \geq 0. \tag{2.7.55}$$

By expressing the probability-diffusion equations in polar coordinates, Seshadri and Lindenberg [118] derive the probability distribution of $Z(t)$, the maximum distance from the origin in the interval $(0, t)$, and the probability distribution of T_b, the time to first exceedance of a prescribed distance b from the origin.

2.8 New Quantum-Physics-Based Probability Distributions

We now present a class of probability models [139; 140; 141] that have a basis in quantum physics; they characterize the inherent uncertainty of properties of single energy quanta emitted by a "perfect radiator" or "blackbody" with a given temperature. The starting point of the derivation of the various probability distributions is the Planck spectrum [107] which expresses how the energy absorbed or emitted in blackbody radiation is distributed among quanta of radiation (photons) with different energies. When matter particles are involved, the blackbody radiation spectrum is consistent with the class of particles (bosons) obeying Bose-Einstein statistics. The focus in this section is on deriving various new probability distributions and their statistical parameters.

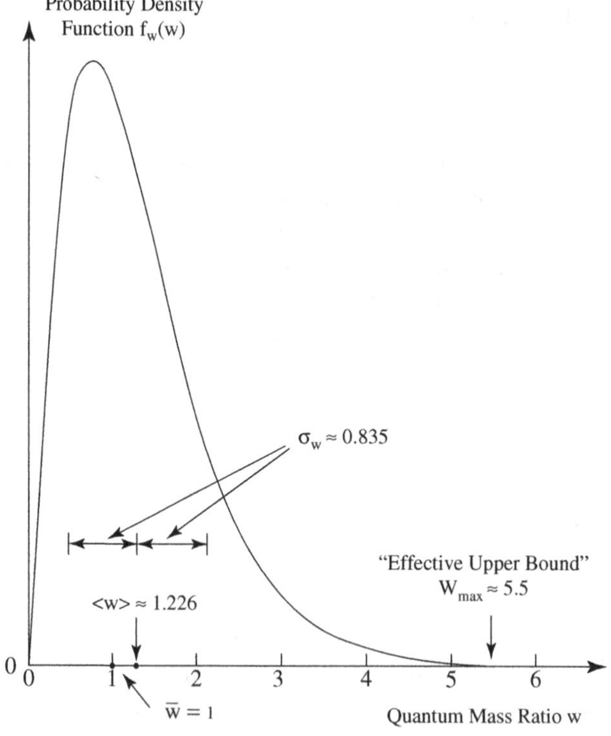

Fig. 2.12 P.d.f. of the mass ratio W. In this figure, $<W> \equiv E[W]$ denotes the arithmetic mean, and $\overline{W}$ is the geometric mean.

Statistics of the "Quantum Mass Ratio" and its Logarithm

A standard way of expressing the Planck radiation function [1; 107], characterizing a large number of energy quanta in local thermal equilibrium, is in terms of the mean number of quanta per unit volume, dN_λ, having wavelengths within the infinitesimal range between λ and $\lambda + d\lambda$,

$$dN_\lambda = \frac{8\pi}{\lambda^4}\left[\exp\left\{\frac{h_P c}{k_B T\lambda}\right\} - 1\right]^{-1} d\lambda, \tag{2.8.1}$$

where T is the radiation temperature; c is the speed of light; k_B is Boltzmann's constant; and h_P is Planck's constant, the "quantum of action". (The numerical values of c, k_B, and h_P are readily available elsewhere.) Integrating over wavelengths λ ranging from zero to infinity yields the mean total number of energy quanta per unit volume as a function of temperature:

$$N_{\text{Total}} = \int_0^\infty dN_\lambda = 16\pi\zeta(3)\left(\frac{k_B T}{h_P c}\right)^3 \approx 20.288\, T^3\ (cm^{-3}), \tag{2.8.2}$$

in which $\zeta(u) = 1 + 2^{-u} + 3^{-u} + \cdots$ is Riemann's Zeta function.

We interpret the Planck radiation spectrum as characterizing the random properties of single energy quanta "sampled" from a collection of quanta in local thermal equilibrium with temperature T. Each quantum possesses a set of functionally-related random properties: its energy $\mathcal{E}$ or (equivalent at-rest) mass $\mathcal{E}/c^2$, wavelength $\lambda = h_P c/\mathcal{E}$, frequency $\nu = \mathcal{E}/h_P$, and momentum $p = \mathcal{E}/c$. Each of these attributes can be expressed, free of units, in terms of a random variable W, referred to herein the "quantum mass ratio", defined as follows:

$$W \equiv \frac{\mathcal{E}}{\overline{\mathcal{E}}} = \frac{\overline{\lambda}}{\lambda}, \tag{2.8.3}$$

where $\overline{\mathcal{E}} \equiv \exp\{E[\ln \mathcal{E}]\}$, the geometric mean [see Eq. (2.3.37)] of the quantum energy, may be expressed as a multiple of $k_B T$:

$$\overline{\mathcal{E}} = c_0 k_B T, \tag{2.8.4}$$

in which $c_0 \simeq 2.134$. The natural logarithm of W is

$$L \equiv \ln W = \ln \mathcal{E} - \ln \overline{\mathcal{E}} = \ln\{\mathcal{E}/(k_B T)\} - \ell_0, \tag{2.8.5}$$

where $\ell_0 \equiv E[\ln\{\overline{\mathcal{E}}/(k_P T)\}] = \ln c_0 \approx 0.758$.

Combining Eqs. (2.8.1) through (2.8.5) yields the probability density functions of the quantum mass ratio W and its natural logarithm, $L \equiv \ln W$, respectively [140]:

$$f_W(w) = \frac{c_0}{2\zeta(3)} \frac{(c_0 w)^2}{e^{c_0 w} - 1}, \qquad w \geq 0, \tag{2.8.6}$$

and

$$f_L(\ell) = \frac{1}{2\zeta(3)} \frac{e^{3(\ell_0+\ell)}}{e^{e^{(\ell_0+\ell)}} - 1}, \qquad -\infty \leq \ell \leq \infty, \tag{2.8.7}$$

where $c_0[2\zeta(3)]^{-1} \approx 0.8877$ and $[2\zeta(3)]^{-1} \approx 0.416$. The p.d.f.'s of the random variables W and L are plotted, respectively, in Figs. 2.12 and 2.13.

The random mass ratio W has a unit geometric mean [see Eq. (2.3.40)]:

$$\overline{W} \equiv \exp\{E[\ln W]\} = 1, \tag{2.8.8}$$

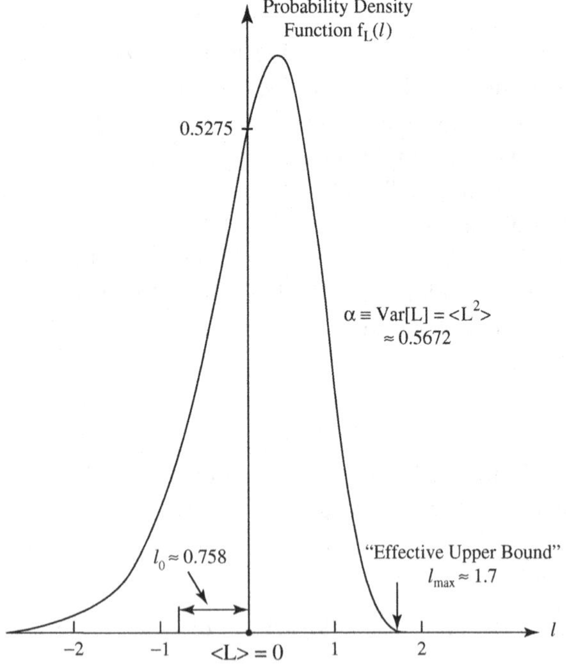

Fig. 2.13 P.d.f. of the logarithm of the mass ratio, $L = \ln W$. The quantities $<L>$ and $<L^2>$ correspond, resp., to the mean $E[L]$ and the mean square $E[L^2]$.

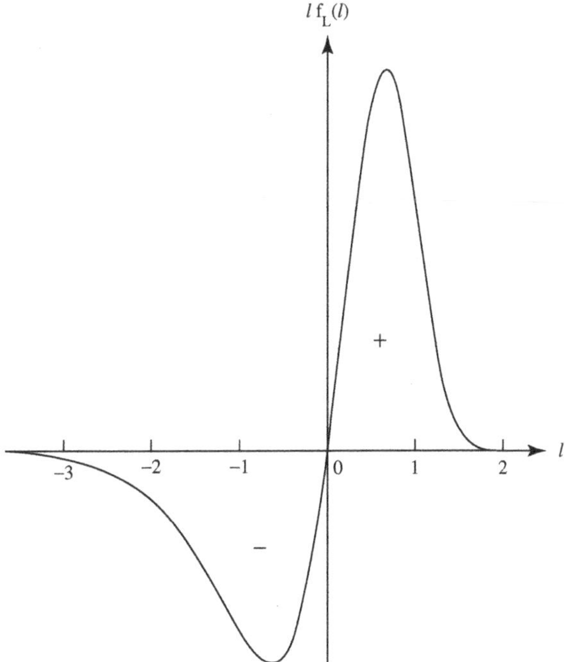

Fig. 2.14 The function $lf_L(l)$ whose integral is $E[L] = 0$.

implying $E[\ln W] = E[L] = 0$; the random variable L has mean zero, owing to the introduction of the shift $l_0 = \ln c_0$. The inherent uncertainty of the properties of single energy quanta in thermal equilibrium is measured by the logarithmic variance of the quantum mass ratio W, or the variance of the mass-ratio logarithm $L = \ln W$, namely:

$$\sigma_L^2 \equiv \mathrm{Var}[\ln W] = \mathrm{Var}[L] = E[L^2] = \int_0^\infty \ell^2 f_L(\ell) d\ell \approx 0.5672. \qquad (2.8.9)$$

(In Figs. 2.13, 2.15 and 2.16, this quantity is also denoted by α.) Expressing the uncertainty in the energy (or wavelength or momentum) of single quanta "sampled" from a population of quanta in thermal equilibrium, for any temperature $0 \leq T_{\text{Planck}} \approx 10^{32} K$, the quantity $\sigma_L^2 \equiv \mathrm{Var}[L] \approx 0.5672$ is a basic constant. It expresses the Uncertainty Principle in a form that differs from Heisenberg's [72]; the latter, proposed in the context of atomic physics, takes the form of an *inequality* that sets a bound (proportional to Planck's constant, a tiny number) on the precision of joint measurements of

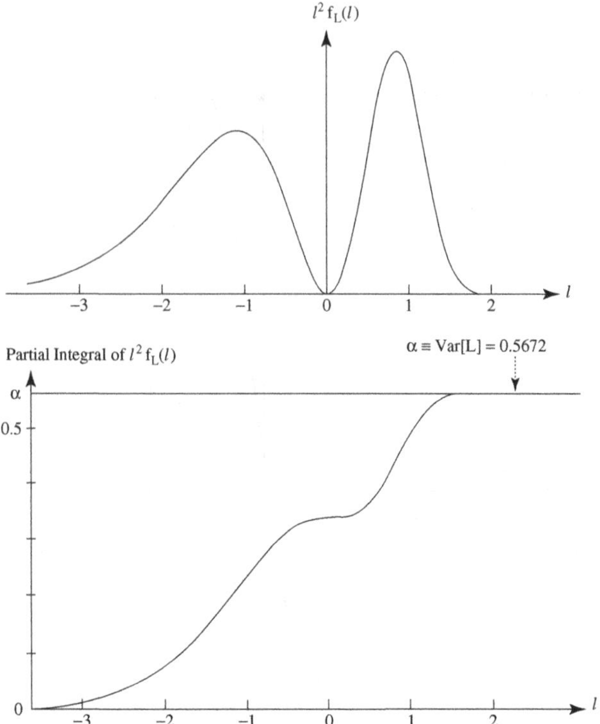

Fig. 2.15 The function $l^2 f_L(l)$ and its partial integral. The complete integral is $\alpha \equiv \mathrm{Var}[L] = E[L^2] \approx 0.5672$.

paired attributes, such as position and momentum, of an elementary particle. The functions $\ell f_{_L}(\ell)$ and $\ell^2 f_{_L}(\ell)$ are plotted in Figs. 2.14 and 2.15, respectively. Their respective complete integrals are the mean $E[L] = 0$ and the variance $\sigma_{_L}^2 = E[L^2] \approx 0.5672$ of the quantum–mass-ratio logarithm.

The statistical moments of the quantum mass ratio W, for any real value $s > -2$, not necessarily integer, are given by

$$E[W^s] = \int_0^\infty w^s f_W(w)dw = \frac{\Gamma(3+s)\zeta(3+s)}{2(c_0)^s\zeta(3)}, \qquad s > -2, \qquad (2.8.10)$$

where $\Gamma(\cdot)$ denotes the Gamma function (which, for positive integers obeys $\Gamma(m+1) = m!$). The quantity $E[W^s]$ and its logarithm are plotted versus s in Fig. 2.16. Taking $s = 0$ confirms that the p.d.f. of W is properly normalized. The case $s = 1$ yields the arithmetic mean $E[W] \approx 1.226$, while the moment of order $s = -2$ is infinite, since $\zeta(1) = \infty$. Other statistics of the mass ratio are: its mean square $E[W^2] \approx 2.31$; its variance $\mathrm{Var}[W] =$

$E[W^2] - E[W]^2 \approx 0.697$; its standard deviation $\sigma_W \equiv (\text{Var}[W])^{1/2} \approx 0.835$; and its coefficient of variation $V_W \equiv \sigma_W / E[W] \approx 0.66$. Also, the quantum-mass-ratio p.d.f. $f_W(w)$ is positively skewed, with coefficient of skewness $\gamma_{1,W} \approx 1.18$. From the definition of the (quantum-mass-ratio-logarithm) cumulant function $K_L(u) = \ln E[e^{iuL}]$, where $i = \sqrt{-1}$, taking into account $W = e^L$, we can infer that Eq. (2.8.10) in effect expresses $E[W^{iu}] = \exp\{K_L(u)\}$, with iu replacing the exponent s. It follows that cumulant function of L is given by

$$\begin{aligned} K_L(u) &= \ln E[W^{iu}] \\ &= \ln\{\Gamma(3+iu)\} + \ln\{\zeta(3+iu)\} - \ell_0 iu - \ln[2\zeta(3)], \end{aligned} \tag{2.8.11}$$

where $\ell_0 \ln c_0 \approx 0.758$ and $\ln[2\zeta(3)] \approx 0.877$; also, as expected, $K_L(0) = \ln E[W^0] = 0$. Since the first three cumulants of L are known, $K_{1,L} = E[L] = 0$, $K_{2,L} = \text{Var}[L] = \sigma_L^2 \approx 0.5672$ and $K_{3,L} = \gamma_{1,L}\sigma_L^3 \approx -0.3755$ (in which $\gamma_{1,L} \approx -0.879$ is the skewness coefficient), the first few terms in the series expansion (Eq. (2.4.15)) of the cumulant function are:

$$\begin{aligned} K_L(u) &= -\sigma_L^2 \frac{u^2}{2} + \gamma_{1,L}\sigma_L^3 \frac{(iu)^3}{3!} + \cdots \\ &\approx -0.5672\frac{u^2}{2} - 0.3755\frac{(iu)^3}{3!} + \cdots . \end{aligned} \tag{2.8.12}$$

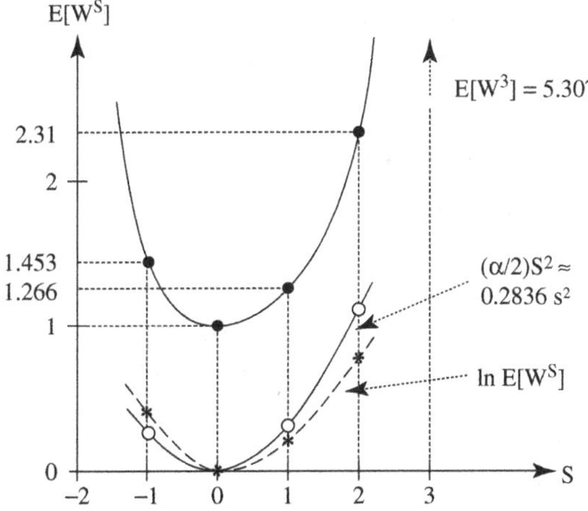

Fig. 2.16 Expectation of mass ratio W raised to the power s, and its logarithm $\ln E[W^s]$. In this figure, $<W^s> \equiv E[W^s]$ and $\alpha \equiv \text{Var}[L] = E[L^2]$.

If the random variable L had a normal (Gaussian) distribution with mean zero and variance σ_L^2, its cumulant function would obey $K_L(u) = -\sigma_L^2 u^2/2$, so the higher-order terms in Eq. (2.8.12) represent physically significant deviations from symmetry and normality. The first-term approximation, $\ln E[W^s] = K_L(s/i) \approx \sigma_L^2 s^2/2 \approx 0.2836 s^2/2$, is shown, along with exact values, in Fig. 2.16.

The quantum mass ratio W possesses the following geometric statistics: geometric mean $\overline{W} = 1$; logarithmic variance $\mathrm{Var}[\ln W] \equiv \sigma_L^2 \approx 0.5672$; and characteristic value (equal to the "stochasticity factor") $\widetilde{W} = \exp\{\sigma_L^2/2\} \approx 1.328$. (The *lognormal* formula for the skewness coefficient would yield, $3V_W + V_W^3 \approx 2.27$, almost twice the true value, $\gamma_{1,W} \approx 1.18$; the difference reflects the fast decay of the p.d.f. of W in the high-energy range compared to a lognormal p.d.f. with the same geometric mean and logarithmic variance). The p.d.f. of W is positively skewed ($\gamma_{1,W} \approx 1.18$) and that of $L = \ln W$ negatively, with skewness coefficient $\gamma_{1,L} \approx -0.879$. Note in Figs. 2.12 and 2.13 the lack of symmetry in the tails of either distribution. The fast decay in the high-energy ("Wien") range of the Planck spectrum means that very-high-energy quanta tend to occur much less frequently than very-low-energy ("Jeans-Rayleigh") quanta. The probability that the mass ratio W (of a randomly chosen quantum of energy, *given* the temperature T) exceeds the value w drops very rapidly for w-values exceeding $w \approx 5.5$, corresponding to $\ell = \ln 5.5 \approx 1.70$ [see Figs. 2.12 and 2.13]. For the same reason, the Gaussian approximation for $f_L(\ell)$, with zero mean and variance σ_L^2, significantly overestimates the probability that L is larger than this "apparent upper bound" $\ell_{\max} \equiv \ln(w_{\max}) = \ln(5.5) \approx 1.70 \approx 2.26\sigma_L$.

Statistics of the "Quantum Size Ratio"

The wavelengths associated with individual energy quanta may be thought of as representing the sizes (radii) of elementary "volumes". The relative sizes of these quantum-specific volumes, *given* the (local-thermal-equilibrium) temperature T, can be measured by the "quantum size ratio", $D \equiv \lambda/\overline{\lambda} = 1/W$, defined as the reciprocal of the "quantum mass ratio". The monotonic one-to-one relationship between D and W means that probability density function of the quantum size ratio D is given by $f_D(d) = (1/d^2) f_W(1/d)$, $d \geq 0$. Inserting Eq. (2.8.6) yields:

$$f_D(d) = \frac{(c_0)^3}{2\zeta(3)} \frac{1}{d^4(e^{c_0/d} - 1)}, \quad d \geq 0, \tag{2.8.13}$$

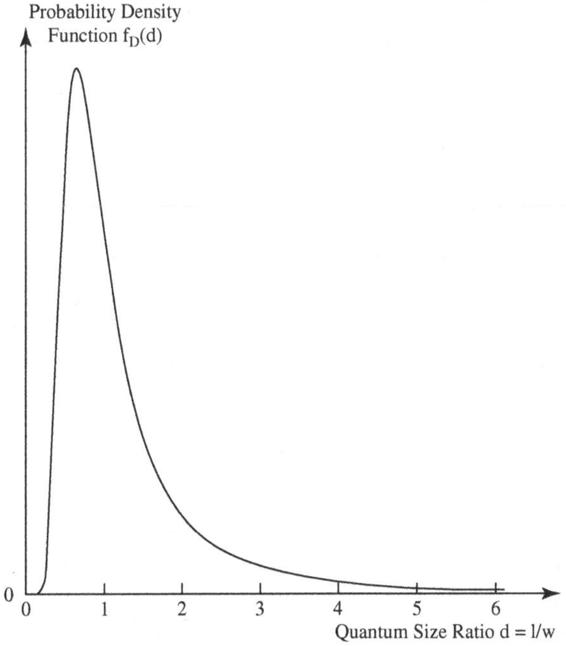

Fig. 2.17 P.d.f. of the "size ratio" $D = 1/W$.

which matches the *shape* of the Planck blackbody radiation spectrum in Eq. (2.8.1). The probability density function of D is plotted in Fig. 2.17.

The probability density function $f_{L'}(\ell')$ of the *logarithm* of the quantum size ratio, $L' \equiv \ln D = -\ln W = -L$, is similar to that of the quantum-mass-ratio logarithm in Eq. (2.8.7). We may write:

$$f_{L'}(\ell') = f_L(-\ell') = \frac{1}{2\zeta(3)} \frac{e^{3(\ell_0 - \ell')}}{e^{e^{(\ell_o - \ell')}} - 1}, \qquad -\infty \leq \ell' \leq \infty. \tag{2.8.14}$$

The two cumulant functions are likewise related: $K_{L'}(u) = K_L(-u)$. The means of L and L' are both zero, their variances are equal ($\sigma_L^2 \approx 0.5672$), and their coefficients of skewness have the same absolute value but the opposite sign: $\gamma_{1,L'} = -\gamma_{1,L} \approx +0.879$.

The "geometric mean" and related statistics of $D = e^{-L}$ are the same as those of the quantum mass ratio ($W = e^L$), namely: geometric mean $\overline{D} = (\overline{W})^{-1} = 1$; characteristic value $\widetilde{D} \approx 1.328$; and logarithmic variance $\mathrm{Var}[\ln D] \equiv 2\ln(\widetilde{D}/\overline{D}) = \sigma_L^2 \approx 0.5672$. The quantum size ratio's arithmetic mean, $E[D] = E[W^{-1}] \approx 1.453$, is obtained by integrating the function

$df_D(d)$. The second moment of the quantum size ratio, $E[D^2] = E[W^{-2}]$, and its variance, $\mathrm{Var}[D] = E[D^2] - E^2[D]$, are both *infinite* (corresponding to $s = -2$ in Fig. 2.16) due to the slow decay of the upper tail of $f_D(d)$. Closer examination of the upper tail's asymptotic expression, for $d \gg 1$,

$$f_D(d) \to \frac{c_0^2}{2\zeta(3)} \frac{1}{d^3} \approx \frac{1.895}{d^3}, \tag{2.8.15}$$

indicates that the "truncated" or "conditional" mean square, $E[D^2|D < d_0]$, defined as the integral from 0 to d_0 of the function $d^2 f_D(d)$ has a term proportional to $\ln d_0$. The latter controls how the "unconditional" variance of the quantum size ratio D approaches infinity. A further consequence is that the conditional or "sample" variance, $\mathrm{Var}[D|D < d_0]$, where d_0 represents the size of any actual (large) volume will be finite.

Families of Related Probability Models

Starting from a set of quantum-physics-based random variables that are statistically independent and identically distributed (i.i.d.), one can readily obtain families of derived probability distributions for *sums* or *products* or *extreme values*, with the total number of i.i.d. random variables in the set serving as a model parameter. Consider, in particular, a set of ν statistically independent quantum-mass-ratio logarithms, their common probability density function $f_L(\ell)$ given by Eq. (2.8.7). One way to derive the probability distribution of the sum, $L_\nu = L_{01} + L_{12} + \cdots + L_{\nu-1,\nu}$, is by iterative convolution , starting with the p.d.f. of $L_2 = L_{01} + L_{12}$, next that of $L_3 = L_2 + L_{23}$, and so on. The p.d.f. of the corresponding product of i.i.d. quantum mass ratios, $W_\nu \equiv \exp\{L_\nu\} = W_{01}W_{12}...W_{\nu-1,\nu}$, where $W_{i-1,i} = \exp\{L_{i-1,i}\}$ for $i = 1, 2..., \nu$, can then be found by a single monotonic (one-to-one) transformation. The p.d.f. of the product of ν quantum size ratios $D_\nu = 1/W_\nu = \exp\{-L_\nu\}$ can be obtained in the same way. The p.d.f.'s of the arithmetic mean L_ν/ν and the geometric means $(W_\nu)^{1/\nu}$ and $(D_\nu)^{1/\nu}$, for $\nu = 1, 2, \ldots$, are also easy to derive.

An elegant alternative to convolution is to express the cumulant function of L_ν as a sum of cumulant functions, $K_{L_\nu}(u) = \nu K_L(u)$, where $K_L(u)$ denotes the cumulant function of the quantum-mass-ratio logarithm L (given by Eq. (2.8.12)).

Scaling $L_\nu \equiv \ln W_\nu$ with respect to its standard deviation, $\sqrt{\nu}\sigma_L$, yields a family of "standardized" random variables, $U^{(\nu)} = L_\nu/(\sqrt{\nu}\sigma_L)$, $\nu = 1, 2, \ldots$, with zero mean and unit variance, whose probability distribution, as ν increases, will tend toward the Standard Normal (Gaussian) distribution. The coefficient of skewness of L_ν can be expressed in terms of its "cumulants", $\gamma_{1,L_\nu} \equiv \nu K_{3,L}/(\nu K_{2,L})^{3/2} = \gamma_{1,L}/\sqrt{\nu} \approx -0.879/\sqrt{\nu}$, indicating how the skewness diminishes as ν grows (consistent with the Central Limit Theorem), but also how negative skewness persists. Finally, a more complete set of probabilistic models can be obtained from the distribution of the "standardized" random variable $U^{(\nu)}$, by shifting its zero mean and scaling its unit variance, yielding a quantum-physics-based family of models characterized by their mean, variance, and skewness coefficient.

Chapter 3

Second-Order Analysis of Homogeneous Random Fields

The only thing that is certain is that nothing is certain.

Pliny the Elder, Historia Naturalis

3.1 Preview, Definitions, and Notation

Second-Order Description of Random Fields

A real, scalar random field $X(\mathbf{t})$ is a collection of random variables at points with coordinates $\mathbf{t} = (t_1, \ldots, t_n)$ in an n-dimensional "parameter space." As described in Sec. 3.2, second-order information about point-to-point variation is contained in the covariance function $B(\mathbf{t}, \mathbf{t}')$, the covariance between values of the random field at two locations $\mathbf{t}$ and $\mathbf{t}'$. If the random field is *homogeneous* (or, for $n = 1$, "stationary"), the covariance function will depend only on the differences $\tau_i = t_i - t'_i$, $i = 1, \ldots, n$, the components of the lag vector $\boldsymbol{\tau}$. The spectral representation is introduced for one-dimensional processes in Sec. 3.3, and for two- and n-dimensional random fields in Sec. 3.4. The generalized Wiener-Khinchine equations state that the n-dimensional Fourier transform of $B(\boldsymbol{\tau})$ equals the spectral density function (s.d.f.) $S(\boldsymbol{\omega})$, where $\boldsymbol{\omega}$ identifies a point in the (generalized) frequency domain. Normalizing these functions with respect to the variance σ^2 yields another Fourier transform pair: the correlation function $\rho(\boldsymbol{\tau})$ and the normalized s.d.f. $s(\boldsymbol{\omega})$.

If the *spatial* character of the random field requires emphasis, the problem is usually formulated in terms of Cartesian spatial coordinates $\mathbf{u} = (u_1, u_2, u_3)$. If the random field is specifically a *space-time process*, it will depend on the vector of spatial coordinates $\mathbf{u}$ and on time t. Spatial lags, denoted by $\nu_i = u_i - u'_i$, are transformed into wave numbers k_i.

Random Fields (Processes) of Lower Dimensionality

If one or more of the parameters of a homogeneous n-dimensional random field $X(\mathbf{t})$ is held constant, one obtains a homogeneous field of dimensionality less than n. We use the notation $X(t_i)$ to refer to a stationary one-dimensional random process on a line parallel to the t_i-axis, while $X(t_i, t_j)$ represents random variation on a plane parallel to the axes t_i and t_j. The two-dimensional (2-D) random field $X(t_i, t_j)$ is characterized in second-order terms by the covariance function $B(\tau_i, \tau_j)$ and the spectral density function $S(\omega_i, \omega_j)$.

Quadrant Symmetry

As explained in Sec. 3.2, the correlation structure of a homogeneous random field is *quadrant symmetric* (q.s.) if the covariance function $B(\boldsymbol{\tau})$ is *even* with respect to each component of the lag vector $\boldsymbol{\tau} = (\tau_1, \ldots, \tau_n)$. Both $B(\boldsymbol{\tau})$ and $S(\boldsymbol{\omega})$ then possess quadrant symmetry, and assorted cross-spectral density functions become real (that is, their imaginary part vanishes). The second-order properties of "q.s. processes" are fully characterized by the covariance function $B(\boldsymbol{\tau})$ defined for positive lags only ($\boldsymbol{\tau} \geq \mathbf{0}$), or by the spectral density function $G(\boldsymbol{\omega}) = 2^n S(\boldsymbol{\omega})$ defined for positive frequencies (wave numbers) only.

The correlation structure is quadrant symmetric if it is (1) isotropic, (2) separable, or (3) ellipsoidal (defined in terms of principal coordinates). In an applications context it is often convenient to describe and analyze random fields based on the assumption of quadrant symmetry.

Moments of the Spectral Density and Correlation Functions

Two sets of moments play important roles in theoretical developments throughout the book. The first set, the *spectral moments*, introduced in Chap. 4, are all characterized by the symbol λ with appropriate addscripts. For example, $\lambda_k^{(i)}$ is the kth moment of the direction-dependent s.d.f. $S(\omega_i)$, the integral over frequency of $\omega_i^k S(\omega_i)$. The joint spectral moment $\lambda_{kl}^{(ij)}$ is defined as the integral of $\omega_i^k \omega_j^l S(\omega_i, \omega_j)$ over the plane of frequencies (ω_i, ω_j); for a homogeneous n-dimensional random field these joint moments can be assembled into an n by n matrix $\boldsymbol{\Lambda}_{kl}$.

The second set of *moments*, defined in terms of the *n-dimensional correlation function* $\rho(\boldsymbol{\tau})$, introduced in Chap. 5 (for $n = 1$), Chap. 6 (for

$n = 2$ and Chap. 7 (for $n = 3$), are identified by the symbol θ for $n = 1$, α for $n = 2$, and β for $n = 3$, while the symbol η is used to refer to a space-time process with $n = 4$ or when the field's dimensionality is unspecified. For example, $\theta_2^{(i)}$ is the second moment of the correlation function $\rho(\tau_i)$, while $\alpha_{11}^{(ij)}$ is the joint moment of second order of the two-dimensional (2-D) correlation function $\rho(\tau_i, \tau_j)$.

3.2 Correlation Function of a Homogeneous Random Field

Basic Properties and Symmetries

In second-order analysis, the variation of an n-dimensional random field $X(\mathbf{t})$ at two locations $\mathbf{t} = (t_1, \ldots, t_n)$ and $\mathbf{t}' = (t'_1, \ldots, t'_n)$ is characterized by the *covariance function*

$$\begin{aligned} B(\mathbf{t}, \mathbf{t}') \equiv B_X(\mathbf{t}, \mathbf{t}') &= \mathrm{Cov}[X(\mathbf{t}), X(\mathbf{t}')] \\ &= E[X(\mathbf{t})X(\mathbf{t}')] - m(\mathbf{t})m(\mathbf{t}'), \end{aligned} \tag{3.2.1}$$

or by the corresponding *correlation function*

$$\rho(\mathbf{t}, \mathbf{t}') \equiv \rho_X(\mathbf{t}, \mathbf{t}') = \frac{B(\mathbf{t}, \mathbf{t}')}{\sigma(\mathbf{t})\sigma(\mathbf{t}')}, \tag{3.2.2}$$

where $m(\mathbf{t})$ and $m(\mathbf{t}')$ are the means, and $\sigma^2(\mathbf{t})$ and $\sigma^2(\mathbf{t}')$ the variances, of $X(\mathbf{t})$ and $X(\mathbf{t}')$, respectively.

If the random field is *homogeneous*, its covariance function will depend on the relative position of the points $\mathbf{t}$ and $\mathbf{t}'$ only, and we may write:

$$B(\mathbf{t}, \mathbf{t}') = B(\mathbf{t} - \mathbf{t}') = B(\boldsymbol{\tau}) = B(\tau_1, \ldots, \tau_n), \tag{3.2.3}$$

where $\boldsymbol{\tau} = \mathbf{t} - \mathbf{t}'$ is the *lag vector* whose components are $\tau_k = t_k - t'_k$ $(k = 1, \ldots, n)$. Also,

$$|B(\tau_1, \ldots, \tau_n)| \le B(0, \ldots, 0) = \sigma^2. \tag{3.2.4}$$

Values of the covariance function $B(\boldsymbol{\tau})$ corresponding to two points in the τ-domain (or the "lag space") located symmetrically with respect to the origin $\boldsymbol{\tau} = \mathbf{0}$ are identical. For example,

$$B(\tau_1, \tau_2, \ldots, \tau_n) = B(-\tau_1, -\tau_2, \ldots, -\tau_n), \tag{3.2.5}$$

and also

$$B(\tau_1, -\tau_2, \tau_3, \tau_4) = B(-\tau_1, \tau_2, -\tau_3, -\tau_4).$$

This implies that the *two-dimensional* covariance function of a homogeneous random process can be represented by two functions, both defined for positive lags only:

$$B(\tau_1, \tau_2) = B(-\tau_1, -\tau_2), \quad \tau_1, \tau_2 \geq 0, \tag{3.2.6}$$

and

$$\overline{B}(\tau_1, \tau_2) \equiv B(-\tau_1, \tau_2) = B(\tau_1, -\tau_2), \quad \tau_1, \tau_2 \geq 0.$$

Likewise, for an n-dimensional random field there are 2^{n-1} such functions defined for $\boldsymbol{\tau} \geq \mathbf{0}$ only; and the same can be said, of course, for the *normalized* covariance function $\rho(\boldsymbol{\tau}) = B(\boldsymbol{\tau})/\sigma^2$.

Quadrant Symmetry

Of considerable practical importance is the class of homogeneous random processes whose covariance function is *even* with respect to each component of the lag vector:

$$B(\tau_1, \ldots, \tau_k, \ldots, \tau_n) = B(\tau_1, \ldots, -\tau_k, \ldots, \tau_n), \quad k = 1, 2, \ldots, n. \tag{3.2.7}$$

Random processes whose covariance function obeys Eq. (3.2.7) are called *quadrant symmetric* (q.s.), a term introduced by the writer [137]. Quadrant symmetry implies homogeneity in the weak sense. For q.s. processes the covariance function *defined for positive lags only* fully represents the correlation structure of $X(\mathbf{t})$, since information from one quadrant of the lag space suffices to describe the multi-dimensional correlation structure. For $n = 2$ it implies $B(\tau_1, \tau_2) = \overline{B}(\tau_1, \tau_2)$; in reference to Fig. 3.1, it means that the covariance of values associated with points 1 and 2 is the same as that associated with points 3 and 4.

The quadrant symmetric correlation structure occurs quite naturally in the context of applications where the random field $X(\mathbf{t})$ is defined in terms of a set of principal (usually orthogonal) coordinates $t_1, \ldots, t_n$. The separation distances $\tau_k = t_k - t'_k$ $(k = 1, \ldots, n)$ are then measured on the respective principal axes of the covariance function. If the correlation structure is homogeneous but not quadrant symmetric, it is generally possible to rotate the axes of the coordinate system to achieve quadrant symmetry.

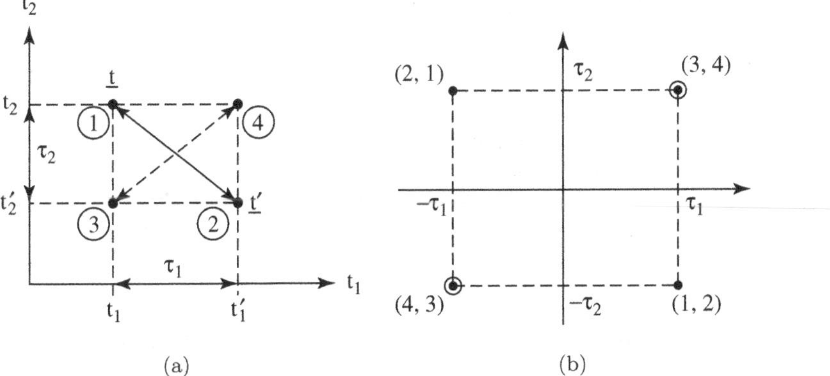

Fig. 3.1 (a) Four-point pattern illustrates the concept of quadrant symmetry; (b) the covariance $B(\tau_1, \tau_2)$ must be the same at all four points in the lag space (τ_1, τ_2).

Important Types of Quadrant Symmetric Covariance Functions

Isotropic Correlation Structure

An isotropic covariance function depends only on the distance τ between the points $\boldsymbol{t}$ and $\boldsymbol{t}'$:

$$\tau = |\boldsymbol{\tau}| = |\mathbf{t} - \mathbf{t}'| = [\tau_1^2 + \tau_2^2 + \cdots + \tau_n^2]^{1/2}. \tag{3.2.8}$$

$B(\boldsymbol{\tau})$ has spherical symmetry, which implies quadrant symmetry. Also,

$$B(\boldsymbol{\tau}) \equiv B(\tau_1, \tau_2, \ldots, \tau_n) = B(\tau, 0, \ldots, 0) = B(\tau) \equiv B^R(\tau), \tag{3.2.9}$$

where $B(\tau) \equiv B^R(\tau)$, the covariance function of random variation *on the line* (in any direction), may be called the "radial covariance function."

The fact that the correlation function is positive definite implies that, for an arbitrary set of coefficients a_k and locations $\boldsymbol{\tau}_k$ in the lag space, the following inequality holds:

$$\sum_{k=1}^{n+1} \sum_{l=1}^{n+1} a_k a_l \rho(\mathbf{t}_k, \mathbf{t}_l) \geq 0. \tag{3.2.10}$$

This gives rise to an interesting lower bound, attributed to Bertil Matern [93], on the value of the correlation function of an n-dimensional *isotropic* random field. For a set of $(n+1)$ *equidistant* locations, taking $a_k = 1$ $(k = 1, \ldots, n+1)$, Eq. (3.2.10) yields

$$(n+1) + [(n+1)^2 - (n+1)]\rho(\boldsymbol{\tau}) \geq 0, \tag{3.2.11}$$

resulting in the lower bound:

$$\rho(\boldsymbol{\tau}) \geq -\frac{1}{n}. \tag{3.2.12}$$

For a two-dimensional isotropic random field ($n = 2$), the lower bound on the correlation function is $-1/2$, while for $n = 3$ it equals $-1/3$.

Ellipsoidal Correlation Structure

By scaling the axes of an isotropic random process, one obtains a process with an ellipsoidal covariance function $B(\boldsymbol{\tau})$ that depends on

$$\tau^2 = \left(\frac{\tau_1}{a_1}\right)^2 + \cdots + \left(\frac{\tau_n}{a_n}\right)^2, \tag{3.2.13}$$

where $a_1, \ldots, a_n$ are (constant) scale factors. The correlation structure can be made isotropic by introducing $\tau'_k = (\tau_k/a_k)$, $k = 1, \ldots, n$. (The bound given by Eq. (3.2.12) also applies to ellipsoidal covariance functions.) In general, the ellipsoidal covariance function depends on

$$\boldsymbol{\tau}^t \mathbf{A}^{-1} \boldsymbol{\tau} = \sum_{k=1}^{n} \sum_{l=1}^{n} \frac{A_{kl}}{|\mathbf{A}|} \tau_k \tau_l, \tag{3.2.14}$$

where $\boldsymbol{\tau}^t$ denotes the transpose of the vector $\boldsymbol{\tau}$ of separation distances; $\mathbf{A}$ is an n by n positive definite matrix; and A_{kl} is the cofactor of the element a_{kl}. The iso-correlation ellipsoids are centered at the origin of the lag space ($\boldsymbol{\tau} = \mathbf{0}$), their size and shape determined by $\mathbf{A}$. In particular, the lengths of the major axes correspond to the eigenvalues of $\mathbf{A}$. Since any positive definite matrix $\mathbf{A}$ can be diagonalized by an appropriate change of variables, Eq. (3.2.14) can always be put into the form of Eq. (3.2.13). In this diagonalized form the covariance function is *quadrant symmetric.*

Separable Correlation Structure

The correlation structure of $X(\mathbf{t})$ is *fully separable* if $B(\boldsymbol{\tau})$ can be expressed as follows:

$$B(\boldsymbol{\tau}) = \sigma^2 \rho(\tau_1)\rho(\tau_2)\cdots\rho(\tau_n), \tag{3.2.15}$$

where $\rho(\tau_k)$ is the covariance function of the one-dimensional random variation $X(t_k)$ along lines parallel to the t_k-axis. Again "quadrant symmetry" is assured. The properties of separability and quadrant symmetry are generally lost upon rotation of the coordinate axes. The correlation structure is *partially separable* if the correlation function $\rho(\boldsymbol{\tau})$ can be expressed as the product of lower-dimension correlation functions, such as in:

$$\rho(\tau_1, \tau_2, \tau_3) = \rho(\tau_1, \tau_2)\,\rho(\tau_3). \tag{3.2.16}$$

In modeling spatial random variation in geology or metereology, in particular, it may be appropriate to hypothesize separable correlation structure for "lateral" variation (t_1, t_2) and "vertical" variation (t_3); and the two-dimensional lateral random variation may presumably be characterized by an isotropic or ellipsoidal correlation function.

In hydrology the space-time correlation structure of the rate of rainfall is assumed by Rodriguez-Iturbe and Mejia [115] to be separable. Likewise, the random field of pressures in a turbulent boundary layer is assumed to have a separable space-time correlation structure by Dyer [53] and Lin [83].

Cross-Covariance Function of Two Coupled Random Fields

The cross-covariance function of two homogeneous random fields $X(\mathbf{t})$ and $Y(\mathbf{t})$ is by definition

$$B_{XY}(\boldsymbol{\tau}) = \frac{\mathrm{Cov}[X(\mathbf{t}), Y(\mathbf{t}+\boldsymbol{\tau})]}{\sigma_X \sigma_Y} = \frac{E[X(\mathbf{t})Y(\mathbf{t}+\boldsymbol{\tau})] - m_X m_Y}{\sigma_X \sigma_Y}. \tag{3.2.17}$$

More generally, for (a vector of) M coupled random fields defined in the same parameter space, one can construct a matrix of cross-covariance functions. Its diagonal elements are covariance functions of the form $B_{XX}(\boldsymbol{\tau}) \equiv B_X(\boldsymbol{\tau})$. For instance, in the study of homogeneous turbulence (Batchelor [11], Lumley [85]), the *vector field* of random velocities (averaged over a time window) can be represented by a 3 by 3 matrix of cross-covariance functions $B_{ij}(\boldsymbol{\tau})$, where the subscripts i and j refer to the spatial coordinates. Space-time correlation structure, a closely related topic (since any n-dimensional random field gives rise to many coupled fields with lower dimensionality), is the subject of Sec. 3.8.

3.3 Spectral Representation of Random Processes on the Line

Introduction

There are several alternate ways of deriving the basic equations that characterize the second-order properties of a stationary random process $X(t)$ in the "frequency domain." The more *formal* approach introduces negative as well as positive frequencies and assumes that the random process has an imaginary as well as a real part, while the more *physical* approach avoids complex-number algebra and considers only nonnegative frequencies. An intermediate approach, the first taken herein, considers both positive and negative frequencies but does not require complex algebra. The stationary random function $X(t)$ is expressed as a sum of its mean m and $2K$ sinusoids with frequencies $\omega_k = \pm[\Delta\omega(2k-1)/2]$, having random amplitudes C_k and phase angles $\Phi_k (k = 1, \ldots, K)$:

$$X(t) = m + \sum_{k=-K}^{K} X_k(t), \tag{3.3.1}$$

where

$$X_k(t) = C_k \cos(\omega_k t + \Phi_k).$$

All the random amplitudes and phase angles are mutually independent, and the phase angles are uniformly distributed over the range $(0, 2\pi)$. Each elementary random function $X_k(t)$ has mean zero and variance

$$\sigma_k^2 = E[X_k^2(t)] = E[C_k^2]E[\cos^2(\omega_k t + \Phi_k)] = \frac{1}{2}E[C_k^2], \tag{3.3.2}$$

where the expectation $E[\cos^2(\omega_k t + \Phi_k)]$ is with respect to the random phase angle Φ_k. The variance of the sum of independent random functions equals the sum of their variances, so that

$$\sigma^2 = \sum_{k=-K}^{K} \sigma_k^2 = \sum_{k=-K}^{K} \frac{1}{2}E[C_k^2]. \tag{3.3.3}$$

This equation indicates that the variance σ^2 can be seen as distributed over the discrete frequencies ω_k $(k = -K, \ldots, -1, 1, \ldots, K)$ and motivates the introduction of a spectral mass function (see Fig. 3.3), as follows;

$$S(\omega_k)\Delta\omega = \frac{1}{2}E[C_k^2]. \tag{3.3.4}$$

Letting $\Delta\omega$ and K approach zero and infinity, respectively, while keeping their product constant, and replacing the summation by an integration, Eq. (3.3.3) becomes

$$\sigma^2 = \sum_{k=-K}^{K} S(\omega_k)\Delta\omega \to \int_{-\infty}^{+\infty} S(\omega)d\omega, \tag{3.3.5}$$

where $S(\omega)$ is the *spectral density function* of $X(t)$. Since it is defined for both positive and negative frequencies, it may be called *two-sided* and, by virtue of its definition, is real and nonnegative.

Wiener-Khinchine Relations

Introducing Eq. (3.3.1) into the definition of the covariance function $B(\tau)$, and accounting for the fact that the random sinusoids $X_k(t)$ are independent, one obtains

$$B(\tau) = \mathrm{Cov}[X(0), X(\tau)] = \sum_{k=-K}^{K} E[X_k(0)X_k(\tau)]. \tag{3.3.6}$$

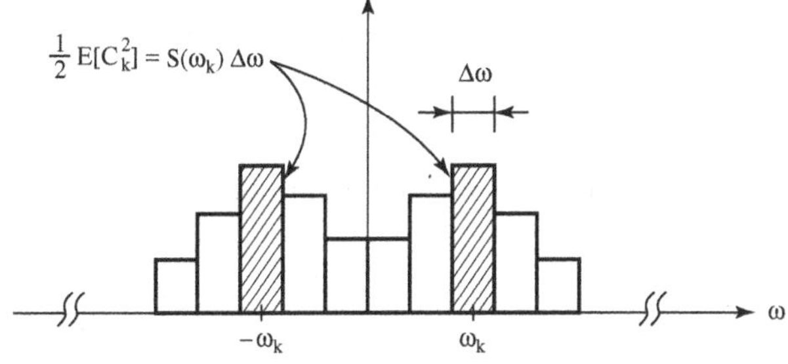

Fig. 3.2 Definition of the two-sided spectral density function $S(\omega)$.

Inserting $X_k(0) = C_k \cos(\Phi_k)$ and $X_k(\tau) = C_k \cos(\omega_k \tau + \Phi_k)$ into Eq. (3.3.6), and accounting for Eq. (3.3.4), leads to

$$B(\tau) = \sum_{k=-K}^{K} E[C_k^2] E\left[\frac{1}{2}[\cos(\omega_k \tau + 2\Phi_k) + \cos \omega_k \tau]\right]$$

$$= \sum_{k=-K}^{K} \frac{1}{2} E[C_k^2] \cos \omega_k \tau = \sum_{k=-K}^{K} S(\omega_k) \Delta\omega \cos \omega_k \tau. \qquad (3.3.7)$$

The limiting form of Eq. (3.3.7) (for $K \to \infty$ and $\Delta\omega \to 0$) is the first relation of the Fourier transform pair that connects $B(\tau)$ and $S(\omega)$:

$$\begin{cases} B(\tau) = \displaystyle\int_{-\infty}^{+\infty} S(\omega) \cos \omega\tau d\omega, & (3.3.8) \\ S(\omega) = \displaystyle\frac{1}{2\pi} \int_{-\infty}^{+\infty} B(\tau) \cos \omega\tau d\tau. & (3.3.9) \end{cases}$$

These equations constitute (one form of) the classical Wiener-Khinchine relations. Setting $\tau = 0$ in Eq. (3.3.8) yields the basic result for the variance, Eq. (3.3.5). Since $B(\tau) = B(-\tau)$, it follows from Eq. (3.3.9) that the function $S(\omega)$ is symmetric about $\omega = 0$:

$$S(-\omega) = S(\omega). \qquad (3.3.10)$$

In other words, $S(\omega)$ is an even function.

One-Sided Spectral Density Function

The symmetry of $S(\omega)$ with respect to $\omega = 0$ motivates the introduction of the *one-sided spectral density function* $G(\omega)$, defined for $\omega \geq 0$ only:

$$G(\omega) = 2S(\omega), \quad \omega \geq 0.$$

In terms of $G(\omega)$, the Wiener-Khinchine relations become

$$\begin{cases} B(\tau) = \displaystyle\int_{0}^{\infty} G(\omega) \cos \omega\tau d\omega, & (3.3.11) \\ G(\omega) = \displaystyle\frac{2}{\pi} \int_{0}^{\infty} B(\tau) \cos \omega\tau d\tau. & (3.3.12) \end{cases}$$

Taking $\tau = 0$ in the first of these equation yields

$$\sigma^2 = B(0) = \int_0^\infty G(\omega)d\omega. \tag{3.3.13}$$

Eqs. (3.3.11) and (3.3.12) can also be obtained *directly* by expressing $X(t)$ as in Eq. (3.3.1), but as a *one-sided* summation of independent random sinusoids, counting only those with positive frequensies.

It is often useful to plot the quantity $\omega G(\omega)$ against $\ln \omega$ (the natural logarithm of ω), since

$$\sigma^2 = \int_{-\infty}^{+\infty} \omega G(\omega)d(\ln \omega). \tag{3.3.14}$$

The units of $G(\omega)$ are X^2-sec [where X symbolizes the units of $X(t)$]. In practice the spectra often depend on the frequency $f = \omega/2\pi$ expressed in cycles per second.

Cumulative and Normalized Spectra

The concept of spectral *density* – by analogy to mass density and probability density – invites consideration of the (cumulative) *spectral distribution function* $F(\omega)$. In the one-dimensional case it is defined as the contribution to the variance from components with frequencies less than or equal to ω and is related to $S(\omega)$ as follows:

$$F(\omega) = \int_{-\infty}^{\omega} S(\omega_1)d\omega_1, \quad -\infty \leq \omega \leq +\infty. \tag{3.3.15}$$

Note that $F(\omega)$ increases monotonically from 0 to σ^2. It can be argued that $F(\omega)$ is a more basic characterization of the frequency content of $X(t)$ than $S(\omega)$ since it permits spectral mass to be concentrated at discrete frequencies. [For example, it permits spectral representation of random functions containing a sinusoid with unknown phase angle.] The derivative of $F(\omega)$, provided it exists, is then

$$S(\omega) = \frac{d}{d\omega}F(\omega), \quad -\infty \leq \omega \leq +\infty. \tag{3.3.16}$$

Normalization with respect to the variance σ^2 leads to the *normalized spectral distribution function*

$$\overline{F}(\omega) = \frac{1}{\sigma^2}F(\omega), \quad -\infty \leq \omega \leq +\infty. \tag{3.3.17}$$

Similarly, the *normalized spectral density functions* (two- and one-sided, respectively) are

$$
\begin{aligned}
s(\omega) &= \frac{1}{\sigma^2} S(\omega), \quad -\infty \leq \omega \leq +\infty, \\
g(\omega) &= \frac{1}{\sigma^2} G(\omega), \quad 0 \leq \omega \leq +\infty.
\end{aligned} \tag{3.3.18}
$$

The one-sided normalized spectrum $g(\omega)$ has mathematical properties similar to those of the probability density function of a nonnegative random variable. Both functions are nonnegative and must have unit area. This analogy facilitates the interpretation of the spectral moments and related parameters introduced in Sec. 4.1.

More Formal Spectral Representation

A zero-mean random process $X(t)$ may be represented as the *real part* of a *complex random process* $X^c(t)$:

$$
\begin{aligned}
X(t) &= \mathrm{Re}\{X^c(t)\} = \mathrm{Re}\left\{\sum_{k=-K}^{K} Z_k e^{i\omega_k t}\right\} \\
&\to \mathrm{Re}\left\{\int_{\mathrm{All}.d\omega} e^{i\omega t} Z(d\omega)\right\}.
\end{aligned} \tag{3.3.19}
$$

It expresses $X^c(t)$ as the sum of many uncorrelated component functions of the form $e^{i\omega t} = \cos\omega t + i\sin\omega t$, each associated with a small interval $d\omega$ on the frequency axis and each multiplied by a (complex) random amplitude $Z_k = Z(d\omega)$ with mean zero. These component functions play the same role as the functions $X_k(t) = C_k \cos(\omega_k t + \Phi_k)$ in Eq. (3.3.1). The new representation differs from Eq. (3.3.1) in that it includes both sine and cosine terms (implying a further decomposition of the spectral content of σ^2). The amplitudes $|Z_k|$ and C_k are related as follows, taking $\Delta\omega = d\omega$,

$$
E[|Z(d\omega)|^2] = E[|Z_k|^2] = \frac{1}{2} E[C_k^2] = S(\omega_k)\Delta\omega, \tag{3.3.20}
$$

where $S(\omega)$ is the *two-sided* spectral density function.

The complex random amplitudes $Z(d\omega)$ constitute an uncorrelated random process in the frequency domain: $E[Z^*(d\omega_1)Z(d\omega_2)] = 0$ if the intervals $d\omega_1$ and $d\omega_2$ do not overlap (the star superscript identifies the complex conjugate). This property makes it easy to evaluate the covariance function of $X(t)$. Inserting $X(0)$ and $X(\tau)$ into the definition, and interchanging the operations of integration and expectation yields

$$B(\tau) = \text{Cov}[X(0), X(\tau)] = E\left[\int_{d\omega_1} Z^*(d\omega_1) \int_{d\omega_2} e^{i\omega\tau} Z(d\omega_2)\right]$$

$$= \int_{d\omega} e^{i\omega\tau} E[|Z(d\omega)|^2]. \tag{3.3.21}$$

The integration is over all (elementary, nonoverlapping) segments $d\omega$ covering the $(-\infty, +\infty)$ frequency axis. Combining Eqs. (3.3.20) and (3.3.21) generates the first of the Wiener-Khinchine (pair of) relations:

$$\begin{cases} B(\tau) = \displaystyle\int_{-\infty}^{+\infty} S(\omega) e^{i\omega\tau} d\omega, & (3.3.22) \\ S(\omega) = \dfrac{1}{2\pi} \displaystyle\int_{-\infty}^{+\infty} B(\tau) e^{-i\omega\tau} d\tau. & (3.3.23) \end{cases}$$

Taking $\tau = 0$ in Eq. (3.3.22) yields the basic formula for σ^2 [Eq. (3.3.5)]. Also, substituting $e^{i\omega\tau} = \cos\omega\tau + i\sin\omega\tau$, and accounting for the fact that $B(\tau)$ and $S(\omega)$ are even functions, leads to the alternate forms of the Wiener-Khinchine relations stated earlier.

Cross-Spectra

Consider *two* stationary random processes $X_1(t)$ and $X_2(t)$, each represented as a sum of $2K$ random sinusoids [as Eq. (3.3.19)]. Their cross-covariance function can be expressed as follows:

$$B_{X_1X_2}(\tau) = \text{Cov}[X_1(0), X_2(\tau)]$$

$$= \sum_{k=-K}^{K} E[X_{k,1}^*(0) X_{k,2}(\tau)] = \sum_{k=-K}^{K} E[Z_{k,1}^* Z_{k,2}] e^{i\omega_k t}. \tag{3.3.24}$$

Define the cross-spectrum $S_{X_1X_2}(\omega)$ as follows:

$$S_{X_1X_2}(\omega_k)\Delta\omega = E[Z_{k,1}^* Z_{k,2}]. \tag{3.3.25}$$

In the limit when $\Delta\omega \to 0$ and $K \to \infty$, Eq. (3.3.24) becomes the first member of the Fourier transform pair linking the cross-covariance function and the cross-spectrum:

$$\begin{cases} B_{X_1X_2}(\tau) = \displaystyle\int_{-\infty}^{+\infty} S_{X_1X_2}(\omega)e^{i\omega\tau}d\omega, & (3.3.26) \\ S_{X_1X_2}(\omega) = \dfrac{1}{2\pi}\displaystyle\int_{-\infty}^{+\infty} B_{X_1X_2}(\tau)e^{-i\omega\tau}d\tau. & (3.3.27) \end{cases}$$

Taking $\tau = 0$ in Eq. (3.3.26) yields the covariance between $X_1(t)$ and $X_2(t)$:

$$\text{Cov}[X_1, X_2] = B_{X_1X_2}(0) = \int_{-\infty}^{+\infty} S_{X_1X_2}(\omega)d\omega. \tag{3.3.28}$$

If the two random processes are in fact identical ($X_1 = X_2 = X$), then

$$S_{XX}(\omega) \equiv S_X(\omega), \quad \text{and} \quad B_{XX}(\tau) \equiv B_X(\tau), \tag{3.3.29}$$

and Eqs. (3.3.26) - (3.3.27) reduce to Eqs. (3.3.22) - (3.3.23).

Spectra of Discrete-Parameter Random Processes

One can argue that it is unnecessary to develop a separate spectral theory for discrete-parameter processes. If a random process is observed intermittently, it can be interpreted as a continuous-parameter process comprised of (randomly sized) delta functions. Nevertheless, two major types of discrete-parameter processes are explicitly considered below: (1) *random series*; and (2) *random impulse processes.*

Random Series

A stationary *random series* $X(t)$ defined at unit time intervals may be represented as a superposition of sinusoids, just as in Eq. (3.3.1). Since t takes only integer values, however, no information is available about high frequency fluctuations, and the spectral components are defined only in the frequency range from zero to the upper frequency limit $\omega = \pi$ or $(-\pi, \pi)$ if two-sided spectra are used. The first of the Wiener-Khinchine relations [Eq. (3.3.8) or (3.3.11)] becomes

$$B(\tau) = \int_{-\pi}^{+\pi} S(\omega)\cos\omega\tau d\omega = \int_{0}^{\pi} G(\omega)\cos\omega\tau d\omega. \tag{3.3.30}$$

The random series may also be represented, as mentioned above, as a special kind of continuous-parameter process consisting of delta functions $X(t_k)\delta(t - t_k)$ located at integer time intervals $t = t_k$:

$$X(t) = \sum_k X(t_k)\delta(t - t_k). \tag{3.3.31}$$

The covariance function $B(\tau)$, equal to zero *except* at the values $\tau = 0, \pm 1, \pm 2, \ldots$, is itself composed of delta functions. The second Wiener-Khinchine relation [Eq. (3.3.9)] may now be expressed as a summation:

$$S(\omega) = \frac{1}{2\pi}\int_{-\infty}^{+\infty} B(\tau)\cos\omega\tau d\tau = \frac{1}{2\pi}\sum_{\tau=-\infty}^{+\infty} B(\tau)\cos\omega\tau d\tau. \tag{3.3.32}$$

The corresponding formula for the one-sided spectral density function, expressed in terms of *nonnegative* values of τ only is:

$$G(\omega) = \frac{2}{\pi}\sum_{\tau=1}^{\infty} B(\tau)\cos\omega\tau + \frac{1}{\pi}B(0). \tag{3.3.33}$$

If $X(t)$ is observed at intervals Δt, it suffices to "scale" the time axis. The effect is that the frequency limit π must be replaced by (the Nyguist frequency) $\pi/\Delta t$ in Eq. (3.3.30), and the expressions for $G(\omega)$ and $S(\omega)$ [Eqs. (3.3.32) and (3.3.33), right side] must be multiplied by Δt.

As an example, consider an *uncorrelated* stationary random series $X(k\Delta t)$ where k is an integer. Equation (3.3.33) leads to the expected result that $G(\omega)$ is constant within the range $0 < \omega < \pi/\Delta t$:

$$G(\omega) = G_0 = \frac{\Delta t}{\pi}B(0) = \frac{\sigma^2\Delta t}{\pi}. \tag{3.3.34}$$

Inserting this expression into Eq. (3.3.30) yields

$$B(\tau) = \frac{\sigma^2\Delta t}{\pi}\int_0^{\pi/\Delta t} \cos\omega\tau d\tau = \begin{cases} \sigma^2, & \tau = 0, \\ 0, & \tau \neq 0. \end{cases} \tag{3.3.35}$$

Note that the variance σ^2 equals $\pi G_0/\Delta t$, where G_0 is the white noise intensity. The implication is that σ^2 must go to infinity as $\Delta t \to 0$.

Random Impulse Processes

A random impulse process $X(t)$, characterized by randomly spaced observation points t_k and random magnitudes $X(t_k)$, can be expressed as a sum

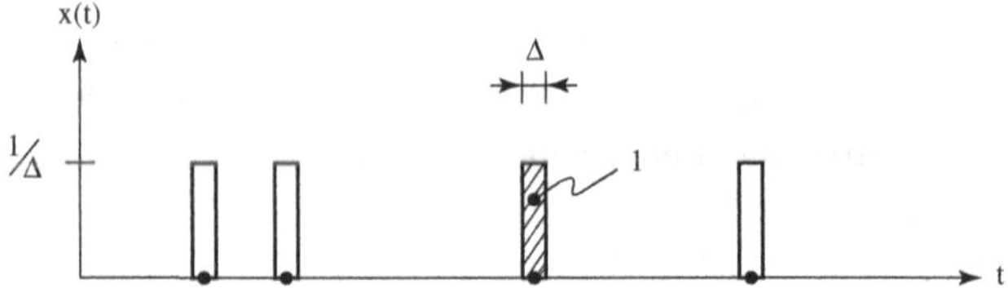

Fig. 3.3 Approximation of a point, or impulse, process.

of delta functions:

$$X(t) = \sum_k X(t_k)\delta(t - t_k). \tag{3.3.36}$$

In general, the impulse "arrival times" t_k and magnitudes $X(t_k)$ may all be correlated. Consider, in particular, a Poisson impulse process ("shot noise") with *unit* magnitudes:

$$X(t) = \sum_k \delta(t - t_k), \tag{3.3.37}$$

where t_k denotes the Poisson-distributed arrival times of the unit impulses $\delta(t - t_k)$. Let λ denote the mean rate of the Poisson arrivals. In analyzing the behavior of this process, it is instructive to replace the delta function by a narrow rectangular window of width Δ and height $1/\Delta$, as shown in Fig. 3.2, and then to take the limit $\Delta \to 0$. Two rectangular windows will overlap if the time between impulse arrivals is less than Δ. Since this time is exponentially distributed, the probability of overlap is

$$1 - e^{-\lambda\Delta} \to \lambda\Delta, \quad \text{for } \Delta \to 0. \tag{3.3.38}$$

Similarly, the probability that the time between Poisson points is less than $\Delta - \tau$ (where $0 \le \tau \le \Delta$) approaches $\lambda(\Delta - \tau)$.

From the definition of expectation (considering that the product $X(t)X(t + \tau)$ is either zero or one), one obtains

$$B(\tau) = E[X(t)X(t+\tau)] = \begin{cases} \lambda(\Delta - |\tau|)/\Delta^2, & |\tau| \le \Delta, \\ 0, & |\tau| \ge \Delta, \end{cases} \tag{3.3.39}$$

which, for $\tau = 0$, yields

$$\sigma^2 = B(0) = \lambda/\Delta. \tag{3.3.40}$$

Note that $B(\tau)$ is in effect the covariance function of a two-valued random process that alternates between $X = 0$ and $X = 1/\Delta$. Its spectral density function, obtained by combining Eqs. (3.3.12) and (3.3.39), equals:

$$G(\omega) = \frac{2}{\pi} \int_0^{\Delta} \frac{1}{\Delta^2} \lambda(\Delta - \tau) \cos \omega\tau d\tau = \frac{\lambda}{\pi} \left[\frac{\sin(\omega\Delta/2)}{(\omega\Delta/2)} \right]^2 . \quad (3.3.41)$$

When $\Delta \to 0$, $\sin(\omega\Delta/2) \to \omega\Delta/2$, and the spectral density function becomes a constant:

$$G_0 \equiv G(\omega) = \frac{\lambda}{\pi}. \quad (3.3.42)$$

Note that, in real-world applications, the variance σ^2 will tend to be arbitrary in the sense that, for physically realizable phenomena, the width Δ will not be zero but small and probably unknown. In the "ideal" case $\Delta = 0$ and $\sigma^2 = \infty$. Evidently, the stable parameters of the Poisson impulse process are the spectral intensity G_0 or the mean rate $\lambda = \pi G_0$.

3.4 Spectral Analysis of Homogeneous Random Fields

Introduction to the Two-Dimensional Case

The extension of Eq. (3.3.1), stated below, expresses a two-dimensional homogeneous random process $X(t_1, t_2)$ with mean m as a *double* summation of uncorrelated sinusoidal functions, each associated with a different point

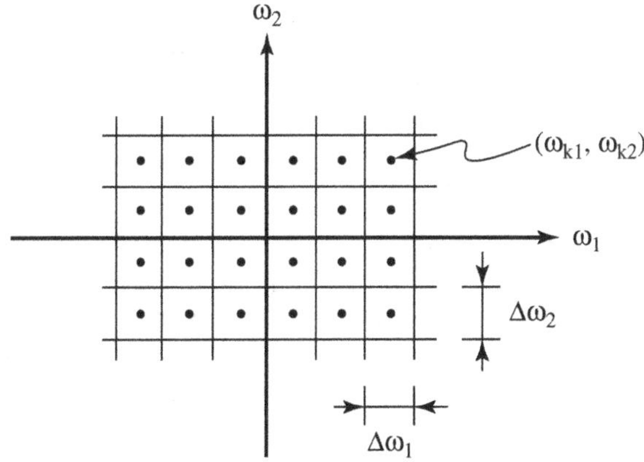

Fig. 3.4 Discretization of the two-dimensional frequency domain.

$(\omega_{k_1}, \omega_{k_2})$ in the two-dimensional frequency domain (see Fig. 3.4):

$$X(t_1, t_2) = m + \sum_{k_1=-K_1}^{K_1} \sum_{k_2=-K_2}^{K_2} X_{k_1 k_2}(t_1, t_2), \tag{3.4.1}$$

where

$$X_{k_1 k_2}(t_1, t_2) = C_{k_1 k_2} \cos(\omega_{k_1} t_1 + \omega_{k_2} t_2 + \Phi_{k_1 k_2}).$$

$C_{k_1 k_2}$ is the random amplitude, and $\Phi_{k_1 k_2}$ the random phase angle, of the component with frequencies $\omega_{k_i} = \pm[\Delta\omega_i(2k_i - 1)/2]$ ($i = 1$, 2 and $k_i = 1, \ldots, K_i$) shown in Fig. 3.4. All amplitudes and phase angles are mutually independent, and each phase angle is uniformly distributed over the range $(0, 2\pi)$. The variance of $X(t_1, t_2)$ is the sum of the variances of all the component functions:

$$\sigma^2 = E[X^2] = \sum_{k_1=-K_1}^{K_1} \sum_{k_2=-K_2}^{K_2} \frac{1}{2} E[C_{k_1 k_2}^2], \tag{3.4.2}$$

in which the factor 1/2 corresponds to the expectation of $\cos^2 \Phi$. The two-dimensional spectral density function can be introduced as follows:

$$S(\omega_{k_1}, \omega_{k_2}) \Delta\omega_1 \Delta\omega_2 = \frac{1}{2} E[C_{k_1 k_2}^2]. \tag{3.4.3}$$

In the limit for $K_1, K_2 \to \infty$ and $\Delta\omega_1, \Delta\omega_2 \to 0$, the integral of $S(\omega_1, \omega_2)$ over all frequencies equals the variance

$$\begin{aligned} \sigma^2 &= \sum_{k_1=-K_1}^{K_1} \sum_{k_2=-K_2}^{K_2} S(\omega_{k_1}, \omega_{k_2}) \Delta\omega_1 \Delta\omega_2 \\ &\to \int_{-\infty}^{+\infty} \int_{-\infty}^{+\infty} S(\omega_1, \omega_2) d\omega_1 d\omega_2. \end{aligned} \tag{3.4.4}$$

From its definition [Eq. (3.4.3)] it follows that the 2-D s.d.f. $S(\omega_1, \omega_2)$ is real and nonnegative; its main property is that integration over the 2-D frequency domain yields the variance $\sigma^2 = \text{Var}[X(t_1, t_2)]$.

Wiener-Khinchine Relations

By introducing Eq. (3.4.1) into the definition of the bivariate covariance function $B(\tau_1, \tau_2)$, accounting for the fact that the random sinusoids

$X_{k_1k_2}(t_1, t_2)$ are mutually independent, and inserting Eq. (3.4.3), one obtains

$$
\begin{aligned}
B(\tau_1, \tau_2) &= \mathrm{Cov}[X(0,0), X(\tau_1, \tau_2)] \\
&= \sum_{k_1=-K_1}^{K_1} \sum_{k_2=-K_2}^{K_2} E[X_{k_1k_2}(0,0)X_{k_1k_2}(\tau_1, \tau_2)] \\
&= \sum_{k_1=-K_1}^{K_1} \sum_{k_2=-K_2}^{K_2} \frac{1}{2} E[C_{k_1k_2}^2] \cos(\omega_{k_1}\tau_1 + \omega_{k_2}\tau_2). \qquad (3.4.5)
\end{aligned}
$$

Using Eq. (3.4.3) and taking K_i and $\Delta\omega_i (i = 1, 2)$ to their limits yields the first relation of the (two-dimensional) Wiener-Khinchine transform pair:

$$
\begin{cases}
B(\tau_1, \tau_2) = \displaystyle\int_{-\infty}^{+\infty} \int_{-\infty}^{+\infty} S(\omega_1, \omega_2) \cos(\omega_1\tau_1 + \omega_2\tau_2) d\omega_1 d\omega_2, & (3.4.6) \\
S(\omega_1, \omega_2) = \dfrac{1}{(2\pi)^2} \displaystyle\int_{-\infty}^{+\infty} \int_{-\infty}^{+\infty} B(\tau_1, \tau_2) \cos(\omega_1\tau_1 + \omega_2\tau_2) d\tau_1 d\tau_2. & (3.4.7)
\end{cases}
$$

Taking $\tau_1 = \tau_2 = 0$ in Eq. (3.4.6) leads to the basic expression for the variance [Eq. (3.4.4)]. An alternate representation of the Wiener-Khinchine relations (in terms of exponential instead of cosine functions) is:

$$
\begin{cases}
B(\tau_1, \tau_2) = \displaystyle\int_{-\infty}^{+\infty} \int_{-\infty}^{+\infty} S(\omega_1, \omega_2) e^{i(\omega_1\tau_1 + \omega_2\tau_2)} d\omega_1 d\omega_2, & (3.4.8) \\
S(\omega_1, \omega_2) = \dfrac{1}{(2\pi)^2} \displaystyle\int_{-\infty}^{+\infty} \int_{-\infty}^{+\infty} B(\tau_1, \tau_2) e^{-i(\omega_1\tau_1 + \omega_2\tau_2)} d\tau_1 d\tau_2. & (3.4.9)
\end{cases}
$$

It is instructive to express the integral in Eq. (3.4.9) as a sum of contributions from the four quadrants of the lag space (τ_1, τ_2) and express the exponential function as follows:

$$
\begin{aligned}
e^{-i(\omega_1\tau_1 + \omega_2\tau_2)} &= \cos(\omega_1\tau_1 + \omega_2\tau_2) - i \sin(\omega_1\tau_1 + \omega_2\tau_2) \\
&= \cos\omega_1\tau_1 \cos\omega_2\tau_2 - \sin\omega_1\tau_1 \sin\omega_2\tau_2 \\
&\quad - i[\sin\omega_1\tau_1 \cos\omega_2\tau_2 + \cos\omega_1\tau_1 \sin\omega_2\tau_2]. \qquad (3.4.10)
\end{aligned}
$$

Accounting for the relations $B(\tau_1, \tau_2) = B(-\tau_1, -\tau_2)$, $B(\tau_1, -\tau_2) = B(-\tau_1, \tau_2)$; $\cos(-x) = \cos x$; and $\sin(-x) = -\sin x$, it is easy to confirm that Eqs. (3.4.7) and (3.4.9) are equivalent (as the imaginary terms vanish), and that $S(\omega_1, \omega_2)$ possesses the same symmetries as $B(\tau_1, \tau_2)$, namely: $S(\omega_1, \omega_2) = S(-\omega_1, -\omega_2)$ and $S(-\omega_1, \omega_2) = S(\omega_1, -\omega_2)$. A similar analysis starting from Eq. (3.4.8) proves the equivalence of Eqs. (3.4.6) and (3.4.8).

The Quadrant Symmetric Case

If the covariance function $B(\tau_1, \tau_2)$ is *quadrant symmetric* (q.s.), its spectral density function $S(\omega_1, \omega_2)$ will also be quadrant symmetric. Starting from Eq. (3.4.7), accounting for $B(\tau_1, \tau_2) = B(\tau_1, -\tau_2)$, one obtains

$$S(\omega_1, \omega_2) = \frac{4}{(2\pi)^2} \int_0^\infty \int_0^\infty B(\tau_1, \tau_2) \cos \omega_1 \tau_1 \cos \omega_2 \tau_2 d\tau_1 d\tau_2. \quad (3.4.11)$$

Likewise, Eq. (3.4.6) becomes

$$B(\tau_1, \tau_2) = 4 \int_0^\infty \int_0^\infty S(\omega_1, \omega_2) \cos \omega_1 \tau_1 \cos \omega_2 \tau_2 s d\omega_1 d\omega_2. \quad (3.4.12)$$

It is attractive to introduce the 2-D spectral density function $G(\omega_1, \omega_2)$ defined for *nonnegative* frequencies only:

$$G(\omega_1, \omega_2) = 4S(\omega_1, \omega_2), \quad \omega_1, \omega_2 \geq 0. \quad (3.4.13)$$

The main property of $G(\omega_1, \omega_2)$ follows from Eqs. (3.4.4) and (3.4.13) :

$$\sigma^2 = \int_0^\infty \int_0^\infty G(\omega_1, \omega_2) d\omega_1 d\omega_2. \quad (3.4.14)$$

Also, when the correlation structure is quadrant symmetric, the 2-D Wiener-Khinchine relations may be expressed as:

$$\begin{cases} B(\tau_1, \tau_2) = \displaystyle\int_0^\infty \int_0^\infty G(\omega_1, \omega_2) \cos \omega_1 \tau_1 \cos \omega_2 \tau_2 d\omega_1 d\omega_2, & (3.4.15) \\ G(\omega_1, \omega_2) = \left(\dfrac{2}{\pi}\right)^2 \displaystyle\int_0^\infty \int_0^\infty B(\tau_1, \tau_2) \cos \omega_1 \tau_1 \cos \omega_2 \tau_2 d\tau_1 d\tau_2. & (3.4.16) \end{cases}$$

Direction-Dependent One-Dimensional Random Variation

If one of the coordinates of the (t_1, t_2) plane is fixed (for instance, $t_2 = t_2^*$) the random function of t_1 may be represented in one of several ways:

$$X(t_1) \equiv X(t_1, t_2^*) \equiv X^{(1)}(t_1, t_2^*). \quad (3.4.17)$$

This one-dimensional, direction-dependent process is stationary and, owing to homogeneity, its statistics will not depend of t_2^*. Its variance σ^2 may be expressed either by Eq. (3.4.4) or in terms of the *one-dimensional* s.d.f. $S(\omega_1)$ characterizing $X(t_1)$:

$$\sigma^2 = \int_{-\infty}^{+\infty} \int_{-\infty}^{+\infty} S(\omega_1, \omega_2) d\omega_1 d\omega_2 = \int_{-\infty}^{+\infty} S(\omega_1) d\omega_1. \tag{3.4.18}$$

Equation (3.4.18) implies that the "marginal" s.d.f. $S(\omega_1)$ can be found from the "joint" or "bivariate" s.d.f. $S(\omega_1, \omega_2)$ by integration over all frequencies ω_2:

$$S(\omega_1) = \int_{-\infty}^{+\infty} S(\omega_1, \omega_2) d\omega_2. \tag{3.4.19}$$

Likewise, the s.d.f. characterizing one-dimensional random variation in function of t_2 is

$$S(\omega_2) = \int_{-\infty}^{+\infty} S(\omega_1, \omega_2) d\omega_1. \tag{3.4.20}$$

It the 2-D correlation structure is quadrant symmetric, Eqs. (3.4.18) through (3.4.20) can be restated in terms of spectral density functions that depend on positive frequencies only:

$$\sigma^2 = \int_0^{\infty} \int_0^{\infty} G(\omega_1, \omega_2) d\omega_1 d\omega_2 = \int_0^{\infty} G(\omega_1) d\omega_1, \tag{3.4.21}$$

and

$$G(\omega_i) = \int_0^{\infty} G(\omega_1, \omega_2) d\omega_j, \quad i \neq j \text{ and } i, j = 1, 2. \tag{3.4.22}$$

While the marginal spectra $S(\omega_i)$ or $G(\omega_i)$ can be obtained when $S(\omega_1, \omega_2)$ is known, the reverse is generally not true.

Cumulative and Normalized Spectra

The bivariate spectral distribution function $F(\omega_1, \omega_2)$ is defined as the contribution to the variance σ^2 from components with frequency-plane coordinates less than ω_1 and ω_2, respectively:

$$F(\omega_1, \omega_2) = \int_{-\infty}^{\omega_1} \int_{-\infty}^{\omega_2} S(\omega_1', \omega_2') d\omega_1' d\omega_2'. \tag{3.4.23}$$

This function increases monotonically with ω_i $(i = 1, 2)$ and permits the bivariate "spectral mass" to be concentrated at discrete points in the frequency domain. We have

$$F(-\infty, -\infty) = 0, \quad \text{and} \quad F(\infty, \infty) = \sigma^2. \tag{3.4.24}$$

The "marginal" spectral distribution function characterizing $X(t_1) \equiv X^{(1)}(t_1, t_2)$ may be found by integrating $F(\omega_1, \omega_2)$ over all values of ω_2:

$$F(\omega_1) = \int_{-\infty}^{+\infty} F(\omega_1, \omega_2) d\omega_2. \tag{3.4.25}$$

Furthermore, provided the derivatives exits, we may write

$$S(\omega_1, \omega_2) = \frac{\partial^2 F(\omega_1, \omega_2)}{\partial \omega_1, \partial \omega_2}, \tag{3.4.26}$$

and

$$S(\omega_i) = \frac{dF(\omega_i)}{d\omega_i}, \quad i = 1, 2. \tag{3.4.27}$$

The calculus involved is much the same as in elementary probability analysis of jointly distributed random variables. This becomes even more obvious when the operations are expressed in terms of *normalized* spectral density functions:

$$s(\omega_1, \omega_2) = \frac{1}{\sigma^2} S(\omega_1, \omega_2), \text{ and} \tag{3.4.28}$$

$$s(\omega_i) = \frac{1}{\sigma^2} S(\omega_i), \quad i = 1, 2. \tag{3.4.29}$$

For 2-D random fields with quadrant symmetric correlation structure, all equations in this section [Eqs. (3.4.23)–(3.4.29)] can be restated in terms of the "positive frequency" spectra $G(\omega_1, \omega_2)$ and $G(\omega_i)$. Of particular interest are the normalized spectral density functions:

$$g(\omega_i) = \frac{1}{\sigma^2} G(\omega_i) \equiv 2s(\omega_i), \quad i = 1, 2. \tag{3.4.30}$$

Normalized spectra have properties similar to those of the probability density functions of two (dependent) random variables. This analogy greatly facilitates the interpretation of joint and marginal *spectral moments* and related parameters, introduced in Sec. 4.1.

In applications, the use of the simpler "positive frequency" version of the spectra is preferred whenever the form of the correlation structure permits it, namely when $B(\tau_1, \tau_2)$ is quadrant symmetric. A number of important special cases (all meeting the q.s. condition) are examined next.

Special Cases

The Correlation Structure is Separable

In this case the 2-D correlation function and normalized spectral density function may be expressed in product form:

$$\begin{aligned} \rho(\tau_1, \tau_2) &= \rho(\tau_1)\rho(\tau_2), \\ g(\omega_1, \omega_2) &= g(\omega_1)g(\omega_2). \end{aligned} \tag{3.4.31}$$

(The equivalent condition in elementary probability theory is that two random variables are statistically independent.)

The Correlation Structure is Isotropic

The covariance depends only on the distance between observation points in the parameter plane (t_1, t_2), and hence on the distance $\tau = \sqrt{\tau_1^2 + \tau_2^2}$ in the lag space. We may write

$$B(\tau_1, \tau_2) = B\left(\sqrt{\tau_1^2 + \tau_2^2}\right) \equiv B^R(\tau) = B(\tau), \tag{3.4.32}$$

where $B(\tau)$ is the covariance function characterizing one-dimensional random variation in any direction. Since Fourier transformation preserves the circular symmetry, the 2-D s.d.f. can be expressed in terms of a radial spectrum $G^R(\omega)$, defined as follows:

$$G(\omega_1, \omega_2) = G\left(\sqrt{\omega_1^2 + \omega_2^2}\right) \equiv G^R(\omega), \tag{3.4.33}$$

where $\omega = \sqrt{\omega_1^2 + \omega_2^2}$ is a distance in the frequency plane (ω_1, ω_2). The integral of $G^R(\omega)$ over all positive values of ω is not equal to σ^2. Rather,

$$\begin{aligned} \sigma^2 &= \int_0^\infty \int_0^\infty G(\omega_1, \omega_2) d\omega_1 d\omega_2 \\ &= \int_0^{\pi/2} d\theta \int_0^\infty \omega G^R(\omega) d\omega = \frac{\pi}{2} \int_0^\infty \omega G^R(\omega) d\omega. \end{aligned} \tag{3.4.34}$$

Recall that integration of $G(\omega_1, \omega_2)$ over one of the frequency coordinates yields the spectral density function that characterizes random variation *on the line* (in any direction, in the isotropic case):

$$G(\omega_1) = \int_0^\infty G(\omega_1, \omega_2) d\omega_2. \tag{3.4.35}$$

$G(\omega_1)$ is the one-dimensional Fourier transform of $B(\tau_1)$, and its integral over $\omega_1 \geq 0$ *does* equal σ^2. The third spectrum is the *radial* spectral density function $R(\omega)$ for which the integral over $0 \leq \omega \leq \infty$ is by *definition* equal to σ^2. From Eq. (3.4.34) it follows that

$$R(\omega) = \frac{\pi}{2} \omega G^R(\omega), \tag{3.4.36}$$

which satisfies

$$\sigma^2 = \int_0^\infty R(\omega) d\omega. \tag{3.4.37}$$

The spectral representation of isotropic random fields is discussed in more detail at the end of this section.

The Correlation Structure is Ellipsoidal

Random fields with an ellipsoidal correlation structure can be made isotropic by scaling the coordinate axes. Their two-dimensional covariance function depends only on

$$\tau = [(a_1\tau_1)^2 + (a_2\tau_2)^2]^{1/2}, \tag{3.4.38}$$

while the associated two-dimensional s.d.f. is a function of

$$\omega = \left[\left(\frac{\omega_1}{a_1} \right)^2 + \left(\frac{\omega_2}{a_2} \right)^2 \right]^{1/2}, \tag{3.4.39}$$

where a_1 and a_2 are scaling factors. In case $a_1 = a_2$, the correlation structure becomes isotropic; and by replacing $a_i\tau_i$ by τ_i' and (ω_i/a_i) by ω_i', all the results for isotropic random fields become applicable to fields having an ellipsoidal correlation structure.

Extension to Higher Dimensions

Basic Relations

The analysis just presented will now be extended to homogeneous random fields of higher dimension, with as starting point the basic representation of the random field $X(\mathbf{t}) = X(t_1, \ldots, t_n)$, with mean m, as a sum of independent random sinusoidal components:

$$X(\mathbf{t}) = m + \sum_{\mathbf{k}=-\mathbf{K}}^{\mathbf{K}} X_{\mathbf{k}}(\mathbf{t}), \tag{3.4.40}$$

where

$$X_{\mathbf{k}}(\mathbf{t}) = C_{\mathbf{k}} \cos(\boldsymbol{\omega}_{\mathbf{k}} \cdot \mathbf{t} + \Phi_{\mathbf{k}}).$$

$C_{\mathbf{k}}$ denotes the random amplitude, and $\Phi_{\mathbf{k}}$ the random phase angle, of the component $X_{\mathbf{k}}(\mathbf{t})$ associated with a point in the n-dimensional frequency domain; the cosine function depends on the inner product of the frequency and time vectors, $\boldsymbol{\omega}_{\mathbf{k}} = (\omega_{k_1}, \ldots, \omega_{k_n})$ and $\mathbf{t} = (t_1, \ldots, t_n)$:

$$\boldsymbol{\omega}_{\mathbf{k}} \cdot \mathbf{t} = \omega_{k_1} t_1 + \omega_{k_2} t_2 + \cdots + \omega_{k_n} t_n, \tag{3.4.41}$$

where $\omega_{k_i} = \pm[\Delta\omega_i(2k_i - 1)/2]$ and $k_i = 1, \ldots, K_i$ is the ith coordinate of a typical point in the n-dimensional frequency domain.

The multi-dimensional spectral density function is defined by

$$S(\boldsymbol{\omega}_k)\Delta\boldsymbol{\omega} = \frac{1}{2} E[C_{\mathbf{k}}^2], \tag{3.4.42}$$

in which $\Delta\boldsymbol{\omega} = \Delta\omega_1 \Delta\omega_2 \cdots \Delta\omega_n$ and the factor $1/2$ accounts for the expectation of $\cos^2 \Phi_{\mathbf{k}}$. The variance of $X(\mathbf{t})$ is obtained by summing the spectral masses throughout the frequency domain:

$$\sigma^2 = \sum_{\mathbf{k}=-\mathbf{K}}^{\mathbf{K}} \frac{1}{2} E[C_{\mathbf{k}}^2] \to \int_{-\infty}^{+\infty} S(\boldsymbol{\omega}) d\boldsymbol{\omega}, \tag{3.4.43}$$

where $d\boldsymbol{\omega} = d\omega_1 d\omega_2 \cdots d\omega_n$ represents an elementary region in the frequency domain. Likewise the covariance function is

$$B(\boldsymbol{\tau}) = \sum_{\mathbf{k}=-\mathbf{K}}^{\mathbf{K}} \frac{1}{2} E[C_{\mathbf{k}}^2] \cos \boldsymbol{\omega}_{\mathbf{k}} \cdot \boldsymbol{\tau}. \tag{3.4.44}$$

Inserting Eq. (3.4.42) generates the first relation of the (generalized)

Wiener-Khinchine transform pair:

$$\begin{cases} B(\boldsymbol{\tau}) = \int_{-\infty}^{+\infty} S(\boldsymbol{\omega}) \cos \boldsymbol{\omega} \cdot \boldsymbol{\tau} d\boldsymbol{\omega}, & (3.4.45) \\ S(\boldsymbol{\omega}) = \dfrac{1}{(2\pi)^n} \int_{-\infty}^{+\infty} B(\boldsymbol{\tau}) \cos \boldsymbol{\omega} \cdot \boldsymbol{\tau} d\boldsymbol{\tau}. & (3.4.46) \end{cases}$$

Taking $\boldsymbol{\tau} = \mathbf{0}$ in Eq. (3.4.45) yields Eq. (3.4.43), the basic relation for the variance of the homogeneous random field.

More Formal Representation

A parallel set of results can be obtained by starting with the more formal representation of a zero mean random field $X(\mathbf{t})$ as the *real part* of a *complex random field* $X^*(t)$ [as in Eq. (3.3.19)]:

$$\begin{aligned} X(\mathbf{t}) &= \mathrm{Re}\{X^*(\mathbf{t})\} = \mathrm{Re}\left\{\sum_{\mathbf{k}} Z_{\mathbf{k}} e^{i\omega_k \cdot \mathbf{t}}\right\} \\ &\to \mathrm{Re}\left\{\int_{d\boldsymbol{\omega}} e^{i\boldsymbol{\omega} \cdot \boldsymbol{\tau}} Z(d\boldsymbol{\omega})\right\}. \end{aligned} \tag{3.4.47}$$

This equation states that $X(\mathbf{t})$ is the sum of many elementary oscillations $e^{i\boldsymbol{\omega} \cdot \mathbf{t}}$, each associated with a different small region $d\boldsymbol{\omega}$ in the frequency domain and each multiplied by a complex random amplitude $Z_{\mathbf{k}} = Z(d\boldsymbol{\omega})$. These random amplitudes constitute an uncorrelated random field with mean zero,

$$E[Z(d\boldsymbol{\omega})] = 0. \tag{3.4.48}$$

For any two regions $d\boldsymbol{\omega}_1$ and $d\boldsymbol{\omega}_2$ that do not overlap, we have

$$E[Z^*(d\boldsymbol{\omega}_1) Z(d\boldsymbol{\omega}_2)] = 0. \tag{3.4.49}$$

The variance of each elementary contribution is

$$E[|Z(d\boldsymbol{\omega})|^2] = E[|Z_{\mathbf{k}}|^2] = \frac{1}{2} E[C_{\mathbf{k}}^2] = S(\boldsymbol{\omega}) d\boldsymbol{\omega}. \tag{3.4.50}$$

The covariance function of $X(\mathbf{t})$ is obtained by inserting $X(\mathbf{0})$ and $X(\boldsymbol{\tau})$ into the definition:

$$B(\boldsymbol{\tau}) = \mathrm{Cov}[X(\mathbf{0}), X(\boldsymbol{\tau})] = E\left[\left(\sum_{\mathbf{k}} Z^*_{\mathbf{k}}\right)\left(\sum_{\mathbf{k}} Z_{\mathbf{k}} \exp\{i\boldsymbol{\omega}_k \cdot \boldsymbol{\tau}\}\right)\right]$$

$$= \sum_{\mathbf{k}} E[|Z_{\mathbf{k}}|^2] \exp\{i\boldsymbol{\omega}_k \cdot \boldsymbol{\tau}\}. \tag{3.4.51}$$

Combining Eqs. (3.4.50) and (3.4.51), and replacing summation by integration, yields the first relation of the Wiener-Khinchine transform pair:

$$\begin{cases} B(\boldsymbol{\tau}) = \displaystyle\int_{-\infty}^{+\infty} S(\boldsymbol{\omega}) e^{i\boldsymbol{\omega}\cdot\boldsymbol{\tau}} d\boldsymbol{\omega}, & (3.4.52) \\ S(\boldsymbol{\omega}) = \dfrac{1}{(2\pi)^n} \displaystyle\int_{-\infty}^{+\infty} B(\boldsymbol{\tau}) e^{-i\boldsymbol{\omega}\cdot\boldsymbol{\tau}} d\boldsymbol{\tau}. & (3.4.53) \end{cases}$$

For real random functions, Eqs. (3.4.45) - (3.4.46) and Eqs. (3.4.52) - (3.4.53) are equivalent since the imaginary terms cancel in pairs.

Cross-Spectra

Consider two homogeneous random fields $X_1(\mathbf{t})$ and $X_2(\mathbf{t})$ defined in the same parameter space $\mathbf{t} = (t_1, t_2, \ldots, t_n)$. The cross-spectrum $S_{X_1X_2}(\boldsymbol{\omega})$ may be introduced as follows:

$$S_{X_1X_2}(\boldsymbol{\omega}_k)\Delta\boldsymbol{\omega} = E[Z^*_{\mathbf{k},1} Z_{\mathbf{k},2}]. \tag{3.4.54}$$

The analysis procedure that yields Eq. (3.4.52) also leads to the first equation of the Fourier transform pair linking $B_{X_1X_2}(\boldsymbol{\tau})$ and $S_{X_1X_2}(\omega)$:

$$\begin{cases} B_{X_1X_2}(\boldsymbol{\tau}) = \displaystyle\int_{-\infty}^{+\infty} S_{X_1X_2}(\boldsymbol{\omega}) e^{i\boldsymbol{\omega}\cdot\boldsymbol{\tau}} d\boldsymbol{\omega}, & (3.4.55) \\ S_{X_1X_2}(\boldsymbol{\omega}) = \dfrac{1}{(2\pi)^n} \displaystyle\int_{-\infty}^{+\infty} B_{X_1X_2}(\boldsymbol{\tau}) e^{-i\boldsymbol{\omega}\cdot\boldsymbol{\tau}} d\boldsymbol{\tau}. & (3.4.56) \end{cases}$$

Taking $\boldsymbol{\tau} = \mathbf{0}$ in Eq. (3.4.55) gives the zero-lag covariance:

$$B_{X_1X_2}(\mathbf{0}) = \mathrm{Cov}[X_1(\mathbf{t}), X_2(\mathbf{t})] = \int_{-\infty}^{+\infty} S_{X_1X_2}(\boldsymbol{\omega}) d\boldsymbol{\omega}. \tag{3.4.57}$$

These relations are used in Sec. 3.5 to derive second-order statistics of a random field $X(\mathbf{t})$ and its (partial) derivatives.

Spectra of Isotropic Random Fields

In general, if a random field $X(\mathbf{t})$ is isotropic, its covariance function $B(\boldsymbol{\tau})$ can be expressed as follows:

$$B(\boldsymbol{\tau}) \equiv B(\tau_1, \tau_2, \ldots, \tau_n) = B(\tau, 0, \ldots, 0) = B(\tau) \equiv B^R(\tau), \qquad (3.4.58)$$

where

$$\tau = [\tau_1^2 + \tau_2^2 + \cdots + \tau_n^2]^{1/2}, \qquad (3.4.59)$$

with $B(\tau) \equiv B^R(\tau)$ denoting the covariance function of the random variation on the line (in any direction); $B^R(\tau)$ may be called the "radial" covariance function.

In the frequency domain three different "scalar" spectral density functions can be defined for an n-dimensional isotropic random field. The first one characterizes the random variation on a line in the parameter space:

$$G(\omega_1) = \int_0^{\omega_2} \cdots \int_0^{\omega_n} G(\omega_1, \omega_2, \ldots, \omega_n) d\omega_2 \cdots d\omega_n, \qquad (3.4.60)$$

obtainable from the radial covariance function $B(\tau)$ by one-dimensional Fourier transformation [Eq. (3.3.12)]:

$$G(\omega) = \frac{2}{\pi} \int_0^{\infty} B(\tau) \cos \omega\tau \, d\tau. \qquad (3.4.61)$$

The second representation of the spectral content of $X(\mathbf{t})$ is

$$G(\boldsymbol{\omega}) = G(\omega_1, \omega_2, \ldots, \omega_n) \equiv G^R(\omega), \qquad (3.4.62)$$

where ω is the vector distance in the (n-dimensional) frequency domain:

$$\omega = [\omega_1^2 + \omega_2^2 + \cdots + \omega_n^2]^{1/2}. \qquad (3.4.63)$$

The Fourier transform link between $B(\boldsymbol{\tau})$ and $G(\boldsymbol{\omega})$ guarantees that circular symmetry is preserved in the frequency domain. However, the "radial spectrum" $G^R(\omega)$ does not possess the property that its integral over ω equals σ^2. Instead, the integration must be over the entire frequency domain:

$$\begin{aligned} \sigma^2 &= \int_{\mathbf{0}}^{\infty} G(\boldsymbol{\omega}) d\boldsymbol{\omega} \\ &= \frac{\pi^{n/2}}{2^{n-1}\Gamma(\frac{n}{2})} \int_0^{\infty} \omega^{n-1} G^R(\omega) d\omega, \quad n = 1, 2, \ldots. \end{aligned} \qquad (3.4.64)$$

For $n = 1$, using $\Gamma(1/2) = \sqrt{\pi}$, this equation becomes

$$\sigma^2 = \int_0^\infty G^R(\omega) d\omega. \tag{3.4.65}$$

For $n = 2$, it yields

$$\sigma^2 = \frac{\pi}{2} \int_0^\infty \omega G^R(\omega) d\omega, \tag{3.4.66}$$

and for $n = 3$, since $\Gamma(3/2) = \sqrt{\pi}/2$, one obtains

$$\sigma^2 = \frac{\pi}{2} \int_0^\infty \omega^2 G^R(\omega) d\omega. \tag{3.4.67}$$

The third representation of the spectral content of an isotropic random field is the "radial spectral density function" $R(\omega)$, which obeys

$$\sigma^2 = \int_0^\infty R(\omega) d\omega. \tag{3.4.68}$$

Comparing Eqs. (3.4.64) and (3.4.68) implies that the functions $G^R(\omega)$ and $R(\omega)$ are related as follows:

$$R(\omega) = \frac{\pi^{n/2}}{2^{n-1}\Gamma(\frac{n}{2})} \omega^{n-1} G^R(\omega), \quad n = 1, 2, \ldots, \tag{3.4.69}$$

which for $n = 1$ reduces to $R(\omega) = G^R(\omega)$.

In conclusion, the three spectra are: (1) the one-dimensional spectral density function $G(\omega)$; (2) the radial spectrum $G^R(\omega)$; and (3) the radial spectral density function $R(\omega)$. Obviously, the different spectral representations are functionally related. Of particular interest is the relationship between the spectra $G^R(\omega)$ and $G(\omega)$ for the case $n = 3$ (Tsuji [131]):

$$G(\omega) = \int_\omega^\infty \frac{G^R(k)}{k} dk, \tag{3.4.70}$$

$$G^R(\omega) = -\omega \frac{dG(\omega)}{d\omega}. \tag{3.4.71}$$

The significance of these relations stems from the fact that they imply certain restrictions on the choice of analytical models for the covariance function $B(\tau)$ and the corresponding 1-D spectral density function $G(\omega)$ for 3-D isotropic random fields. Specifically, since $G^R(\omega)$ must be nonnegative by virtue of its definition, Eq. (3.4.71) implies the condition:

$$\frac{dG(\omega)}{d\omega} \leq 0. \tag{3.4.72}$$

In words, only those 1-D spectral density functions that *decrease monotonically with frequency* qualify as models for 3-D *isotropic* random fields.

3.5 Input-Output Relations for Invariant Linear Systems

System Representation

A multi-dimensional linear operator may be described either by a unit impulse response function $h(\mathbf{t})$ or by a complex frequency response function $H(\boldsymbol{\omega})$, in much the same way as in the 1-D case. The first standard excitation in n-dimensional space is the *unit impulse* at the location $\mathbf{t} = \mathbf{t}^*$:

$$x(\mathbf{t}) = \delta(\mathbf{t} - \mathbf{t}^*) \equiv \delta(t_1 - t_1^*)\delta(t_2 - t_2^*) \cdots \delta(t_n - t_n^*). \tag{3.5.1}$$

The system is assumed to be at rest prior to the application of the unit impulse. The response is

$$y(\mathbf{t}) = h(\mathbf{t} - \mathbf{t}^*) = h(\boldsymbol{\tau}), \tag{3.5.2}$$

where $h(\boldsymbol{\tau})$ is the *unit impulse response function* of the system.

The second standard excitation is the unit-amplitude complex (multi-dimensional) sinusoid:

$$x(\mathbf{t}) = e^{i\boldsymbol{\omega}\cdot\mathbf{t}} \equiv e^{i(\omega_1 t_1 + \omega_2 t_2 + \cdots + \omega_n t_n)}. \tag{3.5.3}$$

The steady-state response to this excitation has the form

$$y(\mathbf{t}) = H(\boldsymbol{\omega})e^{i\boldsymbol{\omega}\cdot\mathbf{t}}, \tag{3.5.4}$$

where $H(\boldsymbol{\omega})$ is the complex *frequency response function* or *transfer function* of the linear system. The system functions $h(\mathbf{t})$ and $H(\boldsymbol{\omega})$ form an n-tuple Fourier transform pair:

$$H(\boldsymbol{\omega}) = \int_{\text{All } \boldsymbol{\tau}} h(\boldsymbol{\tau})e^{i\boldsymbol{\omega}\cdot\boldsymbol{\tau}}\, d\boldsymbol{\tau}, \tag{3.5.5}$$

$$h(\mathbf{t}) = \left(\frac{1}{2\pi}\right)^n \int_{\text{All } \boldsymbol{\omega}} H(\boldsymbol{\omega})e^{-i\boldsymbol{\omega}\cdot\mathbf{t}}.d\boldsymbol{\omega} \tag{3.5.6}$$

Homogeneous Stochastic Response of Linear Systems

The response $Y(\mathbf{t})$ to a random excitation field $X(\mathbf{t})$ can be obtained by superposition, as follows:

$$Y(\mathbf{t}) = \int_{\text{All } \mathbf{u}} X(\mathbf{u}) h(\mathbf{t} - \mathbf{u}) d\mathbf{u} = \int_{\text{All } \boldsymbol{\alpha}} h(\boldsymbol{\alpha}) X(\mathbf{t} - \boldsymbol{\alpha}) d\boldsymbol{\alpha}. \tag{3.5.7}$$

This convolution integral expresses the "response" random field $Y(\mathbf{t})$ in terms of the "excitation" field $X(\mathbf{t})$ and the impulse response function $h(\boldsymbol{\tau})$. If the excitation is a homogeneous random field $X(\mathbf{t})$, the "steady-state" response will also be a homogeneous random field. The mean values are related as follows:

$$m_Y = m_X \int_{\text{All } \boldsymbol{\tau}} h(\boldsymbol{\tau}) d\boldsymbol{\tau}, \tag{3.5.8}$$

or, using Eq. (3.5.5),

$$m_Y = m_X H(\mathbf{0}). \tag{3.5.9}$$

We now evaluate the second-order properties of the response field, first in the time (space) domain and then in the frequency (wave number) domain.

Second-Order Input-Output Relations: Time Domain

The relationship between the covariance functions $B_X(\boldsymbol{\tau})$ and $B_Y(\boldsymbol{\tau})$ is obtained by substituting Eq. (3.5.7) into the definition of the covariance function of $B_Y(\boldsymbol{t})$, yielding

$$\begin{aligned} B_Y(\boldsymbol{\tau}) = E[Y(\mathbf{0})Y(\boldsymbol{\tau})] &= \int_{\boldsymbol{\alpha}} \int_{\boldsymbol{\beta}} h(\boldsymbol{\alpha}) h(\boldsymbol{\beta}) E[X(-\boldsymbol{\alpha}) X(\boldsymbol{\tau} - \boldsymbol{\beta})] \, d\boldsymbol{\alpha} \, d\boldsymbol{\beta} \\ &= \int_{\boldsymbol{\alpha}} \int_{\boldsymbol{\beta}} h(\boldsymbol{\alpha}) h(\boldsymbol{\beta}) B_X(\boldsymbol{\tau} + \boldsymbol{\alpha} - \boldsymbol{\beta}) \, d\boldsymbol{\alpha} \, d\boldsymbol{\beta}, \end{aligned} \tag{3.5.10}$$

in which we assume $m_X = 0$ to simplify the derivation. Under white noise input, when $B_X(\boldsymbol{\tau}) = \sigma_X^2 \delta(\boldsymbol{\tau})$, Eq. (3.5.10) reduces to

$$B_Y(\boldsymbol{\tau}) = \sigma_X^2 \int_{\boldsymbol{\alpha}} h(\boldsymbol{\alpha}) h(\boldsymbol{\alpha} + \boldsymbol{\tau}) d\boldsymbol{\alpha}. \tag{3.5.11}$$

A similar derivation yields the covariance between two response random fields to the same input field $X(\mathbf{t})$, namely

$$Y_1(\mathbf{t}) = \int_{\text{All } \boldsymbol{\alpha}} h_1(\boldsymbol{\alpha}) X(\mathbf{t} - \boldsymbol{\alpha}) d\boldsymbol{\alpha}, \tag{3.5.12}$$

and

$$Y_2(\mathbf{t}) = \int_{\mathrm{All}.\boldsymbol{\beta}} h_2(\boldsymbol{\beta}) X(\mathbf{t} - \boldsymbol{\beta}) d\boldsymbol{\beta}.$$

Again assuming $m_X = 0$ to simplify the derivation, one obtains

$$B_{Y_1Y_2}(\boldsymbol{\tau}) = E[Y_1(\mathbf{0})Y_2(\boldsymbol{\tau})] = \int_{\boldsymbol{\alpha}}\int_{\boldsymbol{\beta}} h_1(\boldsymbol{\alpha})h_2(\boldsymbol{\beta})E[X(-\boldsymbol{\alpha})X(\boldsymbol{\tau} - \boldsymbol{\beta})]d\boldsymbol{\alpha}d\boldsymbol{\beta}$$

$$= \int_{\boldsymbol{\alpha}}\int_{\boldsymbol{\beta}} h_1(\boldsymbol{\alpha})h_2(\boldsymbol{\beta})B_X(\boldsymbol{\tau} + \boldsymbol{\alpha} - \boldsymbol{\beta})\, d\boldsymbol{\alpha}\, d\boldsymbol{\beta}. \qquad (3.5.13)$$

If the input is white noise, so $B_X(\boldsymbol{\tau}) = \sigma_X^2 \delta(\boldsymbol{\tau})$, Eq. (3.5.13) reduces to

$$B_{Y_1Y_2}(\boldsymbol{\tau}) = \sigma_X^2 \int_{\boldsymbol{\alpha}} h_1(\boldsymbol{\alpha})h_2(\boldsymbol{\alpha} + \boldsymbol{\tau})d\boldsymbol{\alpha}. \qquad (3.5.14)$$

More generally, if Y_1 and Y_2 constitute the linear responses to different (correlated) input fields, X_1 and X_2, respectively, the relationship between the output and input covariance functions is

$$B_{Y_1Y_2}(\boldsymbol{\tau}) = \int_{\boldsymbol{\alpha}}\int_{\boldsymbol{\beta}} h_1(\boldsymbol{\alpha})h_2(\boldsymbol{\beta})B_{X_1X_2}(\boldsymbol{\tau} + \boldsymbol{\alpha} - \boldsymbol{\beta})\, d\boldsymbol{\alpha}\, d\boldsymbol{\beta}. \qquad (3.5.15)$$

If $X_1 \equiv X_2 = X$ and $h_1 \equiv h_2 = h$, Eq. (3.5.15) reduces to Eq. (3.5.10), and $B_{Y_1Y_2}(\boldsymbol{\tau}) \equiv B_Y(\boldsymbol{\tau})$.

Second-Order Input-Output Relations: Frequency Domain

The first step in stochastic *frequency domain analysis* is to express the input and output processes in terms of their spectral representation:

$$X(\mathbf{t}) = \int_{d\boldsymbol{\omega}} e^{i\boldsymbol{\omega}\cdot\mathbf{t}} Z_X(d\boldsymbol{\omega}), \qquad (3.5.16)$$

and

$$Y(\mathbf{t}) = \int_{d\boldsymbol{\omega}} e^{i\boldsymbol{\omega}\cdot\mathbf{t}} Z_Y(d\boldsymbol{\omega}). \qquad (3.5.17)$$

In general, $X(\mathbf{t})$ and $Y(\mathbf{t})$ may be complex random processes . The notation Re$\{\cdot\}$ (to identify the real part of a complex quantity) could be added throughout this section, but is omitted to simplify the notation. Inserting

Eqs. (3.5.16) and (3.5.5), in that order, into Eq. (3.5.7) yields

$$Y(\mathbf{t}) = \int_{\boldsymbol{\alpha}} h(\boldsymbol{\alpha})X(\mathbf{t}-\boldsymbol{\alpha})d\boldsymbol{\alpha} = \int_{\boldsymbol{\alpha}} h(\boldsymbol{\alpha})d\boldsymbol{\alpha} \int_{d\boldsymbol{\omega}} e^{i\boldsymbol{\omega}\cdot(\mathbf{t}-\boldsymbol{\alpha})} Z_X(d\boldsymbol{\omega})$$

$$= \int_{d\boldsymbol{\omega}} e^{i\boldsymbol{\omega}\cdot\mathbf{t}} Z_X(d\boldsymbol{\omega}) \int_{\boldsymbol{\alpha}} h(\alpha)e^{i\boldsymbol{\omega}\cdot\boldsymbol{\alpha}}d\boldsymbol{\alpha} = \int_{d\boldsymbol{\omega}} e^{i\boldsymbol{\omega}\cdot\mathbf{t}} Z_X(d\boldsymbol{\omega})H(\boldsymbol{\omega}). \quad (3.5.18)$$

Comparing Eqs. (3.5.17) and (3.5.18) makes it clear that the amplitudes of the components of $X(t)$ and $Y(t)$ associated with the same location $\boldsymbol{\omega}$ in the frequency domain must be related as follows:

$$Z_Y(d\boldsymbol{\omega}) = H(\boldsymbol{\omega})Z_X(d\boldsymbol{\omega}). \quad (3.5.19)$$

The expectation of the square of the absolute value of $Z_Y(d\boldsymbol{\omega})$ is

$$E[|Z_Y(d\boldsymbol{\omega})|^2] = |H(\boldsymbol{\omega})|^2 E[|Z_X(d\boldsymbol{\omega})|^2]. \quad (3.5.20)$$

Finally, combining Eqs. (3.5.20) and (3.3.20) yields the second-order "input-output" relations in the frequency domain:

$$S_Y(\boldsymbol{\omega}) = |H(\boldsymbol{\omega})|^2 S_X(\boldsymbol{\omega}). \quad (3.5.21)$$

If the random fields X and Y are quadrant symmetric, we may also write

$$G_Y(\boldsymbol{\omega}) = |H(\boldsymbol{\omega})|^2 G_X(\boldsymbol{\omega}), \quad \boldsymbol{\omega} \geq \mathbf{0}. \quad (3.5.22)$$

Eqs. (3.5.21) and (3.5.22) state that the response s.d.f. equals the product of the excitation s.d.f. and the squared amplification function of the system.

Cross-Spectra

Now consider *two* response fields $Y_1(\mathbf{t})$ and $Y_2(\mathbf{t})$ generated by the same excitation $X(\mathbf{t})$. In terms the complex frequency response functions, $H_1(\boldsymbol{\omega})$ and $H_2(\boldsymbol{\omega})$, the complex amplitudes of Y_1 and Y_2 are, respectively,

$$Z_{Y_1}(d\boldsymbol{\omega}) = H_1(\boldsymbol{\omega})Z_X(d\boldsymbol{\omega}), \quad (3.5.23)$$

and

$$Z_{Y_2}(d\boldsymbol{\omega}) = H_2(\boldsymbol{\omega})Z_X(d\boldsymbol{\omega}).$$

Taking the expectation of the product of $Z_{Y_1}(d\boldsymbol{\omega})$ and the complex conjugate of $Z_{Y_2}(d\boldsymbol{\omega})$ yields the cross-spectrum

$$S_{Y_1Y_2}(\boldsymbol{\omega}) = H_1(\boldsymbol{\omega})H_2^*(\boldsymbol{\omega})S_X(\boldsymbol{\omega}). \quad (3.5.24)$$

In the special case when $Y_1 \equiv Y$ and $Y_2 \equiv X$, one obtains to the cross-spectrum of the excitation and response fields:

$$S_{XY}(\boldsymbol{\omega}) = H(\boldsymbol{\omega})S_X(\boldsymbol{\omega}). \tag{3.5.25}$$

More generally, if Y_1 and Y_2 are homogeneous responses to different (correlated) excitations X_1 and X_2, the cross-spectra of inputs and outputs are related as follows:

$$S_{Y_1Y_2}(\boldsymbol{\omega}) = H_1(\boldsymbol{\omega})H_2^*(\boldsymbol{\omega})S_{X_1X_2}(\boldsymbol{\omega}). \tag{3.5.26}$$

This relationship reduces to Eq. (3.5.21) when $X_1 \equiv X_2$ and $H_1 \equiv H_2$.

Parallel Results for Lattice Systems

The format and the results of the stochastic response analysis change but little when the parameter space is discrete, a lattice of points. In the frequency domain the applicability of Eq. (3.5.21) is now limited to the range $-\pi \leq \omega \leq \pi$. In the time domain the integrations become summations. The relationship between the excitation and response lattice fields,

$$Y(\mathbf{t}) = \sum_{\boldsymbol{\alpha}} h(\boldsymbol{\alpha})X(\mathbf{t} - \boldsymbol{\alpha}), \tag{3.5.27}$$

is analogous to Eq. (3.5.7). Assuming both fields are homogeneous, their means are related by

$$m_Y = m_X \sum_{\boldsymbol{\alpha}} h(\boldsymbol{\alpha}), \tag{3.5.28}$$

and the covariance of $Y(\mathbf{t})$ is

$$B_Y(\boldsymbol{\tau}) = \sum_{\boldsymbol{\alpha}} \sum_{\boldsymbol{\beta}} h(\boldsymbol{\alpha})h(\boldsymbol{\beta})B_X(\boldsymbol{\tau} + \boldsymbol{\alpha} - \boldsymbol{\beta}). \tag{3.5.29}$$

If the excitation field is an *uncorrelated* lattice process, then Eq. (3.5.29) reduces to

$$B_Y(\boldsymbol{\tau}) = \sigma_X^2 \sum_{\boldsymbol{\alpha}} h(\boldsymbol{\alpha})h(\boldsymbol{\alpha} + \boldsymbol{\tau}), \tag{3.5.30}$$

which the discrete-case equivalent of (3.5.11). In the frequency domain the procedure is the same as for continuous-parameter fields. The quantities $X(\mathbf{t})$ and $Y(\mathbf{t})$ are replaced by $e^{i\boldsymbol{\omega}\cdot\mathbf{t}}$ and $H(\boldsymbol{\omega})e^{i\boldsymbol{\omega}\cdot\mathbf{t}}$, respectively, in

Eq. (3.5.27) which defines the system:

$$H(\boldsymbol{\omega})e^{i\boldsymbol{\omega}\cdot\mathbf{t}} = \sum_{\boldsymbol{\omega}} h(\boldsymbol{\alpha})e^{i\boldsymbol{\omega}\cdot(\mathbf{t}-\boldsymbol{\alpha})}. \tag{3.5.31}$$

Hence

$$H(\boldsymbol{\omega}) = \sum_{\boldsymbol{\alpha}} h(\boldsymbol{\alpha})e^{i\boldsymbol{\omega}\cdot\boldsymbol{\alpha}}, \tag{3.5.32}$$

which is the equivalent of Eq. (3.5.5). The relationship between the spectral density functions of the excitation and response fields is

$$S_Y(\boldsymbol{\omega}) = |H(\boldsymbol{\omega})|^2 S_X(\boldsymbol{\omega}), \quad -\boldsymbol{\pi} \leq \boldsymbol{\omega} \leq \boldsymbol{\pi}, \tag{3.5.33}$$

or, is case the fields are quadrant symmetric,

$$G_Y(\boldsymbol{\omega}) = |H(\boldsymbol{\omega})|^2 G_X(\boldsymbol{\omega}), \quad \mathbf{0} \leq \boldsymbol{\omega} \leq \boldsymbol{\pi}. \tag{3.5.34}$$

It is not uncommon to have mixed treatment of the coordinates, for instance, when time is continuous and space discretized, or *vice versa.* Clearly, this poses no fundamental problems.

3.6 Derivatives and Local Integrals of Random Fields

Statistics of Partial Derivatives

The statistics of the partial derivatives of a random field $X(\mathbf{t})$ contain important information about fluctuations, level excursions, and extremes. In the first part of this section we derive the first- and second-order statistics of the partial derivatives of $X(\mathbf{t})$, tacitly assuming that the random field satisfies the necessary conditions of differentiability. These conditions are explicitly stated and discussed in the second part. The final subsection deals with local integration (or averaging) of random fields.

Spectra of Partial Derivatives

Consider an n-dimensional homogeneous random field $X(\mathbf{t}) = X(t_1, \ldots, t_n)$ with spectral density function $S_X(\boldsymbol{\omega})$ and covariance function $B_X(\boldsymbol{\tau})$. The partial derivative of $X(\mathbf{t})$ with respect to t_j, for $j = 1, \ldots, n$, is

$$\dot{X}_j(\mathbf{t}) = \frac{\partial X(\mathbf{t})}{\partial t_j}. \tag{3.6.1}$$

To obtain the complex transfer function $H(\boldsymbol{\omega})$ associated with this linear operation, we replace $X(\mathbf{t})$ by $e^{i\boldsymbol{\omega}\cdot\mathbf{t}}$ and $\dot{X}_j(\mathbf{t})$ by $H(\boldsymbol{\omega})e^{i\boldsymbol{\omega}\cdot\mathbf{t}}$ in Eq. (3.6.1). The result is

$$H(\boldsymbol{\omega})e^{i\boldsymbol{\omega}\cdot\mathbf{t}} = i\omega_j e^{i\boldsymbol{\omega}\cdot\mathbf{t}}. \tag{3.6.2}$$

Hence

$$H(\boldsymbol{\omega}) = H(\omega_1, \ldots, \omega_n) = i\omega_j, \tag{3.6.3}$$

and the squared amplification function is

$$|H(\boldsymbol{\omega})|^2 = \omega_j^2. \tag{3.6.4}$$

Combining Eqs. (3.6.3) and (3.5.9) confirms that the mean of $\dot{X}_j(\mathbf{t})$ is zero. The spectral density function of $\dot{X}_j(\mathbf{t})$ is [from Eq. (3.5.21)]:

$$S_{\dot{X}_j}(\boldsymbol{\omega}) = \omega_j^2 S_X(\boldsymbol{\omega}). \tag{3.6.5}$$

Covariance Function and Variance of Partial Derivatives

The corresponding impulse response function $h(\mathbf{t})$ is the partial derivative of the "delta function" $\delta(\mathbf{t})$ with respect to t_j. Application of Eq. (3.5.10) yields the following "input-output" relationship in terms of the covariance functions:

$$B_{\dot{X}_j}(\boldsymbol{\tau}) = -\frac{\partial^2 B_X(\boldsymbol{\tau})}{\partial \tau_j^2}. \tag{3.6.6}$$

Provided it exists, the mean square value of the field of partial derivatives (with respect to t_j) is then, for $j = 1, \ldots, n$,

$$\begin{aligned} \sigma_{\dot{X}_j}^2 = B_{\dot{X}_j}(\mathbf{0}) &= -\left[\frac{\partial^2 B_X(\boldsymbol{\tau})}{\partial \tau_j^2}\right]_{\boldsymbol{\tau}=\mathbf{0}} \\ &= \int_{\text{All}\,\boldsymbol{\omega}} \omega_j^2 S_X(\boldsymbol{\omega})d\boldsymbol{\omega} = \int_{-\infty}^{+\infty} \omega_j^2 S_X(\omega_j)d\omega_j, \end{aligned} \tag{3.6.7}$$

where $S_X(\omega_j)$ is the s.d.f. of the one-dimensional random process obtained by keeping all but one (t_j) of the random field's coordinates constant. (These equations reduce, of course, to the well-known results for $n = 1$.)

Results for Second-Order Partial Derivatives

For an n-dimensional continuous-parameter homogeneous random field, properly differentiable, it is possible to define an $n \times n$ matrix of *second-order* partial derivatives:

$$\ddot{X}_{ij}(\mathbf{t}) = \frac{\partial^2}{\partial t_i \partial t_j} X(\mathbf{t}). \tag{3.6.8}$$

The appropriate transfer function characterizing this linear operation is

$$H(\boldsymbol{\omega}) = -\omega_i \omega_j, \tag{3.6.9}$$

and the spectral density function of $\ddot{X}_{ij}(\mathbf{t})$ becomes

$$S_{\ddot{X}_{ij}}(\boldsymbol{\omega}) = \omega_i^2 \omega_j^2 S_X(\boldsymbol{\omega}). \tag{3.6.10}$$

The covariance functions are related as follows:

$$B_{\ddot{X}_{ij}}(\boldsymbol{\tau}) = \frac{\partial^4}{\partial \tau_i^2 \partial \tau_j^2} B_X(\boldsymbol{\tau}). \tag{3.6.11}$$

The mean of $\ddot{X}_{ij}(\mathbf{t})$ is zero, and its variance, *provided it exists*, equals

$$\begin{aligned} \sigma^2_{\ddot{X}_{ij}} &= \left[\frac{\partial^4}{\partial \tau_i^2 \partial \tau_j^2} B_X(\boldsymbol{\tau}) \right]_{\boldsymbol{\tau}=\mathbf{0}} = \int_{\text{All.}\boldsymbol{\omega}} \omega_i^2 \omega_j^2 S_X(\boldsymbol{\omega}) d\boldsymbol{\omega} \\ &= \int_{-\infty}^{+\infty} \int_{-\infty}^{+\infty} \omega_i^2 \omega_j^2 S_X(\omega_i, \omega_j) d\omega_i d\omega_j, \end{aligned} \tag{3.6.12}$$

where $S_X(\omega_i, \omega_j)$ is the s.d.f. characterizing random variation on a plane parallel to the coordinate axes t_i and t_j.

Cross-Correlation Between Partial Derivatives

The cross-spectrum of $\dot{X}_i(\mathbf{t})$ and $\dot{X}_j(\mathbf{t})$ is, from Eq. (3.5.24),

$$S_{\dot{X}_i \dot{X}_j}(\boldsymbol{\omega}) = (i\omega_i)(-i\omega_j) S_X(\boldsymbol{\omega}) = \omega_i \omega_j S_X(\boldsymbol{\omega}). \tag{3.6.13}$$

The cross-covariance function, expressing the covariance of $\dot{X}_i(t)$ and $\dot{X}_j(t + \tau)$, may be obtained from Eq. (3.5.13):

$$B_{\dot{X}_i \dot{X}_j}(\boldsymbol{\tau}) = -\frac{\partial^2 B_X(\boldsymbol{\tau})}{\partial \tau_i \partial \tau_j}, \tag{3.6.14}$$

and the covariance of $\dot{X}_i(\mathbf{t})$ and $\dot{X}_j(\mathbf{t})$ is

$$E[\dot{X}_i\dot{X}_j] = B_{\dot{X}_i\dot{X}_j}(0) = \int_{\text{All.}\boldsymbol{\omega}} \omega_i\omega_j S_X(\boldsymbol{\omega})d\boldsymbol{\omega}$$

$$= \int_{-\infty}^{+\infty}\int_{-\infty}^{+\infty} \omega_i\omega_j S_X(\omega_i,\omega_j)d\omega_i d\omega_j. \tag{3.6.15}$$

For $i \neq j$, in case the correlation structure of $X(\mathbf{t})$ is quadrant symmetric (with respect to t_i and t_j) the covariance $E[\dot{X}_i\dot{X}_j]$ vanishes.

Covariance between a Random Field and its Partial Derivative

Finally, the cross-spectrum of $X(\mathbf{t})$ and $\dot{X}_j(\mathbf{t})$ is

$$S_{\dot{X}_j}(\boldsymbol{\omega}) = i\omega_j S_X(\boldsymbol{\omega}), \tag{3.6.16}$$

which leads to the important result that a homogeneous random field $X(\mathbf{t})$ and its partial derivative $\dot{X}_j(\mathbf{t})$ are uncorrelated:

$$E[X\dot{X}_j] = \int_{-\infty}^{+\infty} S_{\dot{X}_j}(\boldsymbol{\omega})d\boldsymbol{\omega} = i\int_{-\infty}^{+\infty} \omega_j S_X(\omega_j)d\omega_j = 0. \tag{3.6.17}$$

(The proof relies on the fact that $S_X(\omega_j)$ is an even function.)

Derivative statistics are difficult to estimate in actual applications; they are very sensitive to the choice of analytical model for the correlation structure of $X(\mathbf{t})$, in particular to the behavior of $B_X(\boldsymbol{\tau})$ at near-zero values of the lag components, or to the behavior $S_X(\boldsymbol{\omega})$ at very high frequencies. Yet, it is practically impossible to validate models with respect to their behavior in these limiting ranges. The approach taken in Chaps. 5 - 7 seeks to circumvent this problem by permitting a small amount of local averaging of the random field, sufficient to smoothen microscale fluctuations and thereby stabilize estimates of statistics of derivatives.

Conditions for Differentiability in the Mean Square Sense

One-Dimensional Case

It is instructive to consider the one-dimensional case first. Suppose a random function $X(t)$ is sampled at equidistant time intervals Δt. The slope of the line connecting consecutive observations is the random variable

$$\dot{X} = \frac{X(t+\Delta t) - X(t)}{\Delta t}. \tag{3.6.18}$$

The mean of $\dot{X}$ is zero, and its variance is

$$\sigma_{\dot{X}}^2 = \frac{1}{(\Delta t)^2}\{2E[X^2(t)] - 2E[X(t)X(t+\Delta t)]\}$$
$$= \frac{2}{(\Delta t)^2}[\sigma_X^2 - B_X(\Delta t)]. \quad (3.6.19)$$

As the interval vanishes, $\Delta t \to 0$, the mean square derivative will exist if and only if $B_X(\Delta t)$ has the following limiting form as $\Delta t \to 0$:

$$B_X(\Delta t) = \sigma_X^2 - \frac{1}{2}\sigma_{\dot{X}}^2(\Delta t)^2. \quad (3.6.20)$$

This further implies that the first derivative of $B_X(\tau)$ must be zero at $\tau = 0$:

$$\dot{B}_X(0) \equiv \left[\frac{dB_X(\tau)}{d\tau}\right]_{\tau=0} = 0. \quad (3.6.21)$$

Hence, for random processes to possess a mean square derivative, $\dot{B}_X(0)$ must be zero, and the covariance function must have the form of Eq. (3.6.20) near the lag origin. $B_X(\tau)$ then has a true maximum at $\tau = 0$, in the sense that the derivative is zero and the second derivative negative. An alternate way to reach this conclusion is based on the Wiener-Khinchine relation, Eq. (3.3.11). For τ close to zero, using $\sin\omega\tau \simeq \omega\tau$, the first derivative of $B_X(\tau)$ is

$$\dot{B}_X(\tau) = -\int_0^\infty \omega . G_X(\omega)\sin\omega\tau d\omega$$
$$\simeq -\tau\int_0^\infty \omega^2 G_X(\omega)d\omega \simeq -\tau\sigma_{\dot{X}}^2. \quad (3.6.22)$$

Therefore, provided $\sigma_{\dot{X}}^2$ is finite, $\dot{B}_X(0) = 0$ and, of course, $\ddot{B}_X(0) = -\sigma_{\dot{X}}^2$. The implication is that the condition $\dot{B}_X(0) = 0$ is necessary and sufficient for $X(t)$ to be "mean square differentiable," that is, for $\sigma_{\dot{X}}^2$ to be finite.

Multi-Dimensional Case

To extend these arguments to the case of an n-dimensional homogeneous random field $X(\mathbf{t})$, consider the (MacLaurin) series expansion the covariance function near $\tau_1 = \cdots = \tau_n = 0$:

$$B_X(\boldsymbol{\tau}) = a + \sum_{j=1}^{n} b_j|\tau_j| + \sum_{i=1}^{n}\sum_{j=1}^{n} c_{ij}\tau_i\tau_j + \cdots, \quad (3.6.23)$$

where $a = \sigma_X^2$. The covariance function $B_X(0, \ldots, \tau_j, \ldots, 0)$ characterizes one-dimensional random variation along lines parallel to the t_j-axis. The existence of the mean square partial derivatives requires, for $j = 1, \ldots, n$,

$$\begin{aligned} b_j &= \left[\frac{\partial B_X(\boldsymbol{\tau})}{\partial \tau_j}\right]_{\boldsymbol{\tau}=\mathbf{0}} = 0, \\ c_{jj} &= \frac{1}{2}\left[\frac{\partial^2 B_X(\boldsymbol{\tau})}{\partial^2 \tau_j}\right]_{\boldsymbol{\tau}=\mathbf{0}} = \frac{1}{2}\sigma_{\dot{X}_j}^2 . \end{aligned} \tag{3.6.24}$$

Also, from Eqs. (3.6.14) and (3.6.15), the sum of the coefficients preceding $\tau_i \tau_j$, for $i \neq j$, is

$$c_{ij} + c_{ji} = \left[\frac{\partial^2 B_X(\boldsymbol{\tau})}{\partial \tau_i \partial \tau_j}\right]_{\boldsymbol{\tau}=\mathbf{0}} = -E[\dot{X}_i \dot{X}_j]. \tag{3.6.25}$$

Hence for an n-dimensional random field to be *differentiable in the mean square sense*, $B_X(\boldsymbol{\tau})$ must be expressible in the following form as $\boldsymbol{\tau} \to \mathbf{0}$:

$$B_X(\boldsymbol{\tau}) = \sigma_X^2 - \frac{1}{2}\sum_{i=1}^{n}\sum_{j=1}^{n} E[\dot{X}_i \dot{X}_j]\tau_i \tau_j + \cdots . \tag{3.6.26}$$

The necessary and sufficient conditions for the variance of the first-order partial derivatives of $X(\mathbf{t})$ to be finite are

$$\left[\frac{dB_X(\tau_j)}{d\tau_j}\right]_{\tau_j=0} = 0, \quad j = 1, \ldots, n, \tag{3.6.27}$$

or

$$\int_{\text{All}\,\boldsymbol{\omega}} \omega_j^2 S_X(\boldsymbol{\omega}) d\boldsymbol{\omega} = \int_{-\infty}^{+\infty} \omega_j^2 S_X(\omega_j) d\omega_j < \infty, \quad j = 1, \ldots, n, \tag{3.6.28}$$

where $B_X(\tau_j)$ and $S_X(\omega_j)$ characterize one-dimensional random variation along lines parallel to the t_j-axis. If the correlation structure of $X(\mathbf{t})$ is quadrant symmetric, $E[\dot{X}_i \dot{X}_j] = 0$ for $i \neq j$, and the cross-terms involving $\tau_i \tau_j$ will vanish in Eq. (3.6.26).

Local Integration of Random Fields

Consider the *local integral* of the n-dimensional homogeneous random field $X(\mathbf{t}) = X(t_1, t_2, \ldots, t_n)$ over an interval of length T along coordinate axis t_1:

$$I_T(t_1, t_2, \dots, t_n) = \int_{t_1}^{t_1+T} X(u_1, t_2, \dots, t_n) du_1. \tag{3.6.29}$$

We evaluate below various statistics of this derived random field.

Generalized Time Domain Approach

Eq. (3.6.29) describes a linear operation for which the unit impulse function $h(\mathbf{t})$ is

$$\begin{aligned} h(\mathbf{t}) = h(t_1, t_2, \dots, t_n) &= \int_{t_1}^{t_1+T} \delta(u_1)\delta(t_2)\cdots\delta(t_n)\, du_1 \\ &= \begin{cases} \delta(t_2)\cdots\delta(t_n), & -T \le t_1 \le 0, \\ 0, & \text{elsewhere.} \end{cases} \end{aligned} \tag{3.6.30}$$

The principal statistics of the response field may be found from Eqs. (3.5.8) and (3.5.10). The mean of $I_T(\mathbf{t})$ is

$$E[I_T] = \int_0^T E[X(\mathbf{t})]\, dt_1 = T m_X, \tag{3.6.31}$$

where m_X denotes the mean of $X(\mathbf{t})$. The covariance function of $I_T(\mathbf{t})$ is

$$B_{I_T}(\tau_1, \tau_2, \dots, \tau_n) = \int_0^T \int_0^T B_X(\tau_1 + \alpha_1 - \beta_1, \tau_2, \dots, \tau_n)\, d\alpha_1\, d\beta_1. \tag{3.6.32}$$

Setting $\tau_1 = \cdots = \tau_n = 0$ yields the variance of I_T:

$$\begin{aligned} \mathrm{Var}[I_T] &= B_{I_T}(0, \dots, 0) \\ &= \int_0^T \int_0^T B_X(\alpha_1 - \beta_1, 0, \dots, 0)\, d\alpha_1\, d\beta_1. \end{aligned} \tag{3.6.33}$$

Generalized Frequency Domain Approach

The complex frequency response function may be obtained either by calculating the Fourier transform of $H(\mathbf{t})$, or directly from Eq. (3.6.29) which defines the system:

$$H(\boldsymbol{\omega}) e^{i(\omega_1 t_1 + \cdots + \omega_n t_n)} = \int_{t_1}^{t_1+T} e^{i(\omega_1 u_1 + \omega_2 t_2 + \cdots + \omega_n t_n)}\, du_1, \tag{3.6.34}$$

leading to

$$H(\boldsymbol{\omega}) = \frac{1}{i\omega_1}(e^{i\omega_1 T} - 1). \tag{3.6.35}$$

The squared transfer function is

$$|H(\boldsymbol{\omega})|^2 = \left[\frac{1}{i\omega_1}(e^{i\omega_1 T} - 1)\right]\left[\frac{1}{(-i\omega_1)}(e^{-i\omega_1 T} - 1)\right]$$
$$= \frac{1}{\omega_1^2}[2 - (e^{i\omega_1 T} + e^{-i\omega_1 T})] = \frac{4}{\omega_1^2}\sin^2\left(\frac{\omega_1 T}{2}\right). \tag{3.6.36}$$

The spectral density function of $I_T(t)$ is then

$$S_{I_T}(\boldsymbol{\omega}) = \left[\frac{\sin(\omega_1 T/2)}{\omega_1/2}\right]^2 S_X(\boldsymbol{\omega}). \tag{3.6.37}$$

To obtain the variance of $I_T(\mathbf{t})$, we need to integrate over the whole frequency domain.

One-Dimensional Case

For $n = 1$ the random process $I_T(t)$ represents the local integral, over a "time window" T, of the process $X(t)$. From Eq. (3.6.32), the covariance function of $I_T(t)$ is

$$B_{I_T}(\tau) = \int_0^T \int_0^T B_X(\tau + \alpha - \beta)\, d\alpha\, d\beta, \tag{3.6.38}$$

and, from Eq. (3.6.37), its spectral density function is

$$S_{I_T}(\omega) = \left[\frac{\sin(\omega T/2)}{\omega/2}\right]^2 S_X(\omega). \tag{3.6.39}$$

The variance of $I_T(t)$, from Eqs. (3.6.38) and (3.6.39), is

$$\sigma_{I_T}^2 = \int_{-\infty}^{\infty}\left[\frac{\sin(\omega T/2)}{\omega/2}\right]^2 S_X(\omega) d\omega$$
$$= \int_0^T \int_0^T B_X(\alpha - \beta)\, d\alpha\, d\beta. \tag{3.6.40}$$

The double integral above may be simplified by a change of variables, as depicted in Fig. 3.5. Taking $\tau = \alpha - \beta$, integrating with respect to α, and

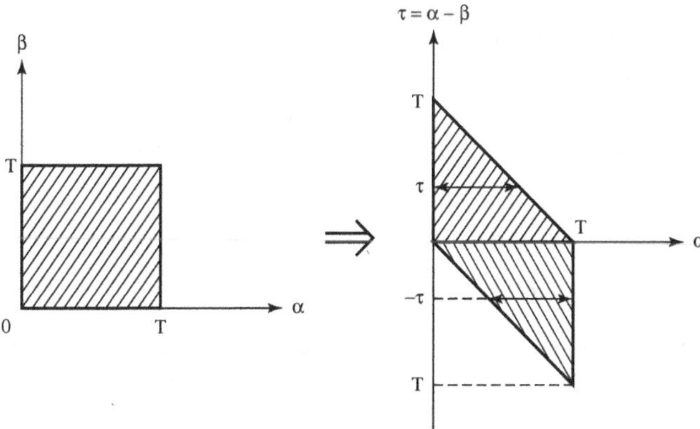

Fig. 3.5 Transformation of the domain of integration.

accounting for the fact that $B_X(\tau)$ is an even function, leads to

$$\sigma^2_{I_T} = 2\int_0^T (T-\tau)B_X(\tau)\,d\tau. \tag{3.6.41}$$

Dividing both sides of Eq. (3.6.41) by $1/T^2$ yields the variance of the *local average* $X_T(t) = (1/T)I_T(t)$, over time interval T, of the process $X(t)$.

Integration over a (2-D) Rectangular Window

A similar analysis yields the second-order statistics of the *integral over a rectangular area* $A = T_1T_2$ in the parameter space:

$$I_A(t_1, t_2, t_3, \ldots) = \int_{t_1}^{t_1+T_1}\int_{t_2}^{t_2+T_2} X(u_1, u_2, t_3, \ldots)\,du_1\,du_2, \tag{3.6.42}$$

where T_1 and T_2 are the dimensions of "moving rectangle." Our interest focuses on the variance of this process, which may be obtained either from the spectral density function

$$S_{I_A}(\boldsymbol{\omega}) = \left[\frac{\sin(\omega_1T_1/2)}{\omega_1/2}\right]^2\left[\frac{\sin(\omega_2T_2/2)}{\omega_2/2}\right]^2 S_X(\boldsymbol{\omega}), \tag{3.6.43}$$

or the covariance function

$$B_{I_A}(\boldsymbol{\tau}) = \int_{\tau_1}^{T_1+\tau_1} d\alpha_1 \int_0^{T_1} d\beta_1 \int_{\tau_2}^{T_1+\tau_2} d\alpha_2$$

$$\cdot \int_0^{T_2} B_X(\alpha_1-\beta_1, \alpha_2-\beta_2, \tau_3, \ldots)\,d\beta_2. \tag{3.6.44}$$

The variance of $I_A(\mathbf{t})$ is obtained from Eq. (3.6.43) by integration over the (two-dimensional) frequency domain, or from Eq. (3.6.44) by quadruple integration (after setting $\boldsymbol{\tau} = \mathbf{0}$). Actually, the quadruple integral can be converted into a double integral [in the same way as the double integral in Eq. (3.6.40) is changed into a single integral, Eq. (3.6.41)].

Integration over a Generalized "Rectangular Box"

These results can be generalized for the case when an n-dimensional homogeneous random field $X(\mathbf{t})$ is locally integrated over a rectangular domain $D = T_1T_2\cdots T_n$. The variance of the local integral $I_D = \int_D X(\mathbf{t})d\mathbf{t}$ is

$$\sigma^2_{I_D} = \int_{-T_1}^{+T_1}\cdots\int_{-T}^{+T_n}(T_1-|\tau_1|)\cdots(T_n-|\tau_n|) \cdot B_X(\tau_1,\ldots,\tau_n)d\tau_1\cdots d\tau_n. \quad (3.6.45)$$

The variance of the *local average* $X_D = D^{-1}I_D$ over the domain D is obtained by dividing $\sigma^2_{I_D}$ by $D^2 = (T_1T_2\cdots T_n)^2$.

These results appear in publications on random function theory (such as Pugachev [109], pp. 302–303) in the context of the derivation of the necessary and sufficient condition for a general homogeneous random field to be *ergodic in the mean.* Ergodicity with respect to the mean is insured if $\sigma^2_{X_D} = \sigma^2_{I_D}/D^2 \to 0$ when $T_1,\ldots,T_n \to \infty$.

Parallel Results for Lattice Fields

When $X(\mathbf{t})$ is a lattice random field, the integration operations become summations over a finite number of adjacent points in the parameter space. Each of the preceding equations (in this section) has its discrete counterpart. For example, consider the random series $I_T(t)$ defined as a sum of $T = m+1$ consecutive values of the stationary random series $X(t)$:

$$I_T(t) = X(t) + X(t-1) + \cdots + X(t-m) = \sum_{\alpha=0}^{m} X(t-\alpha). \quad (3.6.46)$$

The mean of $I_T(t)$ is

$$m_{I_T} = (m+1)m_X, \quad (3.6.47)$$

and the covariance of $I_T(t)$ is

$$B_{I_T}(\tau) = \sum_{\alpha=0}^{m} \sum_{\beta=0}^{m} B_X(\tau + \alpha - \beta). \tag{3.6.48}$$

Its variance may be expressed as follows:

$$\sigma_{I_T}^2 = \sum_{\tau=-m}^{m} (m + 1 - |\tau|) B_X(\tau). \tag{3.6.49}$$

These results parallel Eqs. (3.6.38) and (3.6.41), respectively. The "transfer function" between the input $X(t)$ and the output $I_T(t)$ implied by Eq. (3.6.46) has the form of a geometric progression:

$$H(\omega) = \sum_{\alpha=0}^{m} e^{-i\omega\alpha} = \frac{1 - e^{-i\omega(m+1)}}{1 - e^{-i\omega}}. \tag{3.6.50}$$

The squared modulus of the transfer function is

$$|H(\omega)|^2 = \frac{2 - e^{i\omega(m+1)} - e^{-i\omega(m+1)}}{2 - e^{i\omega} - e^{-i\omega}} = \left[\frac{\sin[\omega(m+1)/2]}{\sin(\omega/2)}\right]^2. \tag{3.6.51}$$

Finally, the spectral density function of $I_T(t)$, where $T = m + 1$, is

$$S_{I_T}(\omega) = \left[\frac{\sin[\omega(m+1)/2]}{\sin(\omega/2)}\right]^2 S_X(\omega), \quad -\pi \leq \omega \leq \pi. \tag{3.6.52}$$

In the limit when $\omega \to 0$, taking $T = m + 1$, Eqs. (3.6.51) and (3.6.52) converge to Eqs. (3.6.35) and (3.6.39), respectively, since $\sin(\omega/2) \to \omega/2$ and $(1 - e^{i\omega}) \to i\omega$. Analogous extension of the multi-dimensional results to *discrete*-parameter processes is quite straightforward.

3.7 Moving Average and Autoregressive Models

Linear system theory provides a powerful tool for generating analytical models for the correlation structure of random processes. Throughout this section it is assumed that a linear equation with constant coefficients defines the relation between an input white noise process and an output process whose second-order statistics are sought. We first consider autoregressive (AR) models, then moving average (MA) models, and lastly the family of autoregressive moving-average (ARMA) models.

Autoregressive (Markovian) Correlation Models in One Dimension

The systems considered in this section are defined by differential or finite-difference equations. The excitation is an uncorrelated (white noise) random field. The problem in each case is to find the second-order statistics of the system response. Consider first the autoregressive (AR) random series $X(t)$ defined by the first-order difference equation

$$X(t+1) = aX(t) + U(t), \quad t = \cdots -1, 0, 1, 2, \ldots, \tag{3.7.1}$$

where $U(t)$ represents an uncorrelated random series. Subtracting $X(t)$ from both sides and rearranging terms yields

$$[X(t+1) - X(t)] + (1-a)X(t) = U(t). \tag{3.7.2}$$

This finite difference equation relates the "input" noise $U(t)$ to the response series $X(t)$. The associated first-order differential equation is

$$\frac{dX(t)}{dt} + \alpha X(t) = U(t). \tag{3.7.3}$$

In the limit, when the time step in Eq. (3.7.2) becomes negligible, $\alpha = 1 - a$. The transfer function characterizing Eq. (3.7.3) is obtained from

$$(i\omega + \alpha)H(\omega)e^{i\omega t} = e^{i\omega t}, \tag{3.7.4}$$

yielding

$$H(\omega) = \frac{1}{i\omega + \alpha}. \tag{3.7.5}$$

Therefore, if G_0 is the constant spectral density of $U(t)$, the s.d.f. of $X(t)$ is

$$G_X(\omega) = G_0|H(\omega)|^2 = \frac{G_0}{\omega^2 + \alpha^2}. \tag{3.7.6}$$

The associated covariance function is

$$B_X(\tau) = \sigma_X^2 e^{-\alpha|\tau|}, \quad |\tau| \geq 0, \tag{3.7.7}$$

in which the variance σ_X^2 is given by:

$$\sigma_X^2 = B_X(0) = \int_0^\infty \frac{G_0}{\omega^2 + \alpha^2} d\omega = \frac{\pi G_0}{2\alpha}. \tag{3.7.8}$$

The covariance function of the corresponding random series, obeying the original finite difference equation [Eq. (3.7.1)], exhibits a geometric decay law, the "discrete-parameter equivalent" of Eq. (3.7.7):

$$B_X(\tau) = \sigma_X^2 a^{|\tau|}, \quad \tau = 0, 1, 2, \ldots. \tag{3.7.9}$$

These correlation functions characterize first-order autoregressive or *Markovian* processes. Eq. (3.7.1) clearly implies that past values of the process $X(t-1)$, $X(t-2)$, etc., do not affect future values $X(t+1)$, $X(t+2)$, etc., if the present, in the form of $X(t)$, is known. Eq. (3.7.3) is in fact the Langevin equation arising in the analysis of Brownian motion , the parameter α corresponding to the "friction coefficient" β in Eq. (2.7.52).

First-Order Autoregressive Models in Space

The Markovian property, as commonly stated, connotes directionality of the parameter: time moves forward only. No such directional preference exists in space. For spatial processes on the line, a self-generating mechanism implying a similar kind of limited memory, but without directional preference, is as follows (with t now representing a spatial coordinate):

$$X(t) = a[X(t-1) + X(t+1)] + U(t). \tag{3.7.10}$$

Subtracting $[-2aX(t)]$ from both sides and reordering the terms yields the second-order difference equation:

$$a[X(t-1) - 2X(t) + X(t+1)] - (1-2a)X(t) = -U(t). \tag{3.7.11}$$

The associated second-order differential equation is

$$\frac{d^2X(t)}{dt^2} - \alpha^2 X(t) = U(t). \tag{3.7.12}$$

In the limit, when the time step becomes negligible, $\alpha^2 = (1-2a)/a$. Only the continuous-time analysis is pursued here. The equation for the system transfer function is

$$(-\omega^2 - \alpha^2)H(\omega)e^{i\omega t} = e^{i\omega t}. \tag{3.7.13}$$

The spectral density function of the response to white noise input is

$$G_X(\omega) = G_0|H(\omega)|^2 = \frac{G_0}{(\omega^2+\alpha^2)^2} = \frac{4\alpha^3\sigma_X^2}{\pi(\omega^2+\alpha^2)^2}, \tag{3.7.14}$$

implying the expression $\sigma_X^2 = \pi G_0/(4\alpha^3)$ for the variance. The associated covariance function is

$$B_X(\tau) = \sigma_X^2(\alpha|\tau| + 1)e^{-\alpha|\tau|}. \tag{3.7.15}$$

Isotropic Multi-Dimensional Case

Whittle [151] extended this type of spatial Markovian analysis to an isotropic n-dimensional field with coordinates t_i $(i = 1, 2, \ldots, n)$. The extension of Eq. (3.7.12) is

$$(\nabla^2 - \alpha^2)X(\mathbf{t}) = U(\mathbf{t}), \tag{3.7.16}$$

where ∇^2 is the Laplace operator:

$$\nabla^2 = \frac{\partial^2}{\partial t_1^2} + \frac{\partial^2}{\partial t_2^2} + \cdots + \frac{\partial^2}{\partial t_n^2} = \sum_{i=1}^{n} \frac{\partial^2}{\partial t_i^2}. \tag{3.7.17}$$

The squared transfer function now has the form

$$|H(\boldsymbol{\omega})|^2 = \left[\alpha^2 + \sum_{i=1}^{n} \omega_i^2\right]^{-2}, \tag{3.7.18}$$

leading to the following s.d.f. of the response to white noise excitation:

$$G_X(\boldsymbol{\omega}) = G_0\left[\alpha^2 + \sum_{i=1}^{n} \omega_i^2\right]^{-2} \equiv G_X^R(\omega), \tag{3.7.19}$$

where $G_X^R(\omega)$ is the "radial spectrum" that depends on $\omega = [\omega_1^2 + \cdots + \omega_n^2]^{1/2}$. Repeated Fourier transformation yields the n-dimensional covariance function which, owing to circular symmetry, depends only on $\tau = [\tau_1^2 + \cdots + \tau_n^2]^{1/2}$, the argument of the radial covariance function

$$B_X(\boldsymbol{\tau}) \equiv B_X^R(\tau) = \frac{G_0}{2}\left(\frac{\pi}{2}\right)^{n/2}\left(\frac{\tau}{\alpha}\right)^p K_p(\alpha\tau), \tag{3.7.20}$$

where $K_p(x)$ is the modified Bessel function of the second kind of order $p = 2 - (n/2)$. The Bessel functions are widely tabulated (see, e.g., [17]). In the three-dimensional case ($n = 3$ and $p = 1/2$), the covariance function takes a simple exponential form:

$$B_X(\tau_1, \tau_2, \tau_3) \equiv B_X^R(\tau) = \frac{\pi G_0}{8\alpha} e^{-\alpha|\tau|}. \tag{3.7.21}$$

Hence, for $n = 3$ we have

$$\sigma_X^2 = \frac{\pi G_0}{8\alpha} \quad \text{and} \quad \rho_X^R(\tau) = e^{-\alpha|\tau|}. \tag{3.7.22}$$

For $n = 1$ ($p = 3/2$) the solution matches that given by Eq. (3.7.15). For $n = 2$ ($p = 1$) the results are

$$\sigma_X^2 = \frac{\pi G_0}{4\alpha^2} \quad \text{and} \quad \rho_X^R(\tau) = \alpha\tau K_1(\alpha\tau). \tag{3.7.23}$$

With proper scaling, the above results remain valid if the correlation structure of $X(\mathbf{t})$ is *ellipsoidal.* The covariance functions then depend on $\tau = [(\alpha_1\tau_1)^2 + \cdots + (\alpha_n\tau_n)^2]^{1/2}$, where $\alpha_1, \ldots, \alpha_n$ are the direction-dependent coefficients of a modified Laplace equation, a straightforward extension of Eq. (3.7.16).

Higher-Order Autoregressive (Markovian) Models

Another extension of Eq. (3.7.1) is to permit the value $X(t)$ of a time series to depend on *two or more* values in its immediate past. If two-step dependence is hypothesized, a second-order differential equation like Eq. (3.7.12) results in continuous-time analysis, and leads to solutions of the form of Eqs. (3.7.14) and (3.7.15). If a three-step memory length is assumed, the result is a third-order differential equation dependent on a single parameter, say α. The spectral density function then becomes

$$G(\omega) = \frac{G_0}{(\omega^2 + \alpha^2)^3} = \frac{16\alpha^5\sigma^2}{3\pi(\omega^2 + \alpha^2)^3}, \tag{3.7.24}$$

and the associated covariance function is

$$B(\tau) = \sigma^2 e^{-\alpha|\tau|}\left\{1 + \alpha|\tau| + \frac{1}{3}\alpha^2|\tau|^2\right\}. \tag{3.7.25}$$

In general, if the memory extends over m steps, an mth-order differential equation results and the associated spectral density and covariance functions (Gelb [63]) are, respectively,

$$G(\omega) = \frac{G_0}{(\omega^2 + \alpha^2)^m} = \frac{(2\alpha)^{2m-1}[\Gamma(m)]^2}{\pi(2m-2)!}\frac{\sigma^2}{(\omega^2 + \alpha^2)^m}, \tag{3.7.26}$$

and

$$B(\tau) = \sigma^2 e^{-\alpha|\tau|}\sum_{k=0}^{m-1}\frac{\Gamma(m)(2\alpha|\tau|)^{m-k-1}}{(2m-2)!k!\Gamma(m-k)}. \tag{3.7.27}$$

Considering that the case $m = 0$ in Eq. (3.7.26) corresponds, in a crude sense, to "white noise", Eqs. (3.7.26) and (3.7.27) represent a rich family of correlation models; by varying one of its parameters (either α or m) while keeping the other one fixed, conditions ranging from near-perfect correlation to complete lack of correlation can be modeled. Note also that all multi-dimensional correlation models discussed in this section possess the property of quadrant symmetry, justifying the use of the spectral density function $G(\boldsymbol{\omega})$, defined for positive frequencies only.

Linear Oscillator Response to White Noise

A notable second-order differential equation of the autoregressive type has the form

$$\frac{d^2X(t)}{dt^2} + 2\zeta\Omega\frac{dX(t)}{dt} + \Omega^2 X(t) = -U(t). \tag{3.7.28}$$

It describes the response of a linear oscillator, with natural frequency Ω and (nonnegative) damping ratio ζ to white noise excitation $U(t)$; specifically, $U(t)$ represents base excitation and $X(t)$ the relative displacement response of the oscillator. The transfer function associated with Eq. (3.7.28) is

$$H(\omega) = \frac{1}{\omega^2 - 2\zeta\Omega\omega i - \Omega^2}, \tag{3.7.29}$$

and the spectral density function of the (relative displacement) response to ideal white noise excitation with spectral density G_0 is

$$\begin{aligned} G_X(\omega) &= G_0|H(\omega)|^2 \\ &= \frac{G_0}{(\Omega^2 - \omega^2)^2 + 4\zeta^2\Omega^2\omega^2}, \quad \omega \geq 0. \end{aligned} \tag{3.7.30}$$

Integrating $G_X(\omega)$ over all frequencies yields the response variance of the (relative displacement) response

$$\sigma_X^2 = \frac{\pi G_0}{4\zeta\Omega^3}. \tag{3.7.31}$$

In case $\zeta = 1$, when the oscillator's damping becomes "critical", Eq. (3.7.30) reduces to Eq. (3.7.14), provided Ω is set equal to α.

Moving Average Correlation Models on the Line

An uncorrelated random series $U(t)$ gives rise to a moving average (MA) series $X(t)$ that can be expressed as follows:

$$X(t) = b_0 U(t) + \cdots + b_m U(t-m) = \sum_{k=0}^{m} b_k U(t-k), \qquad (3.7.32)$$

where $b_0, \ldots, b_m$ are a set of coefficients defining a discrete system function. The term "moving average" is most appropriate when the coefficients are positive and sum to one, as when all the coefficients are equal, $b_k = 1/(m+1)$, $k = 1, 2, \ldots, m$. In any case, it is understood that coefficients b_k with integer subscripts outside the range $(0, m)$ are zero. Stochastic linear system theory readily provides the first- and second-order statistics of $X(t)$. From Eq. (3.5.30), the covariance function of $X(t)$ is, for $|\tau| \le m$,

$$B_X(\tau) = \sigma_U^2 \sum_{k=0}^{m} b_k b_{k+\tau}, \qquad (3.7.33)$$

and $B_X(\tau) = 0$ for $|\tau| > m$. In the frequency domain, the system response function $H(\omega)$ is found by substituting U_t and X_t by $e^{i\omega t}$ and $H(\omega)e^{i\omega t}$, respectively, in Eq. (3.7.32). The result is

$$H(\omega) = \sum_{k=0}^{m} b_k e^{-i\omega k}. \qquad (3.7.34)$$

Hence

$$G_X(\omega) = G_0 |H(\omega)|^2 = G_0 \left| \sum_{k=0}^{m} b_k e^{-i\omega k} \right|^2, \qquad (3.7.35)$$

where G_0 is the "input" spectral density. Many different functional forms may be chosen for the array of coefficients b_k. Each choice leads to a different correlation model, while the value of m indicates the length of the "memory" of the derived random series. If the coefficients b_k are all equal, $b_k = 1/(m+1)$, $k = 0, \ldots, m$, the covariance function of $X(t)$ becomes

$$B_X(\tau) = \frac{\sigma_U^2 (m+1-|\tau|)}{(m+1)^2}, \quad |\tau| = 0, \ldots, m+1, \qquad (3.7.36)$$

and is zero elsewhere. The variance is $\sigma_X^2 = B_X(0) = \sigma_U^2/(m+1)$; the correlation function $\rho_X(\tau)$ has a triangular shape; and the two-sided spectral density function differs from that given by Eq. (3.6.52) by a factor $1/(m+1)^2$.

The continuous equivalent of Eq. (3.7.32) is

$$X(t) = \int_t^{t+T} b(t_1)U(t_1)dt_1. \tag{3.7.37}$$

In particular, taking $b(t) = 1/T$, Eq. (3.7.37) defines the moving average of the white noise process $U(t)$; its second-order statistics differ by a factor $1/T^2$ from those of the local integral process analyzed in Sec. 3.6; and its covariance function has a triangular shape,

$$B_X(\tau) = \begin{cases} \sigma_X^2 \left(1 - \dfrac{|\tau|}{T}\right), & |\tau| \le T, \\ 0, & |\tau| \ge T, \end{cases} \tag{3.7.38}$$

where $\sigma_X^2 = \pi G_0/T$ and G_0 is the white noise spectral density. The associated one-sided spectral density function is

$$G_X(\omega) = G_0 \left[\frac{\sin(\omega T/2)}{\omega T/2}\right]^2, \quad \omega \ge 0. \tag{3.7.39}$$

Moving Averages over Rectangular, Spherical, and Ellipsoidal Domains

Consider a homogeneous uncorrelated random field $U(t_1, t_2)$ averaged over a rectangular area with sides of length T_1 and T_2 parallel to the respective coordinated axes. If the rectangle is permitted to translate (but not rotate) in the parameter space, one obtains a new 2-D random process $X(t_1, t_2)$ that differs by a factor $A = T_1T_2$ from the integral process $I_A(t_1, t_2)$ discussed in Sec. 3.6. Its covariance function is

$$B_X(\tau_1, \tau_2) = \begin{cases} \sigma_X^2 \left(1 - \dfrac{|\tau_1|}{T_1}\right)\left(1 - \dfrac{|\tau_2|}{T_2}\right), & |\tau_1| \le T_1,\ |\tau_2| \le T_2, \\ 0, & \text{elsewhere}, \end{cases} \tag{3.7.40}$$

where $\sigma_X^2 = \pi^2 G_0/(T_1T_2)$. The associated spectral density function is

$$G_X(\omega_1, \omega_2) = G_0 \left[\frac{\sin(\omega_1 T_1/2)}{\omega_1 T_1/2}\right]^2 \left[\frac{\sin(\omega_2 T_2/2)}{\omega_2 T_2/2}\right]^2, \quad \omega_1, \omega_2 \ge 0. \tag{3.7.41}$$

The "product" form of these expressions is a consequence of the fact that the impulse response function and the transfer function of the system are separable,

$$H(\boldsymbol{\omega}) = H(\omega_1)H(\omega_2). \tag{3.7.42}$$

It is easy to see, in light of this, how these equations can be extended to multi-dimensional "purely random" fields locally averaged within a generalized "rectangular box" with dimensions $T_1, T_2, \ldots, T_n$.

Another important class of system functions are those that preserve circular symmetry, such as, in the two-dimensional case:

$$H(\boldsymbol{\omega}) = H\left(\sqrt{\omega_1^2 + \omega_2^2}\right). \tag{3.7.43}$$

If an isotropic "purely random" field $U(t_1, t_2)$ is averaged inside a (moving) circular area, the resulting moving average field $X(t_1, t_2)$ will be isotropic. If averaging occurs instead inside a moving ellipse whose principal axes remain parallel to the coordinate axes (so that the ellipse is translated but not rotated), the response field will have an ellipsoidal correlation structure. Extension of the concepts of circular and ellipsoidal symmetry to systems, averaging regions, and random fields of higher dimension is straightforward.

Autoregressive Moving Average Models

Combining the autoregressive (AR) and moving average (MA) models yields the general one-dimensional ARMA model for which the governing finite difference equation has the form:

$$X(t) = \sum_{k=1}^{m_1} a_k X(t-k) + \sum_{l=0}^{m_2} b_l U(t-l), \tag{3.7.44}$$

giving rise to ARMA-type correlation models. In particular, the spectral density function of the response $X(t)$ to uncorrelated input $U(t)$, with spectral density G_0, can be expressed as follows:

$$G(\omega) = \frac{|B(\omega)|^2}{|A(\omega)|^2} G_0, \tag{3.7.45}$$

where

$$A(\omega) = 1 - \sum_{k=1}^{m_1} a_k e^{-i\omega k}, \qquad B(\omega) = b_0 + \sum_{l=1}^{m_2} b_l e^{-i\omega l}. \tag{3.7.46}$$

The requirements of stationarity (stability) and physical realizability (finite variance) of the response process $X(t)$ impose restrictions on the coefficients a_j and b_k, but this will not be further pursued here.

In extension of Eq. (3.7.44) to multi-dimensional situations, the response field $X(\mathbf{t})$ at a given point $\mathbf{t}$ can be expressed as a linear combination of (1) values $X(\mathbf{t}-\mathbf{j})$ at adjacent locations $(\mathbf{t}-\mathbf{j})$, and (2) values of the purely random excitation field $U(\mathbf{t})$ in the vicinity of $\mathbf{t}$. The analysis, albeit tedious, is straightforward.

3.8 Space-Time Correlation Structure: Basic Relations

This section deals with n-dimensional homogeneous *space-time* processes $X(\mathbf{u}, t)$ that depend on a vector of spatial coordinates $\mathbf{u}$ and on time t. In the notation used in the preceding sections, $(n-1)$ denotes the number of spatial coordinates ($n = 2$, 3, or 4). General second-order analysis, as presented in preceding sections, is of course directly applicable to a space-time process. It is often advantageous, however, to account for the unique role of the time parameter, by transforming the covariance function $B(\boldsymbol{\nu}, \tau)$ with respect to the time lag $\tau = t - t'$ but not with respect to the vector $\boldsymbol{\nu} = \mathbf{u} - \mathbf{u}'$ of separation distances.

Second-Order Statistics

At a given location $\mathbf{u} = \mathbf{u}_0$, the variation with time is represented by the 1-D process $X(\mathbf{u}_0, t)$, and we denote by $S(\omega)$ the two-sided "point" spectral density function of $X(\mathbf{u}, t)$. Its Fourier transform is the temporal covariance function $B(\tau)$. The assumption of homogeneity in space implies that the statistics of $X(\mathbf{u}, t)$ do not depend on $\mathbf{u}$. The variance of $X(\mathbf{u}, t)$ is

$$\sigma^2 = \int_{-\infty}^{+\infty} S(\omega)d\omega = B(\tau)\bigg|_{\tau=0} . \tag{3.8.1}$$

The *unit-area* spectral density function is $s(\omega) = \sigma^{-2}S(\omega)$ and the temporal correlation function is $\rho(\tau) = \sigma^{-2}B(\tau)$. At a given instant, $t = t_0$, the *spatial variation* is represented by a random field $X(\mathbf{u}, t_0)$ that depends on $(n-1)$ spatial coordinates. The second-order properties of this field are described either by the spatial covariance function $B(\mathbf{u}, \mathbf{u}') = B(\mathbf{u} - \mathbf{u}') = B(\boldsymbol{\nu})$ or its Fourier transform $S(\mathbf{k})$, a function of a vector of "wave numbers" $\mathbf{k} = (k_1, k_2, \ldots)$. The symbols ν and k are used throughout this chapter in place of τ and ω, to distinguish clearly between analogous quantities associated with time and space. The variance of $X(\mathbf{u}, t)$ may also be expressed as follows:

$$\sigma^2 = \int_{-\infty}^{+\infty} S(\mathbf{k})d\mathbf{k} = B(\boldsymbol{\nu})\bigg|_{\boldsymbol{\nu}=\mathbf{0}}. \tag{3.8.2}$$

The spatial correlation function is $\rho(\boldsymbol{\nu}) = \sigma^{-2}B(\boldsymbol{\nu})$ and the normalized wave number spectrum is $s(\mathbf{k}) = \sigma^{-2}S(\mathbf{k})$.

The full second-order representation of a homogeneous space-time process may take several essentially equivalent forms, as discussed in Sec. 3.4. The space-time covariance function is by definition the covariance between two observations at different points $(\mathbf{u}, t)$ and $(\mathbf{u}', t')$ in the parameter space:

$$\begin{aligned} B(\mathbf{u}, \mathbf{u}', t, t') &= E[(X(\mathbf{u}, t) - m_X)(X(\mathbf{u}', t') - m_X)] \\ &= B(\mathbf{u} - \mathbf{u}', t - t') = B(\boldsymbol{\nu}, \tau), \end{aligned} \tag{3.8.3}$$

where $\boldsymbol{\nu}$ is the vector of separation distances ($\nu_1 = u_1 - u_1'$, $\nu_2 = u_2 - u_2', \ldots$) and τ is the time lag $t - t'$. By converting the time lag τ into a frequency ω, the *space-time cross-spectral density function* $C(\boldsymbol{\nu}, \omega)$ is obtained. The corresponding Wiener-Khinchine relations are

$$\begin{cases} B(\boldsymbol{\nu}, \tau) = \displaystyle\int_{-\infty}^{+\infty} C(\boldsymbol{\nu}, \omega)e^{i\omega\tau}\, d\omega, & (3.8.4) \\ C(\boldsymbol{\nu}, \omega) = \displaystyle\frac{1}{2\pi}\int_{-\infty}^{+\infty} B(\boldsymbol{\nu}, \tau)e^{-i\omega\tau}\, d\tau. & (3.8.5) \end{cases}$$

If the two locations coincide, that is, $\mathbf{u} = \mathbf{u}'$ or $\boldsymbol{\nu} = \mathbf{0}$, then

$$B(\mathbf{0}, \tau) \equiv B(\tau) \qquad \text{and} \qquad C(\mathbf{0}, \omega) \equiv S(\omega). \tag{3.8.6}$$

Eqs. (3.8.4) and (3.8.5) then reduce to the one-dimensional Wiener-Khinchine relations between $S(\omega)$ and $B(\tau)$. By taking $\tau = 0$ in Eq. (3.8.4), one obtains the spatial covariance function

$$B(\boldsymbol{\nu}, 0) \equiv \sigma^2\rho(\boldsymbol{\nu}) = \int_{-\infty}^{+\infty} C(\boldsymbol{\nu}, \omega)d\omega. \tag{3.8.7}$$

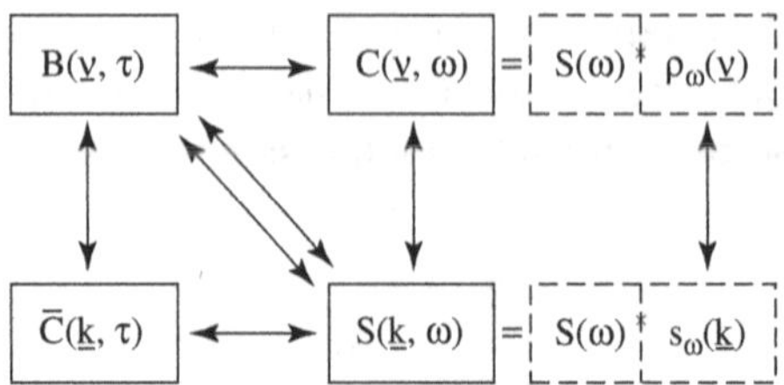

Fig. 3.6 Different but equivalent representations of second-order statistics of a space-time process. Each arrow indicates the existence of a Fourier transform pair.

Frequency-Dependent Spatial Correlation Function

It is useful to normalize the cross-spectral density function with respect to its value at $\boldsymbol{\nu} = \mathbf{0}$. Define

$$\rho_\omega(\boldsymbol{\nu}) = \frac{C(\boldsymbol{\nu}, \omega)}{C(\mathbf{0}, \omega)} = \frac{C(\boldsymbol{\nu}, \omega)}{S(\omega)}. \tag{3.8.8}$$

$\rho_\omega(\boldsymbol{\nu})$ is the *frequency-dependent spatial correlation function*; subscripting by ω is intended to emphasize the dependence on the distance-shift vector $\boldsymbol{\nu}$. At any frequency ω, for $\boldsymbol{\nu} = \mathbf{0}$,

$$\rho_\omega(\mathbf{0}) = 1, \quad \text{any } \omega. \tag{3.8.9}$$

$\rho_\omega(\boldsymbol{\nu})$ has the character of a correlation function as it quantifies the degree of *spatial* correlation associated with individual sinusoidal components of the space-time process $X(\mathbf{u}, t)$. The "composite" spatial correlation function $\rho(\boldsymbol{\nu})$ may be expressed as follows, using Eqs. (3.8.7) and (3.8.8);

$$\rho(\boldsymbol{\nu}) = \int_{-\infty}^{+\infty} \rho_\omega(\boldsymbol{\nu}) s(\omega) d\omega. \tag{3.8.10}$$

In words, the (composite) correlation function $\rho(\boldsymbol{\nu})$ is a weighted combination of frequency-dependent correlation functions $\rho_\omega(\boldsymbol{\nu})$; the weighting function is simply the unit-area spectral density function $s(\omega)$.

If all but one of the spatial coordinates of the space-time process $X(\mathbf{u}, t)$ is held constant, one obtains a direction-dependent process $X(u_i, t)$, $i = 1, \ldots, n-1$, itself a two-dimensional space-time process. The subscript i identifies the "active" spatial coordinate. All the relationships stated for n-dimensional space-time processes can be restated for $X(u_i, t)$.

Four Modes of Second-Order Analysis

The second-order statistics of the space-time process $X(\mathbf{u}, t)$ may be represented in four different ways (see Fig. 3.8):

1. As $B(\boldsymbol{\nu}, \tau) = \sigma^2 \rho(\boldsymbol{\nu}, \tau)$ if neither space nor time coordinates undergo Fourier transformation.
2. As $S(\mathbf{k}, \omega) = \sigma^2 s(\mathbf{k}, \omega)$ if the distance shifts ($\boldsymbol{\nu}$) are transformed into wave numbers ($\mathbf{k}$) and time lag (τ) is transformed into frequency (ω).
3. As $C(\boldsymbol{\nu}, \omega)$ if only the time interval (τ) is transformed into frequency (ω).
4. As $\overline{C}(\mathbf{k}, \tau)$ if only the separation distances ($\boldsymbol{\nu}$) are transformed into wave numbers ($\mathbf{k}$).

Any pair of these functions are related by Fourier transformation, constituting a form of the Wiener-Khinchine relations (see Sec. 3.4). Of particular interest are the following equations:

$$\bar{C}(\mathbf{k}, \tau) = \int_{-\infty}^{+\infty} S(\mathbf{k}, \omega) e^{i\omega\tau} d\omega \tag{3.8.11}$$

and

$$C(\boldsymbol{\nu}, \omega) = \int_{-\infty}^{+\infty} S(\mathbf{k}, \omega) e^{i\boldsymbol{\nu}\cdot\mathbf{k}} d\mathbf{k}. \tag{3.8.12}$$

Taking $\tau = 0$ in Eq. (3.8.11) leads to the wave number spectrum

$$\overline{C}(\mathbf{k}, 0) \equiv S(\mathbf{k}) = \int_{-\infty}^{+\infty} S(\mathbf{k}, \omega) d\omega, \tag{3.8.13}$$

while taking $\boldsymbol{\nu} = \mathbf{0}$ in Eq. (3.8.12) yields the spectral density function

$$C(\mathbf{0}, \omega) \equiv S(\omega) = \int_{-\infty}^{+\infty} S(\mathbf{k}, \omega) d\mathbf{k}. \tag{3.8.14}$$

Frequency-Dependent Wave Number Spectrum

The normalized *frequency-dependent wave number spectrum* may be defined as follows:

$$s_\omega(\mathbf{k}) = \frac{S(\mathbf{k}, \omega)}{S(\omega)} = \frac{s(\mathbf{k}, \omega)}{s(\omega)}. \tag{3.8.15}$$

At any frequency ω, integration over the wave number space yields

$$\int_{-\infty}^{+\infty} s_\omega(\mathbf{k})d\mathbf{k} = 1, \quad \text{any } \omega. \tag{3.8.16}$$

Also,

$$s(\mathbf{k}) = \int_{-\infty}^{+\infty} s_\omega(\mathbf{k})s(\omega)d\omega. \tag{3.8.17}$$

Finally, Eq. (3.8.12) implies that the functions $\rho_\omega(\boldsymbol{\nu})$ and $s_\omega(\mathbf{k})$ form a Fourier transform pair:

$$\begin{cases} \rho_\omega(\boldsymbol{\nu}) = \displaystyle\int_{-\infty}^{+\infty} s_\omega(\mathbf{k})e^{i\mathbf{v}\cdot\mathbf{k}}d\mathbf{k}, & (3.8.18) \\ s_\omega(\mathbf{k}) = \left(\dfrac{1}{2\pi}\right)^{n-1} \displaystyle\int_{-\infty}^{+\infty} \rho_\omega(\boldsymbol{\nu})e^{-i\boldsymbol{\nu}\cdot\mathbf{k}}d\boldsymbol{\nu}. & (3.8.19) \end{cases}$$

The Quadrant Symmetric Case

If the homogeneous space-time process $X(\mathbf{u}, t)$ is real, $B(\boldsymbol{\nu}, \tau)$ will be real, and its complete Fourier transform $S(\mathbf{k}, \omega)$ will also be real. However, the partial transform $C(\boldsymbol{\nu}, \omega)$ will in general be complex. In case $B(\boldsymbol{\nu}, \tau)$ is *quadrant symmetric*, all the Fourier transforms will be real (owing to "evenness" with respect to each component of the space-time lag vector). These comments also apply to the *frequency-dependent* functions $\rho_\omega(\boldsymbol{\nu})$ and $s_\omega(\mathbf{k})$. If the correlation structure of $X(\mathbf{u}, t)$ is quadrant symmetric, $\rho_\omega(\mathbf{v})$ will be real; otherwise, it may be complex. As both $S(\mathbf{k}, \omega)$ and $S(\omega)$ are real, $s_\omega(\mathbf{k})$ will always be real in the quadrant symmetric case.

Since the cross-spectral density function $C(\mathbf{v}, \omega)$ is in general complex, it may be represented as follows:

$$C(\boldsymbol{\nu}, \omega) = R(\boldsymbol{\nu}, \omega) - iQ(\boldsymbol{\nu}, \omega), \tag{3.8.20}$$

where the real and imaginary parts, R and Q, are often referred to as the "cospectrum" and the "quadrature spectrum," respectively. The frequency-dependent spatial correlation function $\rho_\omega(\boldsymbol{\nu})$ may be expressed in the same way. The squared modulus of $\rho_\omega(\boldsymbol{\nu})$ is the *coherence function*:

$$|\rho_\omega(\boldsymbol{\nu})|^2 = \frac{|C(\boldsymbol{\nu}, \omega)|^2}{S^2(\omega)}. \tag{3.8.21}$$

This function is, of course, real and behaves, for a given ω, much like the square of an ordinary correlation function. If the correlation structure is quadrant symmetric, $C(\boldsymbol{\nu}, \omega)$ becomes a real function, and $\rho_\omega(\boldsymbol{\nu}) \equiv |\rho_\omega(\boldsymbol{\nu})|$.

Examples of applications abound in the literature. We mention specifically some references on applications to modeling the space-time variation of earthquake ground motion (Zerva [161], Harichandran and Vanmarcke [70], Boissières and Vanmarcke [19; 20]) and to Monte Carlo simulation of space-time random processes (Vanmarcke *et al.* [143; 146], Roberto *et al.* [114], Frimpong and Achireko [62], Spanos and Zeldin [123]).

Note about Circular and Spherical Harmonics

By restating Laplace's equation [see Eqs. (3.7.16) and (3.7.17)] in spherical coordinates, as is common in particle physics and seismology, solutions can be expressed in terms of Laplace spherical harmonic functions, with coefficients that are analogs of Fourier coefficients. Whereas analysis based on Fourier series applies to (deterministic or random) processes defined in rectangular domains, with trigonometric functions in a Fourier series representing modes of vibration of a string or a rectangular plate, circular or spherical harmonics represent modes of vibration of a circle or a sphere in much the same way. Many results of frequency-domain stochastic analysis presented in this chapter can be generalized by considering expansions in circular or spherical harmonics instead of trigonometric functions. These harmonics have orthogonality properties that enable representation of circular and spherical noise fields by angular power spectra and cross-spectra. The broad topic of stochastic representation and analysis in terms of circular or spherical harmonics is not further pursued herein.

Chapter 4

Spectral Parameters, Level Crossings, and Extremes

Every valley shall be exalted, and every mountain and hill shall be made low; and the crooked shall be made straight, and the rough places plain.

Isaiah, 40:4

4.1 Spectral Moments and Related Parameters

Spectral moments and related parameters of a homogeneous random field $X(\mathbf{t})$ play a vital role in the study of patterns of fluctuation, level crossings and extreme values, those aspects of random variation that are of most practical interest in risk assessment. In this section we introduce the various spectral parameters, first for one-dimensional processes and then for n-dimensional fields, and seek to interpret them in the frequency domain. For notational convenience, the subscript X is dropped (from σ_X^2, $G_X(\omega)$, $\lambda_{k,X}$, and so on) in this section, but it is occasionally needed in later sections to distinguish the statistics of $X(\mathbf{t})$ from those of its envelope $R(\mathbf{t})$ and their respective derivatives.

Definition of One-Dimensional Spectral Parameters

The kth spectral moment of a stationary random process $X(t)$ may be defined as follows:

$$\lambda_k = \int_0^{\infty} \omega^k G(\omega) d\omega = \int_{-\infty}^{+\infty} |\omega|^k S(\omega) d\omega, \quad k = 0, 1, 2, \ldots. \tag{4.1.1}$$

For the moments of *even order* an equivalent definition is

$$\lambda_k = \int_{-\infty}^{+\infty} \omega^k S(\omega) d\omega, \quad k = 0, 2, 4, \ldots. \tag{4.1.2}$$

The moment of order zero equals the variance:

$$\lambda_0 = \int_0^{\infty} G(\omega) d\omega = \int_{-\infty}^{+\infty} S(\omega) d\omega = \sigma^2. \tag{4.1.3}$$

Corresponding to each spectral moment λ_k there exists a characteristic frequency

$$\Omega_k = \left(\frac{\lambda_k}{\lambda_0} \right)^{1/k}, \quad k = 1, 2, \ldots, \tag{4.1.4}$$

and these frequencies form an ordered array, $\Omega_k \leq \Omega_{k+1} (k = 1, 2, \ldots)$. It is useful to note the analogy between the unit-area *spectral* density function $g(\omega)$ and the *probability* density function of a nonnegative random variable. Both functions are nonnegative and must have unit area. As Fig. 4.1 suggests, the analogy leads to the interpretation of the characteristic frequencies Ω_1 and Ω_2 as the "mean" frequency and the "root mean square" frequency, respectively. It also facilitates interpretation of the dimensionless spectral parameter

$$\delta = \left[1 - \frac{\lambda_1^2}{\lambda_0 \lambda_2} \right]^{1/2} = \left[1 - \left(\frac{\Omega_1}{\Omega_2} \right)^2 \right]^{1/2}, \tag{4.1.5}$$

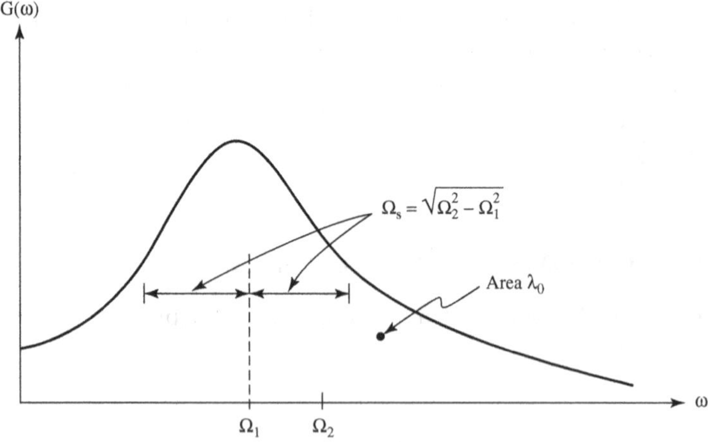

Fig. 4.1 One-sided spectral density function and spectral parameters.

which may be expressed (and interpreted) in terms of the frequency-domain "standard deviation" $\Omega_s = \sqrt{\Omega_2^2 - \Omega_1^2}$, as follows:

$$\delta = \frac{[\Omega_2^2 - \Omega_1^2]^{1/2}}{\Omega_2} = \frac{\Omega_s}{\Omega_2}. \tag{4.1.6}$$

The factor δ must lie between 0 and 1 (since $0 \le \lambda_1^2/\lambda_0\lambda_2 \le 1$, from Schwarz's inequality) and measures the degree of "spread," "dispersion," or "bandwidth" of $G(\omega)$. Specific values for some common spectral density functions are given in Sec. 4.7.

Among the higher-order spectral moments, λ_4 is of most significance. A quantity similar to δ is defined in terms of the moments λ_0, λ_2, and λ_4 as follows (Longuett-Higgins [84]):

$$\varepsilon = \left[1 - \frac{\lambda_2^2}{\lambda_0\lambda_4}\right]^{1/2} = \left[1 - \left(\frac{\Omega_2}{\Omega_4}\right)^4\right]^{1/2}, \tag{4.1.7}$$

which also lies between 0 and 1 and serves as a kind of measure of bandwidth, one that is, however, less direct or accessible than δ. (Higher-order moments tend to be extremely sensitive to details of the high-frequency content of $G(\omega)$ about which information is often scant; see Slepian [120].) The analogy between λ_4 and the 4th moment of a probability density function evokes the concept of "kurtosis," familiar to statisticians.

Bounds on the Spectral Distribution Function

The analogy between the normalized s.d.f. $g(\omega)$ and any probability density function suggests the use of the Chebychev inequality of elementary probability theory to obtain bounds on the normalized one-sided spectral distribution function $\overline{F}(\omega) = \int_0^\omega g(\omega_1)d\omega_1$ in terms of the spectral parameters. For $\omega > \Omega_2$, we may write [133; 134]:

$$\overline{F}(\omega) \ge 1 - \left(\frac{\Omega_s}{\omega - \Omega_1}\right)^2 = 1 - \frac{\delta^2\Omega_2^2}{(\omega - \Omega_1)^2} \ge 1 - \delta^2\left(\frac{\omega}{\Omega_2} - 1\right)^{-2}. \tag{4.1.8}$$

In terms of the lower-order spectral moments, this inequality provides a bound on the fractional contribution to the total variance from frequencies greater than a given value. For example, taking $\delta = 0.2$ and $\omega = 3\Omega_2$, the inequality yields

$$\overline{F}(3\Omega_2) = \int_0^{3\Omega_2} g(\omega)d\omega \ge 1 - \frac{(0.2)^2}{4} = 0.99, \tag{4.1.9}$$

or

$$F(3\Omega_2) = \int_0^{3\Omega_2} G(\omega)d\omega \geq 0.99\sigma^2. \tag{4.1.10}$$

In words, the inequality asserts that for a random process having a bandwidth factor $\delta = 0.2$, less than one percent of the total variance is contributed by components with frequencies above $3\,\Omega_2$. If δ is unknown, replacing it by its upper bound $\delta = 1$ would give $\bar{F} = (3\,\Omega_2) \geq 0.75$. Some additional bounds of this type are mentioned in RF1 [137].

Joint Spectral Moments of $S(\omega_1, \omega_2)$

In terms of the spectral density function $S(\omega_1, \omega_2)$ of a two-dimensional homogeneous random field $X(t_1, t_2)$, we define the joint spectral moment of order k with respect to ω_1 and of order l with respect to ω_2, for $k + l$ *even*, as follows:

$$\lambda_{kl} = \int_{-\infty}^{+\infty}\int_{-\infty}^{+\infty} \omega_1^k \omega_2^l S(\omega_1, \omega_2) d\omega_1 d\omega_2,$$

$$k + l \text{ even}, \;\; k = 0, 1, 2, \ldots, \;\; l = 0, 1, 2, \ldots. \tag{4.1.11}$$

If $k+l$ is odd, the symmetry of $S(\omega_1, \omega_2)$ in relation to $\omega_1 = \omega_2 = 0$ insures that the double integral will vanish. The kth *marginal* spectral moment with respect to ω_1, for k even, is

$$\lambda_{k0} = \int_{-\infty}^{+\infty}\int_{-\infty}^{+\infty} \omega_1^k S(\omega_1, \omega_2) d\omega_1 d\omega_2 = \int_{-\infty}^{+\infty} \omega_1^k S(\omega_1) d\omega_1$$

$$= \int_0^{\infty} \omega_1^k G(\omega_1) d\omega_1 \equiv \lambda_k^{(1)}, \quad k = 0, 2, 4, \ldots, \tag{4.1.12}$$

where the superscript (1) indicates the direction of one-dimensional variation. The *odd* marginal moments are defined, along with the even moments, the same way as in Eq. (4.1.1):

$$\lambda_{k0} = \int_0^{\infty} \omega_1^k G(\omega_1) d\omega_1 \equiv \lambda_k^{(1)}, \quad k = 0, 1, 2, \ldots. \tag{4.1.13}$$

Likewise, the marginal moments with respect to ω_2 are

$$\lambda_{0l} = \int_0^{\infty} \omega_2^l G(\omega_2) d\omega_2 \equiv \lambda_l^{(2)}, \quad l = 0, 1, 2, \ldots. \tag{4.1.14}$$

Note that $\lambda_{00} = \lambda_0^{(1)} = \lambda_0^{(2)} = \sigma^2$. If $S(\omega_1, \omega_2)$ is quadrant symmetric, λ_{kl} may be expressed in terms of the s.d.f. defined for positive frequencies only, as follows:

$$\begin{aligned} \lambda_{kl} &= 0, && k, l \text{ both odd}, \\ \lambda_{kl} &= \int_0^\infty \int_0^\infty \omega_1^k \omega_2^l G(\omega_1, \omega_2) d\omega_1 d\omega_2, && k, l \text{ both even}. \end{aligned} \tag{4.1.15}$$

Matrix of Joint Spectral Moments

All second-order moments can be assembled into a 2 by 2 matrix:

$$\mathbf{\Lambda}_{11} = \begin{bmatrix} \lambda_{20} & \lambda_{11} \\ \lambda_{11} & \lambda_{02} \end{bmatrix} = \begin{bmatrix} \lambda_2^{(1)} & \lambda_{11} \\ \lambda_{11} & \lambda_2^{(2)} \end{bmatrix}. \tag{4.1.16}$$

Likewise, the matrix of fourth-order moments (for k and l even) is

$$\mathbf{\Lambda}_{22} = \begin{bmatrix} \lambda_{40} & \lambda_{22} \\ \lambda_{22} & \lambda_{04} \end{bmatrix} = \begin{bmatrix} \lambda_4^{(1)} & \lambda_{22} \\ \lambda_{22} & \lambda_4^{(2)} \end{bmatrix}. \tag{4.1.17}$$

The Isotropic Case

If $X(t_1, t_2)$ is isotropic, the joint spectral moment λ_{kl} may be computed as follows (for k and l even):

$$\begin{aligned} \lambda_{kl} &= \int_0^\infty \int_0^\infty \omega_1^k \omega_2^l G(\omega_1, \omega_2)\, d\omega_1\, d\omega_2 \\ &= \int_0^{\pi/2} \sin^k \theta \cos^l \theta \, d\theta \int_0^\infty \omega^{k+l+1} G^R(\omega)\, d\omega \\ &= \lambda_{k+l}^R \frac{\Gamma((k+1)/2)\Gamma((l+1)/2)}{\pi \Gamma((k+l+2)/2)}, \quad k, l \text{ even}, \end{aligned} \tag{4.1.18}$$

where λ_{k+l}^R is the spectral moment of order $k+l$ defined in terms of the radial spectral density function $R(\omega) = (\pi/2)\,\omega\, G^R(\omega)$. We may write

$$\lambda_{k+l}^R = \int_0^\infty \omega^{k+l} R(\omega)\, d\omega = \frac{\pi}{2} \int_0^\infty \omega^{k+l+1} G^R(\omega)\, d\omega. \tag{4.1.19}$$

Taking $k = l = 0$ yields the variance; from Eq. (4.1.18) we obtain

$$\sigma^2 = \lambda_{00} = \frac{\lambda_0^R [\Gamma(1/2)]^2}{\pi} = \lambda_0^R. \tag{4.1.20}$$

For $k = l = 2$, Eq. (4.1.18) gives

$$\lambda_{22} = \frac{\lambda_4^R [\Gamma(3/2)]^2}{[\pi\Gamma(3)]} = \frac{1}{8}\lambda_4^R. \tag{4.1.21}$$

The marginal second and fourth moments are, respectively,

$$\lambda_2^{(i)} = \lambda_{02} = \lambda_{20} = \frac{1}{2}\lambda_2^R, \quad i = 1, 2, \tag{4.1.22}$$

and

$$\lambda_4^{(i)} = \lambda_{40} = \frac{3}{8}\lambda_4^R, \quad i = 1, 2. \tag{4.1.23}$$

From Eqs. (4.1.21) and (4.1.23) we deduce

$$\lambda_{22} = \frac{1}{3}\lambda_4^{(i)}, \quad i = 1, 2. \tag{4.1.24}$$

Finally, for a 2-dimensional *isotropic* process, $\lambda_k^{(1)} = \lambda_k^{(2)} = \lambda_k$, and the matrices $\mathbf{\Lambda}_{11}$ and $\mathbf{\Lambda}_{22}$ take, respectively, the following form:

$$\mathbf{\Lambda}_{11} = \begin{bmatrix} \lambda_2 & 0 \\ 0 & \lambda_2 \end{bmatrix}, \quad \mathbf{\Lambda}_{22} = \begin{bmatrix} \lambda_4 & \frac{1}{3}\lambda_4 \\ \frac{1}{3}\lambda_4 & \lambda_4 \end{bmatrix}, \tag{4.1.25}$$

where λ_2 and λ_4 are spectral moments of the direction-dependent one-dimensional random process $X(t_i)$, $i = 1, 2$.

Joint Spectral Moments for n-Dimensional Fields

In specifying joint spectral moments of n-dimensional homogeneous random fields, $X(t_1, \ldots, t_n)$, one needs to identify which pair of coordinates (i, j) is involved. The joint spectral moment of order k with respect to ω_i and of order l with respect to ω_j may be defined in the same way as in Eq. (4.1.11):

$$\begin{aligned}\lambda_{kl}^{(ij)} &= \int_{-\infty}^{+\infty} \cdots \int_{-\infty}^{+\infty} \omega_i^k \omega_j^l S(\boldsymbol{\omega}) d\omega_1 \dots d\omega_n \\ &= \int_{-\infty}^{+\infty} \int_{-\infty}^{+\infty} \omega_i^k \omega_j^l S(\omega_i, \omega_j) d\omega_i d\omega_j, \\ & \quad k+l \text{ even}, \; k = 0, 1, 2, \dots, \; l = 0, 1, 2, \dots, \end{aligned} \tag{4.1.26}$$

where $S(\boldsymbol{\omega})$ is the n-dimensional spectral density function, and $S(\omega_i, \omega_j)$ characterizes random variation on the plane with coordinates t_i and t_j. When k and l are fixed, the joint moments $\lambda_{kl}^{(ij)}$ can be assembled into an $n \times n$ matrix,

$$\mathbf{\Lambda}_{kl} = \begin{bmatrix} \lambda_{kl}^{(ij)} \\ i, j = 1, 2, \dots, n \end{bmatrix}. \tag{4.1.27}$$

If $k = l$, the matrix $\mathbf{\Lambda}_{kk}$ will be symmetric since

$$\lambda_{kk}^{(ij)} = \lambda_{kk}^{(ji)}. \tag{4.1.28}$$

The following relations between the joint and marginal spectral moments are notable:

$$\begin{aligned}\lambda_{k0}^{(ij)} &= \int_{-\infty}^{+\infty} \cdots \int_{-\infty}^{+\infty} \omega_i^k S(\boldsymbol{\omega}) d\omega_1 \dots d\omega_n \\ &= \int_{-\infty}^{+\infty} \omega_i^k S(\omega_i)\, d\omega_i = \lambda_k^{(i)} \end{aligned} \tag{4.1.29}$$

and, more generally,

$$\begin{aligned}\lambda_{kl}^{(ii)} &= \int_{-\infty}^{+\infty} \cdots \int_{-\infty}^{+\infty} \omega_i^{k+l} S(\boldsymbol{\omega}) d\omega_1 \dots d\omega_n \\ &= \int_{-\infty}^{+\infty} \omega_i^{k+l} S(\omega_i)\, d\omega_i = \lambda_{k+l}^{(i)}. \end{aligned} \tag{4.1.30}$$

If $S(\boldsymbol{\omega})$ is quadrant symmetric, we may write

$$\begin{aligned}\lambda_{kl}^{(ij)} &= 0, & & k, l \text{ both odd}, i \neq j, \\ \lambda_{kl}^{(ij)} &= 4 \int_0^{\infty} \int_0^{\infty} \omega_i^k \omega_j^l G(\omega_i, \omega_j) d\omega_i d\omega_j, & & k, l \text{ both even}. \end{aligned} \tag{4.1.31}$$

By far the most important joint spectral moments are those of second order, namely the elements of the matrix $\mathbf{\Lambda}_{11}$. The latter, for quadrant symmetric processes, will be a diagonal matrix.

4.2 Statistics of Partial Derivatives

In the preceding section we introduced various moments (marginal and joint) of the spectral density function $S(\boldsymbol{\omega})$ of a homogeneous random field $X(\mathbf{t})$ and sought to interpret them as measures of variability and correlation characterizing the field's frequency content. In this section it is shown that these spectral moments possess physical meaning in terms of the second-order statistics of the partial (or direction-dependent) derivatives of the random field $X(\mathbf{t})$. The properties of the "envelope" of $X(\mathbf{t})$ are considered in Sec. 4.3, while related results for level excursions and extremes are derived and interpreted in subsequent sections in this chapter.

Consider a homogeneous random field *on the plane* (t_1, t_2) with s.d.f. $S(\omega_1, \omega_2)$. If the coordinate t_2 is kept fixed, the random field reduces (in a sense) to a stationary random function of t_1. We may write alternately

$$X(t_1) \equiv X^{(1)}(\mathbf{t}), \tag{4.2.1}$$

depending upon which aspect of behavior requires emphasis. $X^{(1)}(\mathbf{t})$ may be seen either as a one-dimensional random function of t_1 (with "marginal" statistics, owing to homogeneity, independent of t_2) or as a two-dimensional random field whose coordinate t_2 acts merely as an index. Of course, the same can be said about $X^{(2)}(\mathbf{t})$, *mutatis mutandis*, and in general about the "direction-dependent" process $X^{(i)}(\mathbf{t})$ in relation to an n-dimensional random field $X(\mathbf{t})$.

The derivative of $X(t_1)$ is in effect the *partial* or *direction-dependent* derivative of $X(t_1, t_2)$

$$\dot{X}(t_1) \equiv \frac{d}{dt_1} X(t_1) \equiv \frac{\partial}{\partial t_1} X(\mathbf{t}) \equiv \dot{X}^{(1)}(\mathbf{t}). \tag{4.2.2}$$

This derived random process has mean zero and variance $\lambda_2^{(1)}$. Likewise, $\dot{X}^{(2)}(\mathbf{t})$ has mean zero and variance $\lambda_2^{(2)}$.

First-Order Partial Derivatives

From Eq. (3.6.7), we have, for $i = 1, 2$:

$$\lambda_2^{(i)} = \int_{-\infty}^{+\infty} \int_{-\infty}^{+\infty} \omega_i^2 S(\omega_i) d\omega_i = \mathrm{Var}[\dot{X}^{(i)}] \equiv \sigma_{\dot{X}_i}^2 . \tag{4.2.3}$$

The joint spectral moment of second order λ_{11} is equal to the covariance of the direction-dependent derivatives $\dot{X}^{(1)}(\mathbf{t})$ and $\dot{X}^{(2)}(\mathbf{t})$ at a given point $\mathbf{t} = (t_1, t_2)$ in the parameter space. From Eq. (3.6.15) we may write

$$\lambda_{11} = \int_{-\infty}^{+\infty} \int_{-\infty}^{+\infty} \omega_1 \omega_2 S(\omega_1, \omega_2) d\omega_1 d\omega_2 = E[\dot{X}^{(1)} \dot{X}^{(2)}]. \tag{4.2.4}$$

The dimensionless coefficient

$$v = \frac{\lambda_{11}}{[\lambda_2^{(1)} \lambda_2^{(2)}]^{1/2}} = \frac{E[\dot{X}^{(1)} \dot{X}^{(2)}]}{\sigma_{\dot{X}_1} \sigma_{\dot{X}_2}} \tag{4.2.5}$$

equals the coefficient of correlation between the two direction-dependent derivatives $\dot{X}^{(1)}(t_1, t_2)$ and $\dot{X}^{(2)}(t_1, t_2)$ at the same location in the parameter space. Satisfying $0 \leq v \leq 1$, it is mathematically equivalent to the coefficient of correlation of two random variables (described by a bivariate probability density function).

Higher-Order Partial Derivatives

Similar results can be obtained for higher-order derivatives, provided the moments exist. Consider, in particular,

$$\lambda_{22} = \int_{-\infty}^{+\infty} \int_{-\infty}^{+\infty} \omega_1^2 \omega_2^2 S(\omega_1, \omega_2) d\omega_1 d\omega_2 = E[\ddot{X}^{(1)} \ddot{X}^{(2)}] \tag{4.2.6}$$

and

$$\lambda_4^{(i)} = \int_0^{\infty} \omega_i^4 G(\omega_i) d\omega_i = \mathrm{Var}[\ddot{X}^{(i)}]. \tag{4.2.7}$$

The corresponding dimensionless quantity

$$w = \frac{\lambda_{22}}{[\lambda_4^{(1)} \lambda_4^{(2)}]^{1/2}} = \frac{E[\ddot{X}^{(1)} \ddot{X}^{(2)}]}{\sigma_{\ddot{X}_1} \sigma_{\ddot{X}_2}} \tag{4.2.8}$$

equals the coefficient of correlation between the direction-dependent *second derivatives* $\ddot{X}^{(1)}(t_1, t_2)$ and $\ddot{X}^{(2)}(t_1, t_2)$ at a fixed location (t_1, t_2) in the parameter space. In applications, the higher-order spectral moments are hard to deal with, due to the basic lack of information about, and sensitivity to, the spectral density function at high frequencies.

Parallel Results for n-Dimensional Fields

These results are easily generalized for an n-dimensional field $X(\mathbf{t})$. The covariance between the partial derivatives $\dot{X}^{(i)}(\mathbf{t})$ and $\dot{X}^{(j)}(\mathbf{t})$ is

$$E[\dot{X}^{(i)}\dot{X}^{(j)}] = \lambda_{11}^{(ij)} = \int_{-\infty}^{+\infty}\int_{-\infty}^{+\infty} \omega_i\omega_j S(\omega_i,\omega_j)d\omega_i d\omega_j, \tag{4.2.9}$$

where $S(\omega_i,\omega_j)$ characterizes the frequency content of 2-D random variation on any plane parallel to the t_i and t_j axes. The mean square partial derivative with respect to t_i is [see Eq. (4.2.3)] $\lambda_{11}^{(ii)} = \lambda_2^{(i)} = \sigma_{\dot{X}_i}^2$.

The second-order spectral moments of $X(\mathbf{t})$ can be arranged in matrix form, as follows:

$$\mathbf{\Lambda}_{11} = \begin{bmatrix} \lambda_{11}^{(ij)} \\ i,j=1,\ldots,n \end{bmatrix} = \begin{bmatrix} E[\dot{X}^{(i)}\dot{X}^{(j)}] \\ i,j=1,\ldots,n \end{bmatrix}. \tag{4.2.10}$$

The associated $n \times n$ matrix $\mathbf{V}$ contains the coefficients of correlation between the first-order partial derivatives of $X(\mathbf{t})$ at a given location $\mathbf{t}$,

$$v^{(ij)} = \frac{\lambda_{11}^{(ij)}}{(\lambda_2^{(i)}\lambda_2^{(j)})^{1/2}} = \frac{E[\dot{X}^{(i)}\dot{X}^{(j)}]}{\sigma_{\dot{X}_i}\sigma_{\dot{X}_j}}, \quad i,j=1,\ldots,n. \tag{4.2.11}$$

The diagonal elements are equal to one. For quadrant symmetric random fields, $E[\dot{X}^{(i)}\dot{X}^{(j)}] = 0$ for $i \neq j$, and hence $\mathbf{V}$ becomes an identity matrix. Entirely similar results apply for the covariances $\lambda_{22}^{(ij)}$ of the second-order partial derivatives and the correlation coefficients $w^{(ij)}$ [see Eq. (4.2.8)].

4.3 Basic Envelope Statistics

For a stationary random process $X(t)$ with mean zero, the envelope $R(t)$ may be thought of as a relatively smoothly varying function for which $R(t) \geq |X(t)|$ at all values of t, and $R(t) = |X(t)|$ at or near the peaks of $|X(t)|$. There are a number of one-dimensional envelope definitions in existence, the earliest due to Rice [113]. The literature on multi-dimensional envelopes is scant (Adler [2], Van Dyck [147]) and deals in effect with the envelope of direction-dependent processes such as $X^{(1)}(\mathbf{t})$ rather than with a true n-dimensional field, as considered here. Our starting point is a review of some well-established results for one-dimensional processes.

Alternate Definitions of the 1-D Envelope

The envelope introduced by Rice [113] hypothesizes the existence of a representative midband frequency ω_m. It is based on the fact that a stationary random process $X(t)$ with mean zero can be expressed as follows:

$$X(t) = I_c(t) \cos \omega_m t - I_s(t) \sin \omega_m t, \tag{4.3.1}$$

where $I_c(t)$ and $I_s(t)$ are random functions composed of sinusoids, in much the same way as $X(t)$ itself [see Eq. (3.2.1)]:

$$\begin{aligned} I_c(t) &= \sum_{k=1}^{K} C_k \cos(\omega_k t - \omega_m t + \Phi_k), \\ I_s(t) &= \sum_{k=1}^{K} C_k \sin(\omega_k t - \omega_m t + \Phi_k). \end{aligned} \tag{4.3.2}$$

Note that $I_c(t)$ and $I_s(t)$ are weighted sums of 'cosine' and 'sines', respectively, that share the values of the random amplitudes C_k and phase angles Φ_k. The summation, in this formulation, is over positive frequencies only. The envelope $R(t)$ is then defined by the relationship

$$R^2(t) = I_c^2(t) + I_s^2(t). \tag{4.3.3}$$

In Rice's analysis the midband frequency ω_m possesses physical meaning as the average frequency of the "carrier function" about which the true envelope more or less randomly oscillates. However, the statistics of the envelope process $R(t)$ do not depend on the choice of ω_m. An analytically convenient selection (not considered by Rice) is $\omega_m = 0$. In this case, the component $I_c(t)$ becomes identical to the process itself [see Eq. (4.3.1)], and it is convenient to introduce two "component processes" (the first of which is identical to $X(t)$):

$$\begin{aligned} X_c(t) &= \sum_{k=1}^{K} C_k \cos(\omega_k t + \Phi_k) \equiv X(t), \\ X_s(t) &= \sum_{k=1}^{K} C_k \sin(\omega_k t + \Phi_k). \end{aligned} \tag{4.3.4}$$

The envelope $R(t)$ can now be defined as follows:

$$R^2(t) = X_c^2(t) + X_s^2(t). \qquad (4.3.5)$$

Cramér and Leadbetter [41] define essentially the same envelope by considering the complex random process $X^*(t)$ [see Eq. (3.3.19)]:

$$X^*(t) = \sum_{k=1}^{K} Z_k e^{i\omega_k t} = X_c(t) + iX_s(t). \qquad (4.3.6)$$

(Again the summation is over positive frequencies.) The process $X(t) = X_c(t)$ is the real part of $X^*(t)$, while $X_s(t)$ is its imaginary part (the Hilbert transform of $X(t)$). Their envelope $R(t)$ is defined as the absolute value of the complex random process $X^*(t)$:

$$R(t) = |X^*(t)| = [X_c^2(t) + X_s^2(t)]^{1/2}. \qquad (4.3.7)$$

To establish the equivalence of two interpretations, introduce

$$Z_k = A_k + iB_k \qquad (4.3.8)$$

and

$$e^{i\omega_k t} = \cos\omega_k t + i\sin\omega_k t \qquad (4.3.9)$$

into Eq. (4.3.6). The real and imaginary parts of $X^*(t)$ become

$$\begin{aligned} X_c(t) &= \sum_{k=1}^{K} [A_k \cos\omega_k t - B_k \sin\omega_k t], \\ X_s(t) &= \sum_{k=1}^{K} [A_k \sin\omega_k t - B_k \cos\omega_k t]. \end{aligned} \qquad (4.3.10)$$

Taking $A_k = C_k \cos\Phi_k$ and $B_k = C_k \sin\Phi_k$ renders Eqs. (4.3.4) and (4.3.10) identical.

One-Dimensional Envelope Statistics

It is easy to see that the *associated* random process $X_s(t)$ has the same marginal statistical properties as $X(t) = X_c(t)$ (as only the phase angles of contributing sinusoids differ). Since $m = 0$, the common variance of these processes is

$$E[X_c^2(t)] = E[X_s^2(t)] = \sigma^2. \qquad (4.3.11)$$

The cross-covariance function of $X_c(t)$ and $X_s(t)$ may be obtained by substituting Eq. (4.3.4) into the definition and performing the usual (discrete-to-continouous) limiting operation in the frequency domain:

$$E[X_c(t)X_s(t+\tau)] = \sum_{k=1}^{K} \frac{C_k^2}{2} \sin \omega_k \tau \to \int_0^\infty G(\omega) \sin \omega\tau \, d\tau. \tag{4.3.12}$$

Taking $\tau = 0$ confirms that $X_c(t)$ and $X_s(t)$ are uncorrelated:

$$E[X_c(t)X_s(t)] \equiv E[X_c X_s] = 0. \tag{4.3.13}$$

The same type of analysis leads to

$$E[X_c(t)\dot{X}_s(t)] = \sum_{k=1}^{K} \frac{C_k^2}{2} \omega_k \to \int_0^\infty \omega G(\omega) d\omega = \lambda_1 \tag{4.3.14}$$

and

$$E[\dot{X}_c(t)X_s(t)] = -\lambda_1. \tag{4.3.15}$$

Envelope statistics can now be expressed in terms of the statistics of X_s and X_c. Taking the expectation on both sides of Eq. (4.3.5) leads to the mean square of $R(t)$:

$$E[R^2] = E[X_c^2] + E[X_s^2] = 2\sigma^2. \tag{4.3.16}$$

Taking the (total) derivative of Eq. (4.3.5) with respect to time yields

$$2R\dot{R} = 2X_c\dot{X}_c + 2X_s\dot{X}_s. \tag{4.3.17}$$

The mean slopes are all zero:

$$E[\dot{X}_c] = E[\dot{X}_s] = E[\dot{R}] = 0. \tag{4.3.18}$$

Squaring both sides of Eq. (4.3.17), taking the expectation, and accounting for the fact that the component processes X_c and X_s must have the same statistical properties, one obtains

$$E[(R\dot{R})^2] = 2E[(X_c\dot{X}_c)^2] + 2E[X_s\dot{X}_s X_c\dot{X}_c]. \tag{4.3.19}$$

If $X(t)$ is Gaussian, $X_s(t)$ will also be Gaussian. Owing to stationarity, X_c and $\dot{X}_c$ are uncorrelated (and independent if Gaussian), and so are X_s and $\dot{X}_s$, as well as R and $\dot{R}$. Based on Eq. (2.5.21), $E[X_s\dot{X}_s X_c\dot{X}_c] = E[X_s\dot{X}_c]E[X_c\dot{X}_s]$, and Eq. (4.3.19) becomes

$$E[R^2]E[\dot{R}^2] = 2E[X_c^2]E[\dot{X}_c^2] + 2E[X_s\dot{X}_c]E[X_c\dot{X}_s]. \tag{4.3.20}$$

Finally, inserting Eqs. (4.3.14)–(4.3.16) into Eq. (4.3.20) yields

$$\sigma^2_{\dot{R}} \equiv E[\dot{R}^2] = \sigma^2_{\dot{X}} - \frac{\lambda_1^2}{\sigma^2} = \delta^2\sigma^2_{\dot{X}}, \tag{4.3.21}$$

where [see Eq. (4.1.5)]

$$\delta^2 = 1 - \left[\frac{\lambda_1^2}{\lambda_2\lambda_0}\right]. \tag{4.3.22}$$

The result, $\sigma_{\dot{R}} = \delta\sigma_{\dot{X}}$, may be interpreted as follows: the smaller the spectral bandwidth factor δ, the less intensely the envelope is expected to fluctuate compared to the original process.

The Envelope of an n-Dimensional Gaussian Random Field

Much as in the one-dimensional case, the envelope $R(t)$ of a homogeneous random field $X(\mathbf{t})$ with mean zero can be defined in terms of two component fields $X_c(\mathbf{t}) \equiv X(\mathbf{t})$ and $X_s(\mathbf{t})$:

$$R^2(\mathbf{t}) = X_c^2(\mathbf{t}) + X_s^2(\mathbf{t}), \tag{4.3.23}$$

where the *associated* random field $X_s(\mathbf{t})$ may be thought of as the sum of multi-dimensional sine components, just as $X_c(\mathbf{t})$ is a superposition of cosine components, according to Eq. (3.4.1). The two component fields have identical statistical properties.

By keeping all but one coordinate (t_i) fixed, Eq. (4.3.23) reduces to a relationship defining the one-dimensional envelope $R(t_i) \equiv R^{(i)}(\mathbf{t})$ of the direction-dependent process $X(t_i) \equiv X^{(i)}(\mathbf{t})$:

$$[R^{(i)}(\mathbf{t})]^2 = [X_c^{(i)}(\mathbf{t})]^2 + [X_s^{(i)}(\mathbf{t}]^2, \quad i = 1, \ldots, n. \tag{4.3.24}$$

All the information bearing on one-dimensional envelopes is applicable to $R^{(i)}(\mathbf{t})$, $i = 1, \ldots, n$. In particular, the mean square of the envelope is

$$E[(R^{(i)})^2] = E[R^2] = 2\sigma^2, \tag{4.3.25}$$

where σ^2 is the variance of $X(\mathbf{t})$. The values of the component processes evaluated at the same point are uncorrelated:

$$E[X_c^{(i)}X_s^{(i)}] \equiv E[X_cX_s] = 0. \tag{4.3.26}$$

Also, the covariance between X_c and the partial derivative of X_s with respect to t_i is

$$E[X_c\dot{X}_s^{(i)}] = \lambda_1^{(i)}, \quad i = 1, \ldots, n, \tag{4.3.27}$$

where $\lambda_1^{(i)}$ is the first-order spectral moment of $G(\omega_i)$. Likewise, from Eq. (4.3.15).

$$E[\dot{X}_c^{(i)}X_s] = -\lambda_1^{(i)}, \quad i = 1, \ldots, n. \tag{4.3.28}$$

The derivative of $R(t_i)$ is identical to the partial derivative of $R(\mathbf{t})$ with respect to t_i. Differentiation with respect to t_i of all the terms in Eq. (4.3.23) yields

$$2R\dot{R}^{(i)} = 2X_c\dot{X}_c^{(i)} + 2X_s\dot{X}_s^{(i)}, \tag{4.3.29}$$

which parallels Eq. (4.3.17). Proceeding exactly as in the one-dimensional case leads to the relationship

$$\sigma_{\dot{R}_i} = \delta^{(i)}\sigma_{\dot{X}_i}, \quad i = 1, \ldots, n, \tag{4.3.30}$$

where $\sigma_{\dot{R}_i}$ and $\sigma_{\dot{X}_i}$ denote the standard deviations (r.m.s. values) of the partial derivatives with respect to t_i, of $R(\mathbf{t})$ and $X(\mathbf{t})$, respectively; and $\delta^{(i)}$ is the spectral bandwidth measure characterizing $G(\omega_i)$.

Covariance of Partial Derivatives of $R(\mathbf{t})$

The next step in the analysis is to derive $E[\dot{R}^{(i)}\dot{R}^{(j)}]$, the covariance between partial derivatives of $R(\mathbf{t})$ with respect to two different coordinates t_i and t_j. Equation (4.3.29) may be restated by replacing i by j:

$$R\dot{R}^{(j)} = X_c\dot{X}_c^{(j)} + X_s\dot{X}_s^{(j)}. \tag{4.3.31}$$

Multiplying Eqs. (4.3.29) and (4.3.31), side by side, taking the expectation, and accounting for the fact that $X_c(\mathbf{t})$ and $X_s(\mathbf{t})$ have the same (marginal) statistics, we obtain

$$E[R^2\dot{R}^{(i)}\dot{R}^{(j)}] = 2E[X_c^2\dot{X}_c^{(i)}\dot{X}_c^{(j)}] + 2E[X_c\dot{X}_c^{(i)}X_s\dot{X}_s^{(j)}]. \tag{4.3.32}$$

The following pairs of quantities are uncorrelated: R and $\dot{R}^{(i)}$, R and $\dot{R}^{(j)}$, X_c and $\dot{X}_c^{(i)}$, X_c and $\dot{X}_c^{(j)}$, X_c and X_s, X_s and $\dot{X}_s^{(j)}$. If $X(\mathbf{t})$ is Gaussian, the uncorrelated quantities become independent and, based on Eq. (2.5.21), the fourth-order moment reduces to a product of two second-order moments. The result is

$$E[R^2]E[\dot{R}^{(i)}\dot{R}^{(j)}] = 2E[X^2]E[\dot{X}^{(i)}\dot{X}^{(j)}] + 2E[X_c\dot{X}_s^{(j)}]E[\dot{X}_c^{(i)}X_s]. \quad (4.3.33)$$

Inserting $E[R^2] = 2E[X^2] = 2\sigma^2$, and accounting for Eqs. (4.3.25), (4.3.27) and (4.3.28), yields

$$\begin{aligned} E[\dot{R}^{(i)}\dot{R}^{(j)}] &= E[\dot{X}^{(i)}\dot{X}^{(j)}] - \frac{\lambda_1^{(i)}\lambda_1^{(j)}}{\sigma^2} = \lambda_{11}^{(ij)} - \frac{\lambda_1^{(i)}\lambda_1^{(j)}}{\lambda_0} \\ &= \Delta^{(ij)}[\lambda_2^{(i)}\lambda_2^{(j)}]^{1/2} = \Delta^{(ij)}\sigma_{\dot{X}_i}\sigma_{\dot{X}_j}, \end{aligned} \quad (4.3.34)$$

where $\Delta^{(ij)}$ is a dimensionless parameter defined implicitly in Eq. (4.3.34), either in terms of spectral moments or as the quotient of the covariance $E[\dot{R}^{(i)}\dot{R}^{(j)}]$ and the product $\sigma_{\dot{X}_i}\sigma_{\dot{X}_j}$. Each element $\Delta^{(ij)}$ of the $n \times n$ matrix

$$\mathbf{\Delta} = \begin{bmatrix} \Delta^{(ij)} = \dfrac{E[\dot{R}^{(i)}\dot{R}^{(j)}]}{\sigma_{\dot{X}_i}\sigma_{\dot{X}_j}} \\ \\ i,j = 1,2,\ldots,n \end{bmatrix}, \quad (4.3.35)$$

is proportional to the coefficient of correlation between two partial derivatives of the envelope of $R(\mathbf{t})$; the proportionality factor, considering Eq. (4.3.30), is $\delta^{(i)}\delta^{(j)}$.

For an n-dimensional isotropic random field, we have $\delta^{(i)} = \delta$ for $(i = 1,\ldots,n)$, and the matrix $\mathbf{\Delta}$ takes the following form:

$$\mathbf{\Delta} = \begin{bmatrix} \delta^2 & \cdots & \delta^2 - 1 & \cdots & \delta^2 - 1 \\ & & \delta^2 & \cdots & \delta^2 - 1 \\ \text{(symmetric)} & & & & \vdots \\ & & & & \delta^2 \end{bmatrix}. \quad (4.3.36)$$

Its determinant is

$$|\mathbf{\Delta}| = n\delta^2 - (n-1). \quad (4.3.37)$$

Its elements being correlation coefficients, it can be argued that the matrix $\mathbf{\Delta}$, like the matrix $\mathbf{V}$ [see Eq. (4.2.11)], must be positive semidefinite. The condition $|\mathbf{\Delta}| \geq 0$ then leads to interesting and useful constraints on the one-dimensional spectral bandwidth measures $\delta = \delta^{(i)}$ of isotropic random fields, namely:

$$\delta \geq \sqrt{\frac{n-1}{n}}, \quad n = 1, 2, \ldots. \tag{4.3.38}$$

The lower bound on the bandwidth factor δ of an isotropic random field is $\delta \geq \sqrt{1/2}$ for $n = 2$, and $\delta \geq \sqrt{2/3}$ for $n = 3$. Similar lower bounds, obtained in a same way, exist for the spectral bandwidth measure ε [defined in Eq. (4.1.7)] for isotropic n-dimensional random fields (see RF1 [137]). While Schwarz's inequality requires $0 \leq \varepsilon \leq 1$ for $n = 1$, the additional constraint is $\varepsilon \geq \sqrt{1/3}$ for $n = 2$, and $\varepsilon \geq \sqrt{2/3}$ for $n = 3$.

4.4 Threshold-Crossing Statistics and Extremes

Mean Threshold-Crossing Rate

A classical formula for the mean rate of crossings of the level b by a zero-mean stationary random process $X(t)$ is given by Rice [113]:

$$\nu_b \equiv \nu_{b,X} = \int_{-\infty}^{+\infty} |\dot{x}| f_{X,\dot{X}}(b, \dot{x}) d\dot{x}, \tag{4.4.1}$$

where $f_{X,\dot{X}}(x, \dot{x})$ is the joint probability density function of $X(t)$ and its derivative $\dot{X}(t)$. (Rainal [111] documents and comments on the origin of Rice's formula.) Since $X(t)$ is stationary, the random variables $X(t)$ and $\dot{X}(t)$ are uncorrelated. If they are also independent (which is guaranteed in case $X(t)$ is Gaussian), then

$$\nu_b = f_X(b) \int_{-\infty}^{+\infty} |\dot{x}| f_{\dot{X}}(\dot{x}) d\dot{x} = f_X(b) E[|\dot{X}|], \tag{4.4.2}$$

where $E[|\dot{X}|]$ is the mean of the absolute value of the slope of $X(t)$. Every b-upcrossing (or crossing of b with positive slope) is followed by a b-downcrossing, so the mean rates of up- and down-crossings are given by

$$\nu_b^+ = \nu_b^- = \frac{\nu_b}{2} = \frac{1}{2} f_X(b) E[|\dot{X}|]. \tag{4.4.3}$$

Since differentiation is a linear operation, if $X(t)$ is Gaussian its derivative $\dot{X}$ will also be Gaussian (with mean zero). Hence,

$$E[|\dot{X}|] = 2 \int_0^{\infty} \frac{\dot{x}}{\sqrt{2\pi}\sigma_{\dot{X}}} \exp\left\{ -\frac{\dot{x}^2}{2\sigma_{\dot{X}}^2} \right\} d\dot{x} = \sqrt{\frac{2}{\pi}} \sigma_{\dot{X}}. \tag{4.4.4}$$

Combining Eqs. (4.4.3) and (4.4.4) yields, for a Gaussian random process $X(t)$, the expression:

$$\nu_b^+ = \frac{1}{2\pi}\frac{\sigma_{\dot{X}}}{\sigma_X}\exp\left\{-\frac{b^2}{2\sigma_X^2}\right\}. \tag{4.4.5}$$

Taking $b = 0$ yields the mean rate of zero-crossing with positive slope:

$$\nu_0^+ \equiv \nu_{0,X}^+ = \frac{1}{2\pi}\frac{\sigma_{\dot{X}}}{\sigma_X} = \frac{1}{2\pi}\left(\frac{\lambda_2}{\lambda_0}\right)^{1/2} = \frac{\Omega_2}{2\pi}, \tag{4.4.6}$$

where λ_k equals the kth spectral moment of $X(t)$, and Ω_2 is its second characteristic frequency [see Eq. (4.1.4)].

Mean Rate and Probability Distribution of Local Maxima

The above formulas can also be applied to $\dot{X}(t)$, the derivative of $X(t)$. The zero-crossings with negative slope are of particular interest since for a differentiable process $X(t)$ the conditions $\dot{X}(t) = 0$ and $\ddot{X}(t) < 0$ imply the occurrence of a local maximum. Since the number of local maxima of $X(t)$ is equal to the number of zero-downcrossings of $\dot{X}(t)$, the mean rate of occurrence of local maxima is

$$\nu_{\max} = \nu_{0,\dot{X}}^- = \frac{1}{2\pi}\frac{\sigma_{\ddot{X}}}{\sigma_{\dot{X}}} = \frac{\nu_0^+}{\sqrt{1-\varepsilon^2}}, \tag{4.4.7}$$

where the right side of the equation is obtained by combining Eqs. (4.4.6) and (4.1.7). It may be argued that true narrow-band processes are characterized by the fact that each zero-upcrossing event is followed by a *single* maximum. This condition, $\nu_{\max} = \nu_0^+$, evidently implies $\varepsilon = 0$. At the other extreme, if ε reaches its upper bound ($\varepsilon = 1$), local maxima are predicted to be infinitely more numerous than zero-crossings.

When $\varepsilon = 0$, local maxima are always positive, and their probability density function may be obtained from the expression, Eq. (4.4.5), for the mean level-upcrossing rate ν_b^+. Furthermore, the ratio ν_b^+/ν_0^+ may be interpreted as the *complementary cumulative distribution function* of the local maxima X_m (that is, $X_m = X(t_m)$ at times t_m when $\dot{X}(t_m) = 0$ and $\ddot{X}(t_m) < 0$). Taking $\sigma_X \equiv \sigma$ to simplify the notation, we may write

$$[F_{X_m}^c(b)]_{\varepsilon=0} = \frac{\nu_b^+}{\nu_0^+} = \exp\left\{-\frac{b^2}{2\sigma^2}\right\}, \quad b \geq 0. \tag{4.4.8}$$

Differentiating with respect to b yields the *Rayleigh* probability density function [see Eq. (2.5.40)]:

$$[f_{X_m}(b)]_{\varepsilon=0} = -\frac{d}{db}F^c_{X_m}(b) = \frac{b}{\sigma^2}\exp\left\{-\frac{b^2}{2\sigma^2}\right\}, \quad b \geq 0. \tag{4.4.9}$$

Rice [113] provides an expression similar to Eq. (4.4.1) for the mean number of local maxima above a given level b; see also Cartwright and Longuet-Higgins [84]. The general expression for the probability density function of the local maxima of a stationary Gaussian process (for any value of ε within the range $0 \leq \varepsilon \leq 1$) is

$$f_{X_m(b)} = \frac{\varepsilon}{\sqrt{2\pi}\sigma}\exp\left\{-\frac{b^2}{2\varepsilon^2\sigma^2}\right\} + \frac{b\sqrt{1-\varepsilon^2}}{\sigma^2}\exp\left\{-\frac{b^2}{2\sigma^2}\right\}F_U\left(\frac{b\sqrt{1-\varepsilon^2}}{\sigma\varepsilon}\right), \tag{4.4.10}$$

where $F_U(\cdot)$ is the standard normal c.d.f. For $\varepsilon = 0$ (the "pure" narrow-band case), the first term vanishes and Eq. (4.4.10) reduces to the Rayleigh distribution [Eq. (4.4.9)]. For $\varepsilon = 1$, only the first term remains, and Eq. (4.4.10) becomes the Gaussian p.d.f. with mean zero and variance σ^2, the same p.d.f. that governs $X(t)$ at an arbitrary value of t. This implies that local peaks tend to occur erratically, with equal probability of being below and above the mean of $X(t)$.

The mean rate of occurrence of local maxima above the level b may be obtained from $f_{X_m}(b)$ by partial integration:

$$\nu_{b,m} = \nu_{\max}F^c_{X_m}(b) = \nu_{\max}\int_b^\infty f_{X_m}(b_1)\,db_1, \tag{4.4.11}$$

where $\nu_{\max} = \nu_0^+/\sqrt{1-\varepsilon^2}$ is the mean occurrence rate of local maxima [see Eq. (4.4.7)]. When $b = -\infty$, all maxima are sure to be counted and, as Eq. (4.4.11) confirms, $\nu_{-\infty,m} = \nu_{\max}$.

Using Eq. (4.4.10), it can be shown that excursions above high thresholds b tend to be associated with single local maxima for all but a class of stationary random processes whose bandwidth factor ε is very close to one.

Mean Length of Stay Above or Below a Threshold

Consider the consecutive times $\mathscr{T}_b$ and $\mathscr{T}_b'$, shown in Fig. 4.2, that a stationary random process $X(t)$ spends, respectively, above and below the thresh-

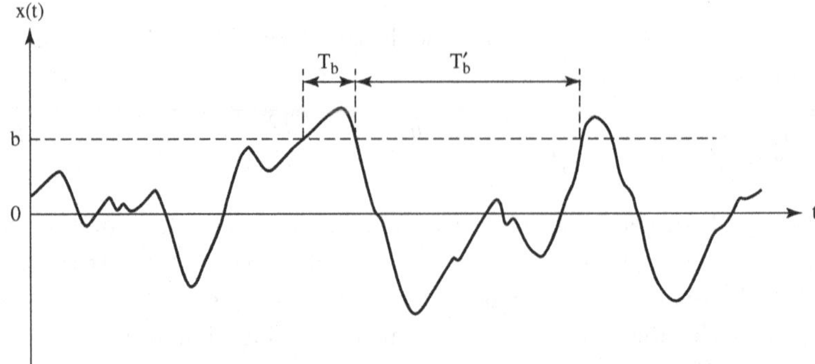

Fig. 4.2 Durations of stay above and below a fixed threshold b.

old b. The mean values $E[\mathscr{T}_b]$ and $E[\mathscr{T}'_b]$ can be obtained (Vanmarcke [132; 135]) from the following two equations that express fundamental results of the theory of recurrent events. The first states that the mean time between successive b-upcrossings is[1]

$$E[\mathscr{T}_b + \mathscr{T}'_b] = \frac{1}{\nu_b^+}. \tag{4.4.12}$$

The second equation expresses the long-run fraction of time during which the random process $X(t)$ is above the level b:

$$\frac{E[\mathscr{T}_b]}{E[\mathscr{T}_b + \mathscr{T}'_b]} = \int_b^\infty f_X(x)dx = F_X^c(b), \tag{4.4.13}$$

where $F_X^c(x)$ is the complementary c.d.f. of $X(t)$. Solving the two equations, Eqs. (4.4.12) and (4.4.13), yields the mean lengths of stay above and below the threshold b, respectively:

$$E[\mathscr{T}_b] = \frac{F_X^c(b)}{\nu_b^+} = \frac{2}{E[|\dot{X}|]} \frac{F_X^c(b)}{f_X(b)}, \tag{4.4.14}$$

$$E[\mathscr{T}'_b] = \frac{F_X(b)}{\nu_b^+} = \frac{2}{E[|\dot{X}|]} \frac{F_X(b)}{f_X(b)}. \tag{4.4.15}$$

Note that the dependence on b is through the first-order probability distribution of $X(t)$. Second-order properties enter only through the quantity $E[|\dot{X}|]$, which is in turn proportional to $\sigma_{\dot{X}}$ [see Eq. (4.4.4)]. If $F_X^c(b)$ is replaced by the series expansion in Eq. (2.5.10) (taking $u = b/\sigma$), we find

[1]According the Feller [56], the averages in Eq. (4.4.12) are (1) over the ensemble and (2) asymptotically over time, for $t \to \infty$.

a simple (asymptotically exact) expression for the mean excursion length:

$$E[\mathscr{T}_b] = \frac{\sigma}{\sqrt{2\pi}\, b\, \nu_0^+}\, \Phi\left(\frac{b}{\sigma}\right) \to \frac{\sigma}{\sqrt{2\pi}\, b\, \nu_0^+}, \qquad b \gg \sigma, \tag{4.4.16}$$

where the function $\Phi(\cdot)$ is given by Eq. (2.5.10) and evaluated in Table 2.2.

Envelope-Crossing Statistics

The crossings of the level b by the envelope $R(t)$ of the stationary random process $X(t)$ may be analyzed in the same way as those of $X(t)$ itself. The mean rate of b-upcrossings by $R(t)$ is given by Rice's formula [Eq. (4.4.2)],

$$\nu_{b,R} = f_R(b) E[|\dot{R}|] = 2\nu^+_{b,R}, \tag{4.4.17}$$

where $f_R(r)$ is the probability density function of $R(t)$, and $E[|\dot{R}|]$ is the mean of the absolute value of the slope of $R(t)$. Equation (4.4.17) assumes that $R(t)$ and $\dot{R}(t)$ are independent, which is true when if $X(t)$ is Gaussian. The expression for the mean value of the envelope excursion time $\mathscr{T}_{b,R}$ is similar to Eq. (4.4.14):

$$E[\mathscr{T}_{b,R}] = \frac{F^c_R(b)}{\nu^+_{b,R}} = \frac{2}{E[|\dot{R}|]}\, \frac{F^c_R(b)}{f_R(b)}, \tag{4.4.18}$$

where $F^c_R(r)$ is the complementary c.d.f. of $R(t)$.

The envelope $R(t)$ of a Gaussian process has a Rayleigh distribution [see Eq. (4.4.9)] with mean square $E[R^2] = 2\sigma^2$:

$$f_R(r) = \frac{r}{\sigma^2} e^{-r^2/2\sigma^2}, \quad r \geq 0. \tag{4.4.19}$$

The associated c.d.f. is

$$F^c_R(r) = \int_r^\infty f_R(r) dr = e^{-r^2/2\sigma^2}, \quad r \geq 0. \tag{4.4.20}$$

The envelope derivative $\dot{R}(t)$ is Gaussian with mean zero and variance $\sigma^2_{\dot{R}} = \delta^2 \sigma^2_{\dot{X}}$. The expected value of $|\dot{R}|$ is

$$E[|\dot{R}|] = \sqrt{2/\pi}\, \sigma_{\dot{R}} = \sqrt{2/\pi}\, \delta\, \sigma_{\dot{X}} = \delta\, E[|\dot{X}|]. \tag{4.4.21}$$

Combining equations readily yields:

$$\nu^+_{b,R} = \sqrt{2\pi}\, \frac{b}{\sigma}\, \nu_b^+\, \delta, \tag{4.4.22}$$

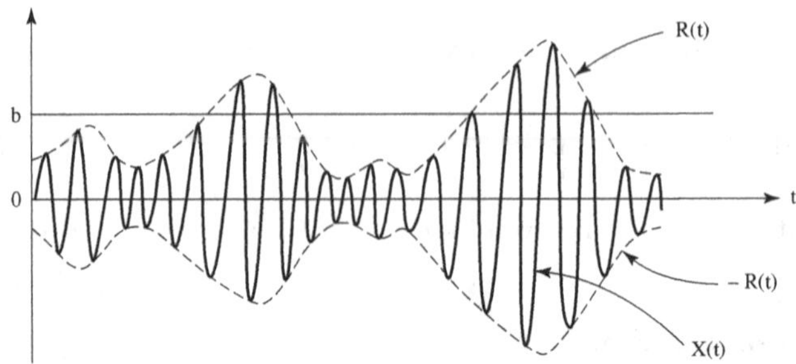

Fig. 4.3 Random function $X(t)$ and its envelope $R(t)$ in relation to a fixed threshold b.

where $\nu_b^+ \equiv \nu_{b,X}^+$. The mean length of envelope excursions above b is then, from Eqs. (4.4.18)–(4.4.22),

$$E[\mathscr{T}_{b,R}] = \frac{\sigma}{\sqrt{2\pi}\, b\, \nu_0^+ \delta}. \tag{4.4.23}$$

Combining Eqs. (4.4.16) and (4.4.23) reveals that the ratio of the mean excursion lengths of $X(t)$ and $R(t)$, respectively, approaches the spectral bandwidth factor δ for high values of b/σ:

$$\frac{E[\mathscr{T}_{b,X}]}{E[\mathscr{T}_{b,R}]} \to \delta, \quad b \gg \sigma, \tag{4.4.24}$$

where $E[\mathscr{T}_{b,X}] \equiv E[\mathscr{T}_b]$, b is the threshold level, and σ is the standard deviation of the Gaussian random process $X(t)$.

Clustering of Threshold Crossings and Local Maxima

Consider the relationship between the patterns of crossings of a given threshold b by a random process (X-crossings) and its envelope (R-crossings). As Fig. 4.3 shows, the process $X(t)$ can exceed b only at times when its envelope $R(t)$ is already above b. The number of X-crossings likely to occur during a single envelope excursion depends primarily on the bandwidth of the process and the height of the threshold. For narrow-band processes and low thresholds, when $(\delta b/\sigma)$ is small, X-crossings tend to occur in clusters or clumps, which immediately follow individual R-crossings. The mean time between consecutive X-crossings in a single clump is the "apparent period" $(\nu_0^+)^{-1}$. Lyon [87] argued that the quotient

$$r_b = \frac{\nu_b^+}{\nu_{b,R}^+} = \frac{\sigma}{\sqrt{2\pi}\delta b} \tag{4.4.25}$$

can be interpreted as the mean value of N_b, the size of a clump of crossings of level b by $X(t)$. (The expression on the right side of Eq. (4.4.25) applies when $X(t)$ is Gaussian.) This concept is very useful when the mean envelope excursion length, $E[\mathscr{T}_{b,R}]$, is significantly greater than $(\nu_0^+)^{-1}$. Note, however, that not all R-crossings are followed by an X-crossing within the next cycle. For relatively wide-band processes and high thresholds, the number of R-crossings may be much larger than the number of X-crossings. Accounting for the fraction of *qualified* R-crossings i.e., those immediately followed by at least one X-crossing, yields an improved approximation for the mean clump size N_b:

$$E[N_b] \simeq \frac{1}{1 - \exp\{-r_b^{-1}\}}. \tag{4.4.26}$$

Note that Eq. (4.4.26) tends toward Lyon's estimate [Eq. (4.4.25)] if $E[N_b] \gg 1$, but has a lower bound of one. For Gaussian processes Eq. (4.4.26) becomes

$$E[N_b] \simeq \left[1 - \exp\left\{-\frac{\sqrt{2\pi}\,\delta\, b}{\sigma}\right\}\right]^{-1}, \tag{4.4.27}$$

which is plotted in Fig. 4.4 as a function of b/σ for different values of δ.

When $b \to 0$, the clump size becomes infinite since $R(t)$ is (by definition) nonnegative. Clearly, the concept of clump size loses its usefulness and meaning when the ratio b/σ is small. Since the mean rate of envelope crossings $\nu_{b,R}$ is proportional to $f_R(b)$ (which attains its peak when $b = \sigma$, the mode of the Rayleigh probability density function), it is appropriate to add the condition $b > \sigma$ to Eqs. (4.4.25)–(4.4.27).

Probability Distribution of Maximum Values

A quantity of central interest in many applications of stochastic process theory to the analysis of natural or engineered systems is the probability that a random field $X(\mathbf{t})$ remains below a prescribed level b within a specified region d_0 in the parameter space, in other words, the probability that the maximum value of $X(\mathbf{t})$ across the region d_0 stays below b. Most of the extensive literature on this topic deals with one-dimensional random processes, or in a random field context, with processes "on the line" (obtained by fixing all but one of the parameters of a random field).

For a one-dimensional random process $X(t)$ that depends on time t,

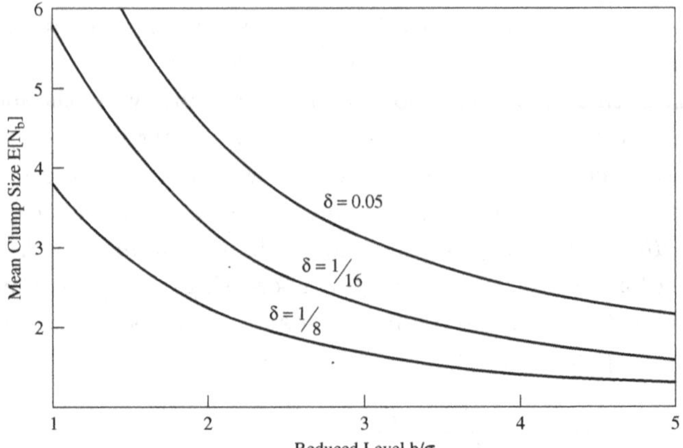

Fig. 4.4 Mean clump size, or the mean number of crossings of $X(t)$ for each crossing of $R(t)$, versus normalized barrier level b/σ. As shown, only upcrossing of $X = b$ are counted.

the problem statement is often: "find the probability $L_b(t_0)$ that the time to first crossing, or first passage, of the level b exceeds t_0." In reliability analysis, the level b may be thought of as a "critical value" not to be surpassed, and $X(t) < b$ defines the "safe region".

Poisson Approximation

A well-known approximation for the "first-passage" probability $L_b(t_0)$ is obtained by assuming that crossings of high levels b occur *independently* according to Poisson counting process $N(t)$ with mean rate ν_b^+. The solution is then

$$L_b(t_0) = P\Big[\operatorname*{Max}_{t_0} X(t) \le b\Big] = P[N(t_0) = 0] = \exp\{-\nu_b^+ t_0\}, \quad (4.4.28)$$

where $\nu_b^+ = \nu_0^+ \exp(-b^2/2\sigma^2)$ for Gaussian processes with mean zero. If the maximum value is expressed as $m + Y\sigma$, where Y is the random *peak factor*, the mean and standard deviation of Y can be expressed as follows (Davenport [48]):

$$E[Y] = \sqrt{2\ln \nu_0^+ t_0} + \frac{0.577}{\sqrt{2\ln \nu_0^+ t_0}}, \quad (4.4.29)$$

$$\sigma_Y = \frac{\pi}{6}\frac{1}{\sqrt{2\ln \nu_0^+ t_0}}. \quad (4.4.30)$$

The complementary c.d.f. of the maximum value can always be bounded as follows (Cramér and Leadbetter [41]):

$$1 - L_b(t_0) \leq P[X(0) > b] + \nu_b^+ t_0 P[X(0) \leq b]. \tag{4.4.31}$$

At relatively high threshold levels, $P[X(0) > b]$ becomes negligible (compared to unity), and the product $\nu_b^+ t_0$ serves as an effective upper bound for $[1 - L_b(t_0)]$. As can be inferred from Eq. (4.4.28), whenever $\nu_b^+ t_0 \ll 1$, $L_b(t_0)$ can often be closely approximated by the expression $[1 - \nu_b^+ t_0]$.

Cramér [40] has shown that for stationary Gaussian processes the Poisson assumption [and hence Eq. (4.4.28)] is asymptotically exact as the level b grows very large. (Cramér's results have been sharpened by Berman [16].) For barrier levels of practical interest, however, its use results in an error whose size and effect depend on the bandwidth of the process and the height of the barrier b (relative to the standard derivation σ). Numerical simulation studies indicate that the error tends to be on the safe side for narrow-band processes (Crandall *et al.* [44]) and slightly on the unsafe side for wide-band processes (Ditlevsen [50]). For wide-band processes and low barriers, the main problem is that the Poisson crossing assumption makes no allowance for the time the process actually spends above the threshold, while for narrow-band processes it neglects statistical dependence between consecutive threshold crossings. As is shown below, both effects can be incorporated into in an approximate analytical solution that converges to the exact (Poisson-process-based) asymptotic solution as $b \to \infty$.

Approximate Expressions for the Reliability Function

The various estimates given below for the "reliability function" all exhibit simple exponential decay,

$$L_b(t_0) = P\Big[\operatorname*{Max}_{t_0} X(t) \leq b\Big] = L_b(0)\exp\{-\alpha_b t_0\}, \quad t_0 \geq 0, \tag{4.4.32}$$

where $L_b(t_0)$ is the probability that $X(t)$ will not exceed the level b during the interval $(0, t_0)$; $L_b(0)$ is the probability of starting in the safe region (implying there is no instantaneous failure); and α_b is the "decay rate" of the reliability function. Simulation studies and theory (Crandall [42], Ditlevsen [50]) support the proposition that the reliability function obeys the exponential decay law at all but very small values of t_0 (of the order of the apparent period $(\nu_0^+)^{-1}$). At very high barrier levels, the quantities $L_b(0)$ and α_b should approach their (asymptotically exact) values, from

Eqs. (4.4.28) and (4.4.32):

$$L_b(0) = 1, \qquad \alpha_b = \nu_b^+. \tag{4.4.33}$$

The first improvement of these asymptotic values accounts for the fact that excursion durations are finite. This becomes significant at relatively low barrier levels. Recall that $\mathscr{T}_b'$ is the time spent below b between consecutive excursions; its mean is given by Eq. (4.4.15). If the pattern of b-crossings is modeled as a two-state Markov process instead of a Poisson process, then $\mathscr{T}_b'$ is exponentially distributed, and the parameters of the reliability function become

$$L_b(0) = \frac{E[\mathscr{T}_b']}{E[\mathscr{T}_b + \mathscr{T}_b']} = F_X(b), \qquad \alpha_b = \frac{1}{E[\mathscr{T}_b']} = \frac{\nu_b^+}{F_X(b)}. \tag{4.4.34}$$

The second correction becomes significant for narrow-band processes, as it accounts for the fact that crossings of the level b may occur in clumps. Clumps "arrivals" are modeled as a Poisson process with reduced mean rate $\nu_b^+ \,/\, E[N_b]$ (whose inverse may be interpreted as the "mean time between occurrences of clumps"). The parameters of $L_b(t_0)$ now become

$$L_b(t_0) = 1, \quad \alpha_b = \nu_b^+[1 - \exp\{-r_b^{-1}\}], \quad b > \sigma. \tag{4.4.35}$$

The two effects may be combined by arguing that only a fraction (equal to $F_X(b)$) of the mean time between clump occurrences, $E[N_b]/\nu_b^+$, is spent below the threshold. The parameter $L_b(0)$ is then the same as in Eq. (4.4.34), while the decay rate Eq. (4.4.35) is reduced by the factor $F_X(b)$:

$$L_b(0) = F_X(b), \qquad \alpha_b = \nu_b^+ \frac{1 - \exp\{-r_b^{-1}\}}{F_X(b)}, \quad b > \sigma. \tag{4.4.36}$$

4.5 Expected Size of Regions of Excursion

Practical questions about the behavior of a random field often focus on its relationship to threshold levels of various heights: How frequently is the threshold exceeded? What is the size of each region of excursion? Much information about excursions of a homogeneous random field $X(\mathbf{t})$ above relatively high levels stems from our knowledge of the behavior of the one-dimensional processes generated by fixing all but one of the coordinates. Actually, the concept of level crossings is no longer adequate in a multi-dimensional setting. We must now refer to regions of excursion above the level $X = b$, or "level sets". These excursion regions are bounded by level curves ($n = 2$) or level surfaces ($n = 3$) which tend to have extremely complex shapes at low threshold levels. As the level b grows, excursion regions become increasingly isolated, each tending to contain a single local maximum above b. In this section and the next, interest focuses on excursions above relatively high levels. We obtain estimates (asymptotically exact for $b \to \infty$) for the expected size and the mean rate of occurrence of isolated regions of excursion, for homogeneous Gaussian random fields and their envelopes. This leads to an estimate for the probability distribution of the extreme value of a random field within a prescribed domain in the parameter space. As throughout this chapter, it is assumed in Secs. 4.5 and 4.6 that the random field satisfies the necessary conditions of differentiability, conditions that are then critically examined in later chapters.

The Two-Dimensional Case

Consider a homogeneous Gaussian random field $X(\mathbf{t})$ on the plane (t_1, t_2) and a fixed level $X = b$. If t_2 is kept constant, one visualizes the stationary random function $X(t_1)$ crossing the level b more or less frequently depending upon the relative level height b/σ. The same is true for $X(t_2)$, the random variation in function of t_2 only. The theory of one-dimensional level excursions presented in Sec. 4.4 provides the following information about the direction-dependent processes $X(t_1)$ and $X(t_2)$:

1. The mean rate of b-crossings by the process $X(t_i)$, $i = 1, 2$, is

$$\nu_b^{(i)} = f(b)E[|\dot{X}^{(i)}|] = \frac{\sqrt{2}}{\pi} f(b)\sigma_{\dot{X}_i}, \tag{4.5.1}$$

   where $\sigma_{\dot{X}_i}$ is the standard deviation of the derivative of $X(t_i)$ (or the partial derivative of the field $X(\mathbf{t})$ with respect to t_i), and $f(x)$ is the p.d.f. of $X(\mathbf{t})$.

2. The mean length (measured along the t_i axis) of excursions above b by the process $X(t_i)$, based on Eq. (4.4.14), is

$$E[\mathscr{T}_b^{(i)}] = F^c(b)\left[\frac{1}{2}\nu_b^{(i)}\right]^{-1} = \sqrt{2\pi}\frac{F^c(b)}{f(b)}\frac{1}{\sigma_{\dot{X}_i}}, \tag{4.5.2}$$

   where $F^c(x)$ is the c.d.f. of $X(\mathbf{t})$.

3. At relatively high threshold levels of practical interest (say, $b/\sigma \geq 2$) b-crossings by $X(t_i)$ tend to be followed by single local maxima unless the associated spectral bandwidth factors $\varepsilon^{(i)}$ and $\delta^{(i)}$ are close to one.

The third statement implies that isolated regions of excursion in the parameter space (regions throughout which $X(\mathbf{t}) \geq b$) tend to have a simple shape and contain only a single point where $\dot{X}(t_i) = 0$ $(i = 1, 2)$, indicative of the existence of a local maximum.

The first statement implies that areas of excursion above the level b tend to become ever more isolated as b increases. Moreover, since b-upcrossing events on a line constitute a Poisson process when $b \to \infty$ (Cramér [40]), it follows that the pattern of occurrence of local maxima above a high level b will tend toward a two-dimensional Poisson process.

The second statement provides information about the size of the isolated excursion areas. Specifically, the quantities $E[\mathscr{T}_b^{(i)}]$, $i = 1, 2$, may be thought of as the "average dimensions" of an area of excursion above b.

Assuming that appropriate conditions of differentiability hold, the 2-D random field will attain a local maximum at location $\mathbf{t}_m$ if and only if

$$\begin{aligned} &\dot{X}^{(1)}(\mathbf{t}_m) = 0, \quad \dot{X}^{(2)}(\mathbf{t}_m) = 0, \\ &\ddot{X}^{(1)}(\mathbf{t}_m) \leq 0, \quad \ddot{X}^{(2)}(\mathbf{t}_m) \leq 0. \end{aligned} \tag{4.5.3}$$

The covariance of the directional (partial) derivatives $\dot{X}^{(1)}(\mathbf{t})$ and $\dot{X}^{(2)}(\mathbf{t})$ equals the joint spectral moment of second order $\lambda_{11}^{(12)} = \lambda_{11}$ [Eq. (4.2.4)]. If the directional derivatives (at $\mathbf{t}_m$) are uncorrelated, that is if $\lambda_{11} = 0$, then the expected size of an isolated area of excursion above b will be approximately equal to the product of the mean excursion lengths $E[\mathscr{T}_b^{(i)}]$, $i = 1, 2$. Thus

$$E[\mathscr{A}_b] \simeq E[\mathscr{T}_b^{(1)}]E[\mathscr{T}_b^{(2)}] = \frac{2\pi}{\sigma_{\dot{X}_1}\sigma_{\dot{X}_2}}\left(\frac{F^c(b)}{f(b)}\right)^2, \quad \text{if } \lambda_{11} = 0. \tag{4.5.4}$$

The product of standard deviations in the denominator of Eq. (4.5.4) may be replaced by the square root of the determinant of the matrix $\mathbf{\Lambda}_{11}$:

$$\sigma_{\dot{X}_1}\sigma_{\dot{X}_2} = [\lambda_2^{(1)}\lambda_2^{(2)}]^{1/2} = |\mathbf{\Lambda}_{11}|^{1/2}, \tag{4.5.5}$$

where

$$\mathbf{\Lambda}_{11} = \begin{bmatrix} \lambda_2^{(1)} & 0 \\ 0 & \lambda_2^{(2)} \end{bmatrix}. \tag{4.5.6}$$

The partial derivatives at $\mathbf{t}_m$ are generally correlated, and rotating the coordinate axes will change λ_{11} as well as the diagonal elements of the matrix $\mathbf{\Lambda}_{11}$. The quantity of interest, however, the expected size of excursion areas above b, should remain invariant to rotation of the coordinate system. The determinant of the matrix $\mathbf{\Lambda}_{11}$ provides just this kind of invariance, and it is the appropriate quantity to replace $[\lambda_2^{(1)}\lambda_2^{(2)}]$ in case the partial derivatives are correlated. In general, we may write

$$E[\mathscr{A}_b] \simeq 2\pi \left(\frac{F^c(b)}{f(b)}\right)^2 |\mathbf{\Lambda}_{11}|^{-1/2} = \frac{2\pi}{\sigma_{\dot{X}_1}\sigma_{\dot{X}_2}} \left(\frac{F^c(b)}{f(b)}\right)^2 |\mathbf{V}|^{-1/2}, \quad (4.5.7)$$

where $\mathbf{V}$ is the matrix of partial-derivative correlation coefficients whose elements are given by Eq. (4.2.11).

A parallel analysis procedure can be applied to the *envelope* process $R(t_1, t_2)$, *mutatis mutandis.* In particular, the equivalent of Eq. (4.5.7) is

$$E[\mathscr{A}_{b,R}] \simeq \frac{2\pi}{\sigma_{\dot{X}_1}\sigma_{\dot{X}_2}} \left(\frac{F_R^c(b)}{f_R(b)}\right)^2 |\mathbf{\Delta}|^{-1/2}. \quad (4.5.8)$$

The elements of the matrix $\mathbf{\Delta}$ are defined, and interpreted in terms of envelope-slope statistics, in Eq. (4.3.35).

The Multi-Dimensional Case

For an n-dimensional homogeneous Gaussian random field $X(\mathbf{t})$, the expected size of an isolated domain of excursion above a high level b is

$$E[\mathscr{D}_{b,X}] \simeq \left[\sqrt{2\pi}\frac{F^c(b)}{f(b)}\right]^n |\mathbf{\Lambda}_{11}|^{-1/2}, \quad n = 1, 2, \ldots. \quad (4.5.9)$$

By defining the *geometric mean* of the root-mean-square directional derivatives $\sigma_{\dot{X}_i} = [\lambda_2^{(i)}]^{1/2}$, as follows:

$$\sigma_{\dot{X}} \equiv [\sigma_{\dot{X}_1}\,\sigma_{\dot{X}_2}\cdots\sigma_{\dot{X}_n}]^{1/n}, \quad (4.5.10)$$

Eq. (4.5.9) may also be expressed as follows:

$$E[\mathscr{D}_{b,X}] \simeq \left(\frac{\sqrt{2\pi}}{\sigma_{\dot{X}}}\frac{F^c(b)}{f(b)}\right)^n |\mathbf{V}|^{-1/2}, \quad (4.5.11)$$

where $\mathbf{V}$ is the matrix of correlation coefficients [see Eq. (4.2.11)]. Likewise, for the multi-dimensional envelope field,

$$E[\mathscr{D}_{b,R}] \simeq \left(\frac{\sqrt{2\pi}}{\sigma_{\dot{X}}}\frac{F^c_{R(b)}}{f_R(b)}\right)^n |\mathbf{\Delta}|^{-1/2}. \tag{4.5.12}$$

For $n = 1$, since $|\mathbf{V}| = 1$ and $|\mathbf{\Delta}| = \delta^2$, the above equations reduce to their one-dimensional equivalents.

If $X(\mathbf{t})$ is Gaussian, $F^c(b)$ may be replaced by the series expansion in terms of $u = (b - m)/\sigma$ [see Eq. (2.5.10)]. For $u \gg 1$, this leads to

$$E[\mathscr{D}_{b,X}] \simeq \left[\frac{\sqrt{2\pi}}{u}\frac{\sigma}{\sigma_{\dot{X}}}\Phi(u)\right]^n \frac{1}{|\mathbf{V}|^{1/2}} \to \left(\frac{\sqrt{2\pi}}{u}\frac{\sigma}{\sigma_{\dot{X}}}\right)^n \frac{1}{|\mathbf{V}|^{1/2}}. \tag{4.5.13}$$

Similarly, for the envelope, using the Rayleigh distribution, one obtains

$$E[\mathscr{D}_{b,R}] \simeq \left(\frac{\sqrt{2\pi}}{u}\frac{\sigma}{\sigma_{\dot{X}}}\right)^n \frac{1}{|\mathbf{\Delta}|^{1/2}}. \tag{4.5.14}$$

Hence, the ratio of the mean sizes of excursion regions above the same threshold b, for a Gaussian random field and its envelope, respectively, for $u = (b - m)/\sigma \gg 1$ converges to

$$\frac{E[\mathscr{D}_{b,X}]}{E[\mathscr{D}_{b,R}]} \to \left(\frac{|\mathbf{\Delta}|}{|\mathbf{V}|}\right)^{1/2}. \tag{4.5.15}$$

For $n = 1$, $|\mathbf{V}| = 1$ and $|\mathbf{\Delta}| = \delta^2$, and the right side of Eq. (4.5.15) reduces to δ, in agreement with the one-dimensional result, Eq. (4.4.24).

We have derived estimates for the mean size of isolated regions of excursion above relatively high threshold levels for homogeneous Gaussian random fields and their envelopes. These estimates appear to be asymptotically exact (for $b \to \infty$) and lead, as is shown in Sec. 4.6, to the asymptotically correct mean rate of occurrence of isolated regions of excursion above b. This result for the asymptotic mean excursion rate for $X(\mathbf{t})$, but not $R(\mathbf{t})$, is attributed to Nosko [100] by Belayev [12; 13]; see also Hasofer [71].

The geometry of the contour lines that characterize the excursion regions above a fixed level, or the "level sets", becomes increasing complex as the (absolute value of the) level, relative to the mean, decreases. Adler [3] investigates more fully the geometry of the excursion regions and explores various topological measures that are only to a limited extent amenable to formal probabilistic treatment.

4.6 Statistics of Level Excursions and Extremes

Knowledge about the frequency and size of isolated regions of excursion above a high threshold b enables one to derive of an asymptotically exact solution for the probability distribution of the maximum value of a homogeneous random field $X(\mathbf{t})$. As the level b rises, regions of excursion containing local maxima become increasingly rare and isolated. Since the pattern of b-upcrossing events on any line in the parameter space tends toward a Poisson process when $b \to \infty$ (Cramér [40]), the pattern of isolated excursions by a random field $X(t_1, t_2)$ above a high level $X = b$ will tend toward a spatial Poisson process when $b \to \infty$.

Mean Rate of Local Maxima above High Barriers

The Two-Dimensional Case

Let μ_b denote the mean number of local maxima of $X(t_1, t_2)$ per unit area in the parameter space, namely the asymptotic mean rate of the spatial Poisson process of excursion events. We can express the expected sizes of the *aggregate excursion region* (the total area of all regions where X exceeds b) within a given area of size a_0 in two different ways, as follows:

$$E[\mathscr{A}_b]\,\mu_b\,a_0 = F^c(b)\,a_0. \tag{4.6.1}$$

The left side is the product of $\mu_b a_0$, the mean number of excursion events, and $E[\mathscr{A}_b]$, the mean area associated with each excursion. (The underlying assumption of independence between the *number* of events and their *size* appears consistent with the asymptotic Poisson behavior.) The right side of Eq. (4.6.1) is the product of the area a_0 and the mean fraction of the parameter space where X is above the level b. The quantity a_0 cancels out of Eq. (4.6.1), yielding following expression for $\mu_{b,X}$, the spatial mean rate of local maxima above b:

$$\mu_b = \mu_{b,X} = \frac{F^c(b)}{E[\mathscr{A}_b]} \simeq \frac{\sigma_{\dot{X}_1}\sigma_{\dot{X}_2}}{2\pi}\frac{[f(b)]^2}{F^c(b)}|\mathbf{V}|^{1/2}. \tag{4.6.2}$$

A similar expression can be obtained for the spatial mean rate $\mu_{b,R}$ of local maxima above the level b for the *envelope* $R(t_1, t_2)$ of the random field $X(t_1, t_2)$. The equation that parallels Eq. (4.6.1) is

$$E[\mathscr{A}_{b,R}]\,\mu_{b,R}a_0 = F_R^c(b)a_0. \tag{4.6.3}$$

Solving for $\mu_{b,R}$ and inserting Eq. (4.5.10) yields

$$\mu_{b,R} = \frac{F_R^c(b)}{E[\mathscr{A}_{b,R}]} \simeq \frac{\sigma_{\dot{X}_1}\sigma_{\dot{X}_2}}{2\pi}\frac{[f_R(b)]^2}{F_R^c(b)}|\boldsymbol{\Delta}|^{1/2}. \tag{4.6.4}$$

The Multi-Dimensional Case

The same methodology leads to analytical expressions for the mean rates of local maxima above high barriers for an n-dimensional homogeneous random field $X(\mathbf{t})$ and its envelope $R(\mathbf{t})$. The sizes of isolated domains of excursion above b are denoted by $\mathscr{D}_b$ and $\mathscr{D}_{b,R}$ for the process X and its envelope R, respectively. For $X(\mathbf{t})$ the expected size of the *aggregate excursion region* in a domain d_0 in the parameter space is again, as in Eq. (4.6.1), expressed in two ways:

$$E[\mathscr{D}_b]\,\mu_{b,X}\,d_0 = F^c(b)\,d_0. \tag{4.6.5}$$

For the envelope $R(\mathbf{t})$ the corresponding equation is

$$E[\mathscr{D}_{b,R}]\,\mu_{b,R}\,d_0 = F_R^c(b)\,d_0. \tag{4.6.6}$$

Inserting into Eqs. (4.6.5) and (4.6.6) the expressions for $E[\mathscr{D}_{b,X}]$ and $E[\mathscr{D}_{b,R}]$, respectively, yields the asymptotic mean rates of excursion,

$$\mu_{b,X} \simeq \left(\frac{\sigma_{\dot{X}}}{\sqrt{2\pi}}f(b)\right)^n (F^c(b))^{1-n}|\mathbf{V}|^{1/2}, \tag{4.6.7}$$

and

$$\mu_{b,R} \simeq \left(\frac{\sigma_{\dot{X}}}{\sqrt{2\pi}}f_R(b)\right)^n (F_R^c(b))^{1-n}|\boldsymbol{\Delta}|^{1/2}, \tag{4.6.8}$$

where $\sigma_{\dot{X}}$ denotes the geometric mean of the r.m.s. directional (partial) derivatives $\sigma_{\dot{X}_i}$ $(i = 1, 2, \ldots, n)$. If $X(\mathbf{t})$ is Gaussian, $F^c(b)$ may be expressed in terms of the normalized threshold, $u = (b - m)/\sigma$, as follows:

$$F^c(b) = \frac{1}{\sqrt{2\pi}u}e^{-u^2/2}\,\Phi(u), \tag{4.6.9}$$

where $\Phi(u)$ is the series expansion given by Eq. (2.5.10). Since $\Phi(u) \to 1$ for $u \gg 1$, the asymptotic mean rate becomes

$$\mu_{b,X} \simeq |\mathbf{V}|^{1/2}\left(\frac{\sigma_{\dot{X}}}{\sigma}\right)^n \frac{u^{n-1}}{(2\pi)^{(n+1)/2}}e^{-u^2/2}. \tag{4.6.10}$$

Likewise, the asymptotic mean excursion rate for the envelope $R(\mathbf{t})$ is

$$\mu_{b,R} \simeq |\mathbf{\Delta}|^{1/2} \left(\frac{\sigma_{\dot{X}}}{\sigma} \frac{u}{\sqrt{2\pi}} \right)^n e^{-u^2/2}. \tag{4.6.11}$$

Measure of the Tendency for Clustering

The ratio of the mean excursion rates, for $u = (b - m)/\sigma \gg 1$, is

$$\frac{\mu_{b,X}}{\mu_{b,R}} \simeq \left(\frac{1}{2\pi} \frac{|\mathbf{V}|}{|\mathbf{\Delta}|} \right)^{1/2} \frac{1}{u}. \tag{4.6.12}$$

Provided it exceeds one, this ratio may be regarded as an estimate of the "mean cluster size", the mean number of excursions by X for each excursion by R. Note that R is always above the level b when X is, while not every excursion by R is accompanied by an excursion by X. The argument based on the concept of "qualified" envelope excursions remains valid in the n-dimensional case. It leads to an approximation for the fraction of "empty" envelope excursions, and an improved estimate for the mean cluster size:

$$E[N_b] \simeq \left[1 - \exp\left(-\frac{\mu_{b,R}}{\mu_{b,X}} \right) \right]^{-1}. \tag{4.6.13}$$

When the ratio $\mu_{b,X}/\mu_{b,R}$ is small, there are many empty envelope excursions and the mean cluster size is close to its lower bound, $E[N_b] = 1$. When the ratio $\mu_b/\mu_{b,R}$ is large, each envelope excursion region tends to contain many local maxima of $X(\mathbf{t})$, and the two estimates, Eqs. (4.6.12) and (4.6.13), for the "mean cluster size" should be close.

Probability Distribution of the Maximum Value of a Random Field

The Two-Dimensional Case

From Sec. 4.5 and the foregoing discussion, we know that local maxima of $X(t_1, t_2)$ above a sufficiently high level b tend occur randomly according to a two-dimensional Poisson process with mean spatial rate $\mu_b = \mu_{b,X}$. Therefore the random number $N_b(a_0)$ of local maxima above b in an area of size a_0 may be described by a Poisson distribution with mean $\mu_b a_0$:

$$P[N_b(a_0) = k] = \frac{(a_0 \mu_b)^k}{k!} e^{-a_0 \mu_b}, \quad k = 0, 1, 2. \tag{4.6.14}$$

The *reliability function* $L_b(a_0)$, defined as the probability that the maximum of $X(t_1, t_2)$ remains below b throughout the area a_0, is

$$L_b(a_0) = P\Big[\operatorname*{Max}_{a_0} X(t_1,t_2) \le b\Big]$$
$$= P[N_b(a_0) = 0] = \exp\{-a_0\,\mu_b\}, \tag{4.6.15}$$

where μ_b is given by Eq. (4.6.2). While exact in the limit when $b \to \infty$, Eq. (4.6.15) does not account for several effects that gain importance when the level b decreases. First, there is a nonzero probability that the field is above b at any point (t_1, t_2), whereas Eq. (4.6.15) implies $L_b(0) = 1$. We have instead

$$L_b(0) = F(b). \tag{4.6.16}$$

Second, the excursion areas $\mathscr{A}_b$ are not "points", but have finite size. An improved estimate for the spatial mean rate, denoted by μ_b, of excursions, can be obtained by modifying Eq. (4.6.1) as follows:

$$E[\mathscr{A}_b]\,\mu_b\,a_0\,F(b) = F^c(b)\,a_0. \tag{4.6.17}$$

The expression on the left side differs from Eq. (4.6.1) in that the area a_0 is reduced by the areal fraction $F(b)$ occupied by the excursion regions. Incorporating the two effects yields the following improved estimate for the reliability function:

$$L_b(a_0) = F(b)\exp\left\{-\frac{\mu_b\,a_0}{F(b)}\right\}, \tag{4.6.18}$$

which converges toward Eq. (4.6.15) as b increases, since $F(b) \to 1$ for $b \to \infty$.

This estimate for the reliability function can be further refined by accounting for possible clustering of local maxima of X above b, the tendency for clustering being measured by the mean clump size $E[N_b]$, given by Eq. (4.6.13). The improved estimate for $L_b(a_0)$ is obtained by replacing the mean rate μ_b in Eq. (4.6.18) by $\mu_b/E[N_b]$, the mean rate of occurrence of clumps of local maxima. The result is presented below [see Eq. (4.6.21)] for n-dimensional homogeneous random fields.

The Multi-Dimensional Case

Extending the preceding analysis to multi-dimensional random fields is straightforward. Denote by $N_b(d_0)$ the number of excursions by $X(\mathbf{t})$ above a high level b in a domain of size d_0. As $b \to \infty$ the probability mass function of $N_b(d_0)$ approaches the Poisson law, for $k = 0, 1, 2, \ldots,$

$$P[N_b(d_0) = k] = (\mu_b\, d_0)^k \frac{1}{k!} \exp\{-\mu_b\, d_0\}. \tag{4.6.19}$$

The c.d.f. of the maximum value of $X(\mathbf{t})$ anywhere in the region d_0 defines the *reliability function* $L_b(d_0)$. For very high barriers b, we may write:

$$L_b(d_0) = P\Big[\operatorname*{Max}_{d_0} X(\mathbf{t}) < b\Big] = P[N_b(d_0) = 0] = \exp\{-\mu_b\, d_0\}. \tag{4.6.20}$$

For lower threshold levels, the estimate of the reliability function given by Eq. (4.6.20) fails to account for the following effects: (1) there is a finite probability of being above the threshold when the domain d_0 becomes very small; (2) the regions of excursion above b are not "points" as is implied by the multi-dimensional Poisson model; and (3) regions of excursion by $X(\mathbf{t})$ above b may be clustered (in a way that is not accounted for by the Poisson model). All three effects are incorporated in the following estimate for the reliability function:

$$L_b(d_0) = F(b) \exp\left\{-\frac{\mu_{b,X}\, d_0}{F(b)} \left[1 - \exp\left(-\frac{\mu_{b,R}}{\mu_{b,X}}\right)\right]\right\}, \tag{4.6.21}$$

where $\mu_{b,X}$ and $\mu_{b,R}$ are given by Eqs. (4.6.7) and (4.6.8), respectively. As expected, Eq. (4.6.21) converges to the simple exponential expression [Eq. (4.6.20)] when $b \to \infty$. Note also that the approximation $L_b(d_0) \approx (1 - \mu_b\, d_0)$ may be sufficiently accurate in many practical applications.

4.7 Spectral Parameters of Common Correlation Models

The results in the preceding sections demonstrate the extent to which the spectral moments λ_k and related parameters (the characteristic frequencies Ω_k and the dimensionless bandwidth measures δ and ε) dominate important aspects of the behavior of stationary random functions. In this section we examine a number of common mathematical models for the correlation structure of one-dimensional random processes, with a focus on evaluating their spectral moments and parameters.

Much of the theory in the preceding sections presumes the existence of the second spectral moment of $X(t)$ or the mean square of $\dot{X}(t)$,

$$\sigma_{\dot{X}}^2 = \lambda_2 = \int_0^\infty \omega^2 G(\omega)\, d\omega = -\left[\frac{d^2 B(\tau)}{d\tau^2}\right]_{\tau=0} = \ddot{B}(0). \tag{4.7.1}$$

In other words, the results are meaningful only if the mean square derivative is finite. Recall that the condition $\dot{B}(0) = 0$ is necessary and sufficient for mean square differentiability. In particular, the following well-known stochastic models do not satisfy this condition:

1. an ideal white noise process ($\lambda_k = \infty$ for $k = 0, 1, 2, \ldots$);
2. the moving average of ideal white noise (whose correlation function is triangular, hence $\dot{B}(0)$ is constant, *not* zero);
3. the first-order Markov process (the response of a first-order linear system to ideal white noise input) whose correlation function is exponential, $B(\tau) = \sigma^2 \exp\{-a|\tau|\}$, for $|\tau| \geq 0$.

If the white noise s.d.f. has a *finite upper limit* ω_1, then the second spectral moments does exist. Similarly, if the "input" into the moving-average operator (case 2) or the first-order linear system (case 3) is band-limited white noise (with a *finite* upper limit on frequency), the resulting stochastic response processes do become mean-square differentiable. However, in all these cases the actual value of $\sigma_{\dot{X}}^2 = \lambda_2$ will be very sensitive to the (often arbitrary) choice of the frequency limit.

We now consider a number of specific mathematical models for the spectral density function. The subscript X is omitted throughout for notational convenience.

Low-Pass White Noise

The one-sided spectral density function $G(\omega)$, shown in Fig. 4.5, is constant for frequencies below ω_1, and zero otherwise:

$$G(\omega) = \begin{cases} G_0, & 0 \leq \omega \leq \omega_1, \\ 0, & \text{elsewhere}. \end{cases} \tag{4.7.2}$$

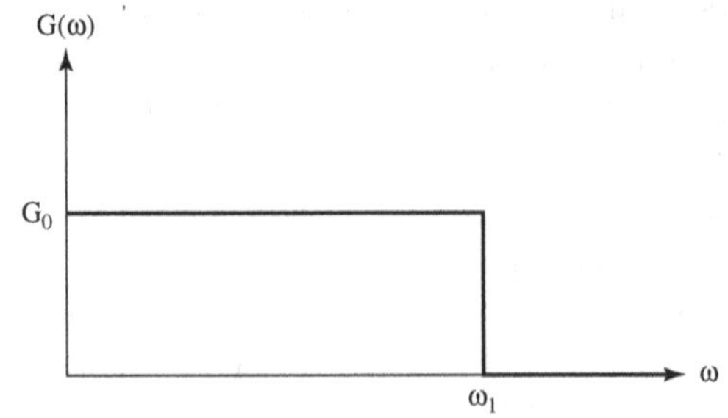

Fig. 4.5 Spectral density function of band-limited (low-pass) white noise.

The spectral moments are

$$\lambda_k = G_0 \int_0^{\omega_1} \omega^k \, d\omega = \frac{1}{k+1}\omega_1^{k+1} G_0, \quad k = 0, 1, 2, \ldots. \tag{4.7.3}$$

The variance, in particular, is $\lambda_0 = \sigma^2 = G_0\,\omega_1$. The characteristic frequencies take the form

$$\Omega_k = \left(\frac{1}{k+1}\right)^{1/k} \omega_1, \quad k = 1, 2, \ldots, \tag{4.7.4}$$

giving $\Omega_1 = 0.5\omega_1$, $\Omega_2 \simeq 0.58\omega_1$, and $\Omega_4 \simeq 0.67\omega_1$. The bandwidth measures δ and ε are constants:

$$\delta = \frac{1}{2}, \qquad \varepsilon = \frac{2}{3}. \tag{4.7.5}$$

The above values also apply in the limit for ideal white noise (as $\omega_1 \to \infty$), even though the spectral moments themselves approach infinity.

Sinusoid with Random Phase Angle

As shown in Fig. 4.6, the spectral mass is concentrated at a single (non-zero) frequency, and the spectral density function is:

$$G(\omega) = \delta(\omega - \omega_1)\sigma^2 = \begin{cases} \sigma^2, & \omega = \omega_1, \\ 0, & \omega \neq \omega_1. \end{cases} \tag{4.7.6}$$

The spectral moments and characteristic frequencies are, respectively,

$$\begin{aligned} \lambda_k &= \omega_1^k \sigma^2, \quad k = 0, 1, 2, \ldots, \\ \Omega_k &= \omega_1, \qquad k = 1, 2, \ldots, \end{aligned} \tag{4.7.7}$$

and both bandwidth measures attain their lowest possible value, $\delta = \varepsilon = 0$.

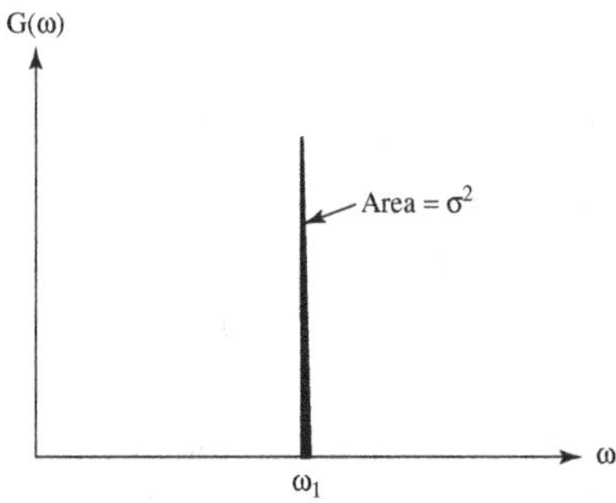

Fig. 4.6 Spectral density function of a pure sinusoid with random phase angle.

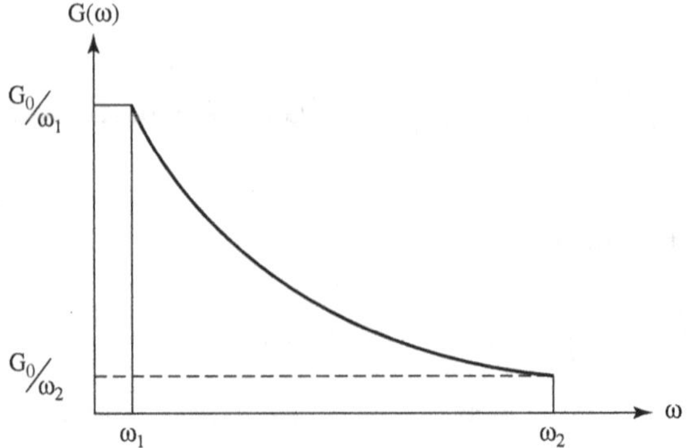

Fig. 4.7 Spectral density function of band-limited fractional noise ($1/f$ noise).

Band-Limited Fractional or "1/*f*" Noise

Assume that the spectral density function varies, as shown in Fig. 4.7, in inverse proportion to ω over a limited band of frequencies:

$$G(\omega) = \frac{G_0}{\omega}, \quad \omega_1 \leq \omega \leq \omega_2. \tag{4.7.8}$$

Define the "geometric average" frequency $\omega_0 = \sqrt{\omega_1 \omega_2}$ and the (dimensionless) factor $b = (\omega_2/\omega_0) = (\omega_0/\omega_1) = \sqrt{\omega_2/\omega_1}$ which measures the width of the frequency band (ω_1, ω_2). The spectral moments of $G(\omega)$ are

$$\lambda_k = G_0 \int_{\omega_1}^{\omega_2} \omega^{k-1}\, d\omega = \begin{cases} 2\, G_0 \ln b, & k = 0, \\ G_0, \omega_0^k\, b^k [1 - b^{-2k}], & k = 1, 2, \ldots. \end{cases} \tag{4.7.9}$$

Hence the characteristic frequencies become

$$\Omega_k = \omega_0\, b \left[\frac{1 - b^{-2k}}{\ln b}\right]^{1/k}, \quad k = 1, 2, \ldots, \tag{4.7.10}$$

and the squared bandwidth factors are

$$\delta^2 = 1 - \frac{1}{\ln b} \frac{(1 - b^{-2})^2}{1 - b^{-4}} \xrightarrow[b \text{ large}]{} 1 - \frac{1}{\ln b}, \tag{4.7.11}$$

$$\varepsilon^2 = 1 - \frac{1}{\ln b} \frac{(1 - b^{-4})^2}{1 - b^{-8}} \xrightarrow[b \text{ large}]{} 1 - \frac{1}{\ln b}. \tag{4.7.12}$$

Clearly, as the frequency band (ω_1, ω_2) widens, $b \to \infty$, the bandwidth factors gradually approach their upper limit $\delta = \varepsilon = 1$. As this happens, the variance σ^2 itself slowly grows without limit, in proportion to $\ln b$.

A Family of Autoregressive Correlation Models

The family of spectral density functions defined below has two parameters, the frequency α and an integer measure of "memory length" m.

$$G(\omega) = \frac{G_0}{(\alpha^2 + \omega^2)^m}, \quad \alpha > 0,\ m = 0, 1, 2, 3, \ldots. \tag{4.7.13}$$

Note that $m = 0$ corresponds to white noise and $m = \infty$ to a process whose correlation function is a constant. The case $m = 1$ is treated in Sec. 3.7; its variance is $\lambda_0 = \pi G_0/2\alpha$ [see Eq. (3.7.8)], but its first and second spectral moments are infinite. The area under $G(\omega)$ is

$$\lambda_0 = \frac{1 \cdot 3 \cdot 5 \cdots (2m-3)}{2 \cdot 4 \cdot 6 \cdots (2m-2)} \frac{\pi G_0}{2\alpha^{2m-1}}, \quad m = 2, 3, \ldots. \tag{4.7.14}$$

The first and second spectral moments are, respectively,

$$\lambda_1 = \frac{1}{2(m-1)} \frac{G_0}{\alpha^{2(m-1)}}, \quad m = 2, 3, \ldots, \tag{4.7.15}$$

and

$$\lambda_2 = \frac{\alpha^2}{(2m-3)} \lambda_0, \quad m = 2, 3, \ldots. \tag{4.7.16}$$

The characteristic frequencies Ω_1 and Ω_2 are, respectively,

$$\begin{aligned} \Omega_1 &= \frac{\alpha}{(m-1)\pi} \frac{2 \cdot 4 \cdot 6 \cdots (2m-2)}{1 \cdot 3 \cdot 5 \cdots (2m-3)}, && m = 2, 3, \ldots, \\ \Omega_2 &= \frac{\alpha}{(2m-3)^{1/2}}, && m = 2, 3, \ldots. \end{aligned} \tag{4.7.17}$$

The bandwidth measure δ is given by

$$\delta = \left[1 - \left(\frac{\Omega_1}{\Omega_2}\right)^2\right]^{1/2}, \quad m = 2, 3, \ldots. \tag{4.7.18}$$

Numerical values of these parameters are listed in Table 4.1 for values of m ranging between 2 and 20.

Response of a Simple Oscillator Excited by White Noise

The spectral density function of the stationary random response of a simple linear oscillator with viscous damping, excited by ideal white noise (with

Table 4.1 Spectral parameters associated with the family of autoregressive correlation models.

m	$\dfrac{\lambda_0 a^{2m-1}}{G_0}$	$\dfrac{\Omega_1}{\alpha}$	$\dfrac{\Omega_2}{\alpha}$	δ
2	0.7857	0.6363	1.0	0.7714
3	0.5893	0.4242	0.5774	0.8296
4	0.4911	0.3394	0.4472	0.8616
5	0.4297	0.2909	0.3779	0.8810
7	0.3545	0.2351	0.3016	0.9037
10	0.2915	0.1906	0.2425	0.9221
15	0.2348	0.1521	0.1925	0.9380
20	0.2021	0.1303	0.1644	0.9470

spectral density G_0), is given by [see Eq. (3.7.30)]:

$$G(\omega) = \frac{G_0}{(\Omega^2 - \omega^2)^2 + 4\,\zeta^2\Omega^2\omega^2}, \quad \omega \geq 0, \tag{4.7.19}$$

where Ω is the undamped natural frequency and ζ the damping ratio of the oscillator. The first three spectral moments are

$$\begin{aligned}
\lambda_0 &= \frac{\pi G_0}{4\,\zeta\,\Omega^3} = \sigma^2, \\
\lambda_1 &= \frac{1}{\sqrt{1-\zeta^2}}\,\frac{\pi G_0}{4\,\zeta\,\Omega^2} \\
&\quad \times \left[1 - \frac{1}{\pi}\tan^{-1}\left(\frac{2\,\zeta\sqrt{1-\zeta^2}}{1-2\,\zeta^2}\right)\right] \quad \text{for } \zeta \neq 1, \\
\lambda_2 &= \frac{\pi G_0}{4\,\zeta\,\Omega}.
\end{aligned} \tag{4.7.20}$$

Higher-order moments $(\lambda_3, \lambda_4, \ldots)$ are infinite owing to diverging high-frequency contributions, and the bandwidth measure ε is undefined. Note that the second characteristic frequency equals Ω:

$$\Omega_2 = \sqrt{\frac{\lambda_2}{\lambda_0}} = \Omega. \tag{4.7.21}$$

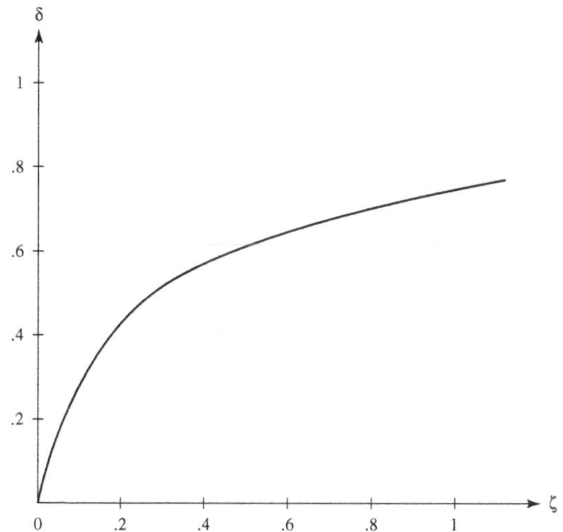

Fig. 4.8 Spectral bandwidth factor δ characterizing the response to white noise of a simple linear oscillator with damping ratio ζ.

The square of the bandwidth factor δ depends only on ζ:

$$\begin{aligned}\delta^2 &= 1 - \frac{1}{1-\zeta^2}\left[1 - \frac{1}{\pi}\tan^{-1}\left(\frac{2\,\zeta\sqrt{1-\zeta^2}}{1-2\,\zeta^2}\right)\right]^2 \\ &\simeq \frac{4\,\zeta}{\pi}[1 - 1.1\,\zeta] \quad \text{for } \zeta \le 0.2. \end{aligned} \tag{4.7.22}$$

Note that δ^2 is nearly linear in ζ for lightly damped systems, that is, $\delta^2 \simeq (4/\pi)\zeta$. For a critically damped system, when $\zeta = 1$, one obtains $\Omega_1 = (2/\pi)\Omega$ and $\delta = (1 - 4/\pi^2)^{1/2} = 0.77$. As the damping ratio ζ increases beyond its critical value ($\zeta = 1$), the bandwidth factor slowly approaches its upper limit, $\delta = 1$ (see Fig. 4.8).

Exponentially Decaying Spectral Density Function

For the negative-exponential spectral density function,

$$G(\omega) = G_0\, e^{-a\omega}, \quad \omega \ge 0, \tag{4.7.23}$$

the spectral moments are:

$$\lambda_k = \frac{G_0\, k!}{a^{k+1}}, \quad k = 0, 1, 2, \ldots, \tag{4.7.24}$$

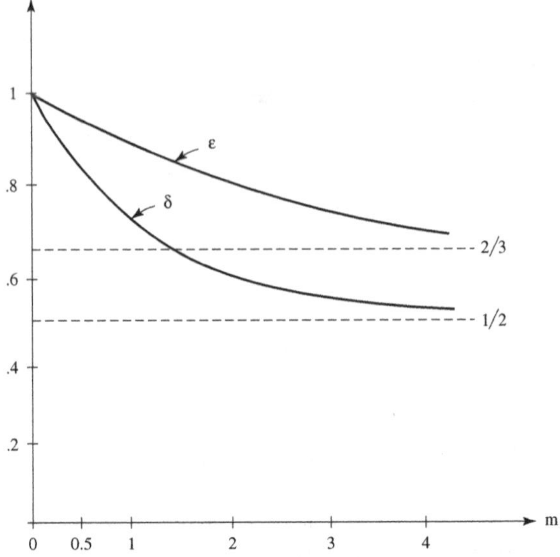

Fig. 4.9 Spectral bandwidth measures δ and ε for a family of models with exponentially decaying spectral density function.

yielding the spectral bandwidth measures:

$$\delta = \sqrt{1 - \frac{1}{2}} \simeq 0.707 \quad \text{and} \quad \varepsilon = \sqrt{1 - \frac{1}{6}} \simeq 0.913. \tag{4.7.25}$$

An extended family of spectral density function exhibiting exponential decay is represented by

$$G(\omega) = G_0 \exp\{-(a\,\omega)^m\}, \quad a, m > 0. \tag{4.7.26}$$

Its spectral moments are

$$\lambda_k = \frac{G_0}{m} \Gamma\left(\frac{k+1}{m}\right) a^{-(k+1)}, \quad k = 0, 1, 2, \ldots, \tag{4.7.27}$$

and the corresponding squared bandwidth factors are

$$\delta^2 = 1 - \frac{\Gamma^2(2/m)}{\Gamma(1/m)\Gamma(3/m)} \quad \text{and} \quad \varepsilon^2 = 1 - \frac{\Gamma^2(3/m)}{\Gamma(1/m)\Gamma(5/m)}. \tag{4.7.28}$$

The factors δ and ε are plotted as a function of m in Fig. 4.9. For $m \to 0$, δ and ε approach their unit upper bound, while for $m \to \infty$ they tend toward the values that characterize low-pass white noise, namely $\delta = 1/2$ and $\varepsilon = 2/3$ [see Eq. (4.7.5)].

4.8 Some Extensions to Nonhomogeneous Random Fields

The limited purpose of this section is to present a brief overview of approaches to analyzing nonhomogeneous random fields and point the reader to some application-oriented literature on this broad and important topic. The scope of this book is otherwise restricted to homogeneous random field theory which in most cases provides the basis and starting point for methodology to deal with problems involving nonhomogeneous fields.

One can often make a nonhomogeneous random field approximately homogeneous with respect to some of its parameters by "standardizing" it, that is, by subtracting from $X(\mathbf{t})$ the mean $m(\mathbf{t})$, and dividing the difference by the standard deviation $\sigma(\mathbf{t})$. The standardized (transformed) random field, defined by

$$X^*(\mathbf{t}) = \frac{X(\mathbf{t}) - m(\mathbf{t})}{\sigma(\mathbf{t})}, \tag{4.8.1}$$

has zero mean and unit variance. The fact that the (transformed-field) first-order statistics m and σ^2 are constant does, of course, not guarantee that the second-order statistics will satisfy the homogeneity condition (that $\rho(\mathbf{t}, \mathbf{t}')$ depends on $|\boldsymbol{\tau}| = |\mathbf{t} - \mathbf{t}'|$ only). An interesting example is the time-dependent "Brownian noise" process considered in Sec. 2.7, whose mean and variance increase linearly with t. Even after standardization, its correlation function retains the form given by Eq. (2.7.26). In a case like this, a further transformation is possible to bring about so-called "weak homogeneity," or homogeneity with respect to first- and second-order statistics. It consists of a transformation of the parameter space; in the case of Brownian noise the substitution $s = \ln t$ or $t = e^s$ accomplishes the purpose (as the transformed correlation function will depend on $\tau = |s - s'|$). If the simple standardization [Eq. (4.8.1)] is also applied to $X(t)$, the resulting doubly transformed process $X^*(s)$ will be weakly homogeneous. If it is also Gaussian – as is the case for Brownian noise – it then becomes homogeneous in the strong sense. The same technique can be applied to multi-dimensional Brownian noise. Each coordinate may be transformed logarithmically, yielding a process whose correlation function depends on $|\boldsymbol{\tau}| = |\mathbf{s} - \mathbf{s}'|$ only.

It is sometimes possible to "average out" the effect of those parameters with respect to which the random field is nonhomogeneous. For example, in metereology and geology, random properties of interest are often nonhomogeneously dependent on elevation (above the ground) or depth (below the surface), respectively. If properties are averaged or integrated over a range of elevations or depths, the resulting two-dimensional random field becomes approximately homogeneous. (Related, one might also consider representing two-dimensional random variation on a spherical surface, using polar coordinates and spherical harmonics.)

In seeking to model random field nonhomogeneity, consider that stochastic model building, in practice, is an art as much as a science. We mention below a number of additional approaches to extending the basic (homogeneous) theory to deal with nonhomogeneity.

1. First, using multi-scale modeling, consider the random field to be a superposition of statistically independent component fields with very different correlation properties. In this way, long-range fluctuations (trends) can be incorporated explicitly in the random field model.

2. Second, permit some of the random field statistics (such as m, σ, δ) to be functions of the (time and/or space) coordinates; the changes may occur intermittently, perhaps in jumps, or relatively slowly and continuously. The variation may, depending on the application: (a) be entirely deterministic, (b) possess a known functional form with uncertain parameters, or (c) be itself (amenable to modeling as) a random field.

3. Third, permit the covariance function or the spectral density function of the random field to be vary slowly with $\mathbf{t}$. The second-order properties thus become "evolutionary", and statistics such as σ^2, Ω and δ, and the mean threshold-excursion rates, also become (relatively slowly varying) functions of $\mathbf{t}$.

Formal definitions of "evolutionary spectra" of one-dimensional (time-dependent) random processes have been proposed by Page [102], Mark [92], and Priestley [108]. Important applications of nonstationary theory arise in random vibration analysis of structures suddenly exposed to excitation, such ground motion caused by an earthquake. Lightly damped structures do not reach a condition of stationary response before the random excitation ceases. The evolutionary spectral density function and the time-dependent statistics of the response of vibratory systems were studied by Caughey and Stumpf [27], Hammond [69], Lin [83], Vanmarcke [133; 136], Chakravorty [28], Corotis and Vanmarcke [39], and Corotis and Marshall [37]. The papers by Corotis *et al.* [38], Yang [159], and Petocz [106] focus on the distribution of extreme values of nonstationary random functions, in particular the random response of linear oscillators.

Chapter 5

Local Average Processes on the Line

All nature is but art, unknown to thee
All chance, direction, which thou canst not see.

Alexander Pope, Essay on Man

5.1 Variance Function and Scale of Fluctuation

This section introduces the scale of fluctuation θ and the variance function (in two alternate forms) of a stationary random process $X(t)$, and states the basic relations between these quantities and the more common descriptions of the correlation structure. The theory is presented assuming $X(t)$ is a continuous-parameter process, but it is shown in Sec. 5.6 that the principal results remain valid in case $X(t)$ is a random series or a point process.

The Variance Function

A continuous-parameter stationary random process $X(t)$ with mean m and variance σ^2 gives rise to a family of *moving average* processes $X_T(t)$, defined as follows:

$$X_T(t) = \frac{1}{T}\int_{t-T/2}^{t+T/2} X(u)\,du, \tag{5.1.1}$$

where T denotes the averaging time. The relationship between the processes $X(t)$ and $X_T(t)$ is illustrated in Fig. 5.1. The mean is not affected by the averaging operation, while the variance of $X_T(t)$ may be expressed as:

$$\mathrm{Var}[X_T] \equiv \sigma_T^2 = \gamma(T)\sigma^2. \tag{5.1.2}$$

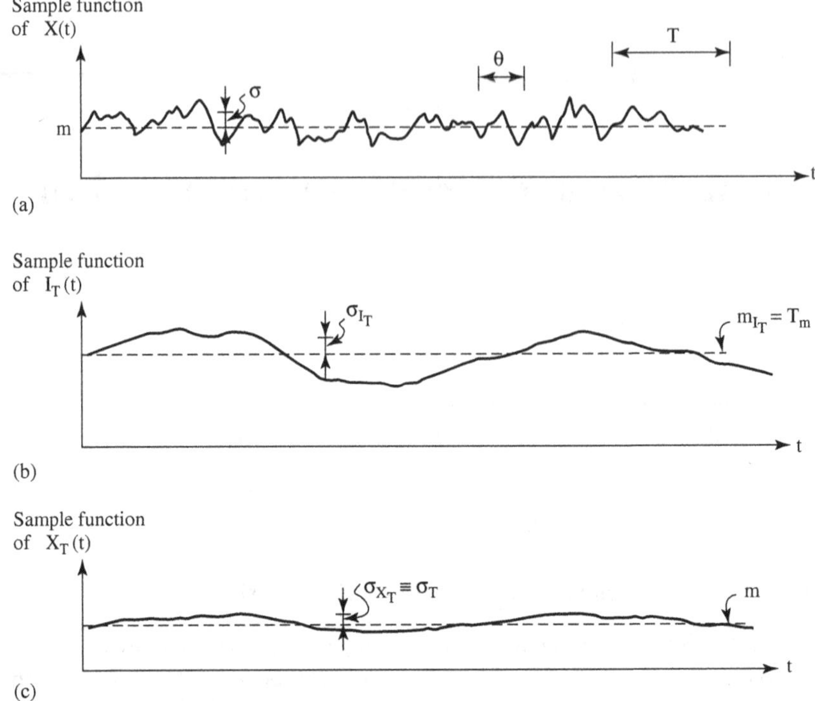

Fig. 5.1 (a) Sample function of a random process $X(t)$ with mean m, standard deviation σ, and scale of fluctuation θ; (b) sample function of the local integral process obtained by integrating $X(t)$ locally over a moving interval of size T; (c) sample function of the local average process $X_T(t) = (1/T)I_T(t)$.

$\gamma(T)$ is by definition the *variance function* of $X(t)$, which measures the reduction of the point variance σ^2 under local averaging. It is a dimensionless function that possesses the following properties:

$$\gamma(T) \geq 0, \tag{5.1.3}$$

$$\gamma(0) = 1, \tag{5.1.4}$$

$$\gamma(-T) = \gamma(|T|) = \gamma(T). \tag{5.1.5}$$

Its square root, $[\gamma(T)]^{1/2}$, is a "reduction factor" to be applied to the point standard deviation σ in order to obtain σ_T, the standard deviation of the local average process $X_T(t)$.

The variance function $\gamma(T)$ is related to the correlation function $\rho(\tau)$ as follows, from Eq. (3.6.41):

$$\gamma(T) = \frac{1}{T^2}\int_0^T\int_0^T \rho(t_1 - t_2)\,dt_1 dt_2 = \frac{2}{T}\int_0^T \left(1 - \frac{\tau}{T}\right)\rho(\tau)\,d\tau. \qquad (5.1.6)$$

To focus the ideas, consider these common analytical models for the correlation function and the corresponding variance functions, obtained using Eq. (5.1.6).

Case 1. The *triangular* correlation function, which decreases linearly from 1 to 0 as $|\tau|$ goes from 0 to a (see Eq. (3.7.38)):

$$\rho(\tau) = \begin{cases} 1 - \dfrac{|\tau|}{a}, & 0 \le |\tau| \le a, \\ 0 & |\tau| \ge a. \end{cases} \qquad (5.1.7)$$

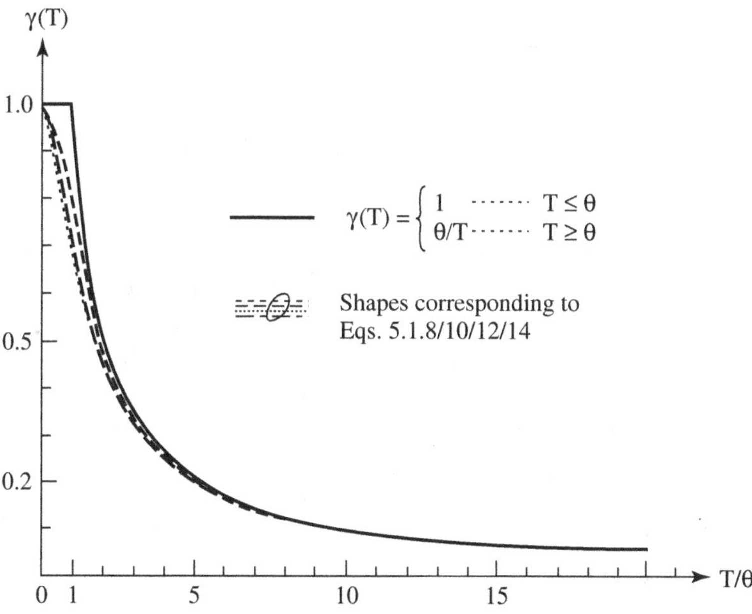

Fig. 5.2 Variance function $\gamma(T)$ plotted against the normalized averaging interval T/θ for several common "wide-band" correlation models.

Its variance function, shown in Fig. 5.2, is:

$$\gamma(T) = \begin{cases} 1 - \dfrac{T}{3\,a}, & 0 \le T \le a, \\ \left(\dfrac{a}{T}\right)\left[1 - \dfrac{a}{3\,T}\right], & T \ge a. \end{cases} \tag{5.1.8}$$

Case 2. The *exponential* correlation function associated with a first-order autoregressive (or Markov) process [see Eq. (3.7.7)]:

$$\rho(\tau) = e^{-|\tau|/b}, \quad |\tau| \ge 0, \tag{5.1.9}$$

$$\gamma(T) = 2\left(\frac{b}{T}\right)^2\left(\frac{T}{b} - 1 + e^{-T/b}\right), \quad T \ge 0. \tag{5.1.10}$$

Case 3. The correlation function associated with a *second-order autoregressive* process [see Eq. (3.7.15)]:

$$\rho(\tau) = \left[1 + \frac{|\tau|}{c}\right] e^{-|\tau|/c}, \quad |\tau| \ge 0, \tag{5.1.11}$$

$$\gamma(T) = 2\frac{c}{T}\left[2 + e^{-T/c} - 3\frac{c}{T}(1 - e^{-T/c})\right], \quad T \ge 0. \tag{5.1.12}$$

Case 4. The *Gaussian* (*squared exponential*) correlation function :

$$\rho(\tau) = e^{-(\tau/d)^2}, \quad |\tau| \ge 0, \tag{5.1.13}$$

$$\gamma(T) = \left(\frac{d}{T}\right)^2\left[\sqrt{\pi}\,\frac{T}{d}\,E\left(\frac{T}{d}\right) + e^{-(T/d)^2} - 1\right], \quad T \ge 0, \tag{5.1.14}$$

where $E(\cdot)$ is the widely tabulated "error function" which increases from 0 to 1 as its argument goes from 0 to ∞. In terms of the Standard Gaussian cumulative distribution function (see Table 2.2), $E(u) = 2[F_U(u) - 0.5]$.

Note that $\gamma(T) \to 0$ when $T \to \infty$ for these four models. This is the condition for "ergodicity in the mean" mentioned in Sec. 3.6. More significant is the fact that $\gamma(T)$ tends to become inversely proportional to T at large values of T. For the above-listed four cases we find:

$$\begin{aligned} &\textbf{Case 1.}\ \ \gamma(T) \to a/T, \\ &\textbf{Case 2.}\ \ \gamma(T) \to 2b/T, \\ &\textbf{Case 3.}\ \ \gamma(T) \to 4c/T, \\ &\textbf{Case 4.}\ \ \gamma(T) \to \sqrt{\pi}d/T. \end{aligned} \tag{5.1.15}$$

This motivates the introduction of a parameter that equals the proportionality constant in the limiting expression of the variance function.

The Scale of Fluctuation

The scale of fluctuation is *defined* as follows:

$$\theta = \lim_{T\to\infty} T\gamma(T), \tag{5.1.16}$$

or

$$\gamma(T) = \theta/T, \quad \text{when } T \gg \theta. \tag{5.1.17}$$

Evidently, for the four models considered above the "scale" θ is simply related to the model parameters, as follows: $\theta = a$ (Case 1); $\theta = 2b$ (Case 2); $\theta = 4c$ (Case 3); and $\theta = \sqrt{\pi}d$ (Case 4). When the four variance functions are plotted against T/θ (see Fig. 5.2), they are seen converging quite rapidly toward the common asymptotic expression $\gamma(T) = \theta/T$.

Closer examination of Eq. (5.1.6) reveals a condition for the existence of the scale of fluctuation, in the sense implied by the definition, Eq. (5.1.16). From Eq. (5.1.6)

$$\gamma(T) = \frac{2}{T}\left[\int_0^T \rho(\tau)\,d\tau - \frac{1}{T}\int_0^T \tau\rho(\tau)\,d\tau\right]. \tag{5.1.18}$$

A necessary condition for θ to exist is

$$\lim_{T\to\infty} \frac{1}{T}\int_0^T \tau\rho(\tau)\,d\tau = 0. \tag{5.1.19}$$

A more stringent condition is that the "first moment" of $\rho(\tau)$ (for $\tau \geq 0$) be finite. Since $\rho(\tau)$ is an even function, its complete "moments" θ_i, $i = 0$, 1, 2, ..., can be defined as follows:

$$\theta_i = 2\int_0^\infty \tau^i \rho(\tau)d\tau = \int_{-\infty}^\infty \tau^i\rho(\tau)d\tau, \quad i = 0, 1, 2\ldots. \tag{5.1.20}$$

The necessary and sufficient condition for the validity of all interpretations of θ involves the "second moment" θ_2 and is stated later in this section. Eq. (5.1.18) implies that the scale of fluctuation, provided it exists, is the integral of the correlation function:

$$\theta = 2\int_0^\infty \rho(\tau)\,d\tau = \int_{-\infty}^{+\infty} \rho(\tau)\,d\tau, \tag{5.1.21}$$

so that $\theta = \theta_0$, the moment of order zero of the correlation function. From the Wiener-Khinchine relation [Eq. (3.3.12)], it further follows that θ is proportional to the ordinate at the frequency origin ($\omega = 0$) of the unit-area (one-sided) spectral density function $g(\omega)$:

$$\theta = \pi g(0). \tag{5.1.22}$$

For example, the unit-area s.d.f. associated with the triangular correlation function [Eq. (5.1.7)] is

$$g(\omega) = \frac{a}{\pi}\left[\frac{\sin(\omega\, a/2)}{\omega\, a/2}\right]^2. \tag{5.1.23}$$

For $\omega \to 0$, $\sin(\omega a/2) \simeq \omega a/2$, and hence

$$\theta = \pi g(0) = \pi\left(\frac{a}{\pi}\right) = a, \tag{5.1.24}$$

in agreement with the result obtained from the correlation function [Eq. (5.1.7)] or the variance function [Eq. (5.1.8)]. Whichever formula is most convenient can serve to evaluate θ:

$$\theta = \begin{cases} T\,\gamma(T), & \text{as } T \to \infty, \\ 2\displaystyle\int_0^\infty \rho(\tau)\,d\tau, & \\ \pi g(\omega), & \text{as } \omega \to 0. \end{cases} \tag{5.1.25}$$

It must be understood, however, that the *first* formula serves as the definition. (If the first formula yields a finite value for θ, the other two formulas are guaranteed to give the same value.)

Alternate Form for the Variance Function

The variance function may also be defined in terms of the *local integral* of the random process $X(t)$:

$$I_{T(t)} = \int_{t-T/2}^{t+T/2} X(u)\,du \equiv T X_T(t). \tag{5.1.26}$$

Since the random functions $X_T(t)$ and $I_T(t)$ differ only by the factor T, their respective variances are simply related, namely:

$$\text{Var}[I_{T(t)}] = T^2\sigma_T^2 = T^2\sigma^2\gamma(T). \tag{5.1.27}$$

It is notationally convenient (especially in Sec. 5.3) to introduce the "variance function of the local integral process", as follows:

$$\Delta(T) = T^2\gamma(T), \tag{5.1.28}$$

which yields $\mathrm{Var}[I_T]$ when multiplied by the point variance σ^2:

$$\mathrm{Var}[I_{T(t)}] = \sigma^2\Delta(T). \tag{5.1.29}$$

When T approaches zero, $\Delta(T) \to T^2$, while the asymptotic expression when $T \to \infty$ is $\Delta(T) \to \theta T$. Note from Fig. 5.3 that the function $\Delta(T)/T \equiv T\gamma(T)$ is zero at $T = 0$ and converges toward θ as the ratio T/θ grows. The dotted-line curve corresponds to the exact expression for the case when the correlation function $\rho(\tau)$ is triangular [Case 1 above; see Eq. (5.1.7)].

Some Analytical Models for the Variance Function

Simplest (but Overly Idealized) Model

Fig. 5.2 suggests that the asymptotic expression for the variance function, $\gamma(T) = \theta/T$, provides an approximation as well as an upper bound for the true variance function of typical wide-band random processes when T

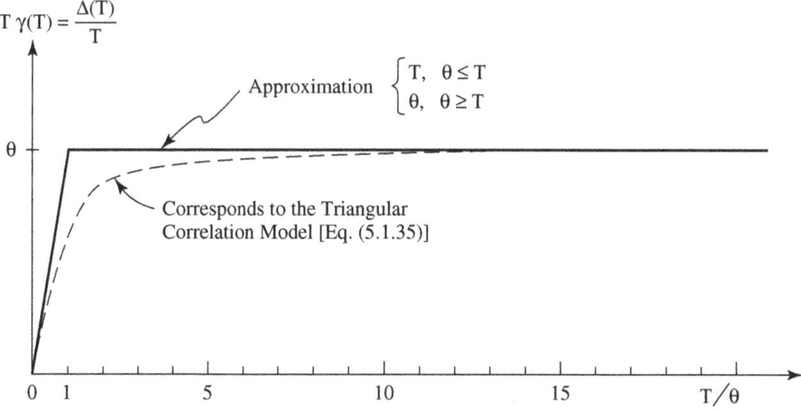

Fig. 5.3 Alternate representations of the variance functions $\gamma(T)$ and $\Delta(T)$.

exceeds θ. Indeed, the simple analytical model

$$\gamma(T) = \begin{cases} 1, & 0 \le T \le \theta, \\ \dfrac{\theta}{T}, & T \ge \theta, \end{cases} \tag{5.1.30}$$

is exact for $T \to 0$ and $T \to \infty$. A significant drawback, however, is that, according to Eq. (5.1.6), it implies a boxcar-shaped correlation function,

$$\rho(\tau) = \begin{cases} 1, & 0 \le |\tau| \le \theta, \\ 0, & |\tau| > \theta, \end{cases} \tag{5.1.31}$$

that is unacceptable because it fails to meet the requirement that θ be equal to the area under $\rho(\tau)$; instead, the area equals 2θ. A similar, but proper, correlation function is considered next.

Based on "Boxcar" Correlation Function

A simple model for the correlation function, consistent with the definition of θ and having a "boxcar" shape similar to Eq. (5.1.31), is

$$\rho(\tau) = \begin{cases} 1, & 0 \le |\tau| \le \dfrac{\theta}{2}, \\ 0 & |\tau| \ge \dfrac{\theta}{2}. \end{cases} \tag{5.1.32}$$

The corresponding variance function, computed using Eq. (5.1.6), is

$$\gamma(\tau) = \begin{cases} 1, & 0 \le T \le \dfrac{\theta}{2}, \\ \dfrac{\theta}{T}\left(1 - \dfrac{\theta}{4T}\right), & T \ge \dfrac{\theta}{2}. \end{cases} \tag{5.1.33}$$

Although the sudden drop from "perfect correlation" below $\theta/2$ to "complete lack of correlation" above $\theta/2$ may be unrealistic, this simple model is one of a number of wide-band (single-scale) stochastic models appropriate for use in situations where available information about the correlation structure of $X(t)$ consists of θ only.

Based on Triangular Correlation Function

To set the stage for what follows, we briefly reconsider the triangular correlation function ("Case 1" above):

$$\rho(\tau) = \begin{cases} 1 - \dfrac{|\tau|}{\theta}, & |\tau| \leq \theta, \\ 0, & |\tau| \geq \theta. \end{cases} \tag{5.1.34}$$

Its variance function, with the parameter a now replaced by the scale of fluctuation θ, has the following form:

$$\gamma(T) = \begin{cases} 1 - \dfrac{T}{3\,\theta}, & 0 \leq T \leq \theta, \\ \dfrac{\theta}{T}\left(1 - \dfrac{\theta}{3T}\right), & T \geq \theta. \end{cases} \tag{5.1.35}$$

Recommended Approximation for the Variance Function

By definition, the variance function $\gamma(T)$ is inversely proportional to T when T is much larger than θ. To see how $\gamma(T)$ approaches its limiting shape, it is useful to examine Eq. (5.1.6) more closely. Note that it depends on the partial first moment of the correlation function $\rho(\tau)$:

$$2\int_0^T \tau\rho(\tau)\,d\tau = \int_{-T}^{+T} |\tau|\rho(\tau)\,d\tau, \tag{5.1.36}$$

which approaches the (complete) first moment θ_1 [given by Eq. (5.1.20), with $i = 1$] when $T \to \infty$.

For typical simple correlation functions for which θ_2 is finite (including all those previously presented in Sec. 5.1), the partial first moment $\int_{-T}^{T} \tau\rho(\tau)\,d\tau$ will approach its limit θ_1 relatively rapidly, and the partial integral $\int_{-T}^{T} \rho(\tau)\,d\tau$ approaches its own limit $\theta \equiv \theta_0$ even faster. Therefore, it follows from Eq. (5.1.6) that $\gamma(T)$ can be approximated as:

$$\gamma(T) \to \frac{\theta}{T} - \frac{\theta_1}{T^2} = \frac{\theta}{T}\left(1 - k_1\frac{\theta}{T}\right), \quad T \gg \theta, \tag{5.1.37}$$

in which

$$k_1 = \theta_1/\theta^2 \tag{5.1.38}$$

is a dimensionless parameter of the correlation function $\rho(\tau)$. For the four cases considered in Sec. 5.1, the respective values are: $k_1 = 1/3$ (Case 1); $k_1 = 1/2$ (Case 2); $k_1 = 6/16$ (Case 3); and $k_1 = 1/\pi$ (Case 4). The asymptotic expression, Eq. (5.1.37), happens to be exact for the case of the triangular correlation function (since there are no further contributions, beyond $T = \theta$, to the partial first moment $\int_{-T}^{+T} \tau\rho(\tau)\,d\tau$).

In conclusion, a very useful approximation for the variance function of any single-scale wide-band stationary random process is:

$$\gamma(\tau) = \begin{cases} 1 - k_1 \dfrac{T}{\theta}, & 0 \le T \le \theta, \\[2ex] \dfrac{\theta}{T}\left(1 - k_1 \dfrac{\theta}{T}\right), & T \ge \theta. \end{cases} \tag{5.1.39}$$

One can either insert the value of k_1 given by Eq. (5.1.38) or use a "default value" such as $k_1 = 1/3$, corresponding to the triangular correlation function Eq. (5.1.7). As Fig. 5.2 indicates, this provides a very good approximation for the variance functions of typical wide-band processes. The approximation is plotted in a different way, for $T\gamma(T) = \Delta(T)/T$, as the dotted-line curve in Fig. 5.3.

Historical Note

The integral correlation measure defined by Eq. (5.1.21) was first proposed by Taylor [127]. In the theory of turbulence it serves primarily as a guide for determining the time interval required to obtain stable (low variance) estimates of the (ensemble) mean of fluctuating quantities such as fluid velocities and pressures. In particular, the ratio T/θ is interpreted as the "equivalent number of independent observations" contained in a sampling interval T. In the analysis of isotropic turbulence (Batchelor [11], Monin and Yaglom [98], Tennekes and Lumley [128]), multi-dimensional correlation measures likewise play a limited and auxiliary role; in essence, they help to quantify the condition for ergodicity with respect to the mean. Of course, the interpretation of equivalent independent observations also provides quantitative support for the argument that sample averages and integrals of stationary random processes with finite scale θ tend to become Gaussian when $T/\theta \to \infty$, by virtue of the Central Limit Theorem (Lumley [85]).

In this book the measures $\gamma(T)$ and θ are introduced as operational quantities, directly useful (and often necessary) for characterizing second-order properties, evaluating level-crossing statistics, and in general interpreting, estimating, and analyzing patterns of random variation.

5.2 Scale of Fluctuation: Frequency-Domain Interpretation

Relationship between $\gamma(T)$ and $g(\omega)$

The variance function can be related to $g(\omega)$, the unit-area spectral density function of $X(t)$. From Eq. (3.6.39), the s.d.f. of $X_T(t)$ is given by:

$$G_T(\omega) = G(\omega)\left[\frac{\sin(\omega T/2)}{\omega T/2}\right]^2. \tag{5.2.1}$$

Integrating over frequency and dividing by σ^2 yields

$$\gamma(T) = \int_0^\infty g(\omega)\left[\frac{\sin(\omega T/2)}{\omega T/2}\right]^2 d\omega. \tag{5.2.2}$$

For $T \to 0$, $\sin(\omega T/2) \to (\omega T/2)$, and Eq. (5.2.2) becomes

$$\gamma(0) = \int_0^\infty g(\omega)\, d\omega = 1. \tag{5.2.3}$$

To examine what happens to Eq. (5.2.2) when $T \to \infty$, we introduce a change of variables, $u = (\omega T/2\pi)$, so that only contributions to the integral associated with $u \to 0$ matter. We may write:

$$\lim_{T\to\infty} \gamma(T) = g(0)\frac{2\pi}{T}\int_0^\infty \left(\frac{\sin \pi u}{\pi u}\right)^2 du. \tag{5.2.4}$$

The integral on the right side of Eq. (5.2.5) equals 1/2, and hence

$$\lim_{T\to\infty} \gamma(T) = \frac{\pi}{T} g(0) = \frac{\theta}{T}. \tag{5.2.5}$$

Eqs. (5.2.3) and (5.2.5) confirm, by frequency domain analysis, the behavior of $\gamma(T)$ at the extremes when $T \ll \theta$ and $T \gg \theta$, respectively.

Behavior of $g(\omega)$ near the Frequency Origin

The behavior of $g(\omega)$ near $\omega = 0$ is of considerable interest. The formula $\theta = \pi g(0)$ would have little practical value unless $g(0)$ can be estimated, say, by $g(\omega_0)$ where ω_0 represents a small (non-zero) frequency. Related, in deriving Eq. (5.2.5) from Eq. (5.2.4), it was tacitly assumed that $g(\omega)$ does not vary appreciably near $\omega = 0$.

Consider the MacLaurin series for $g(\omega)$, the polynomial expansion of $g(\omega)$ near $\omega = 0$:

$$g(\omega) = g(0) + \dot{g}(0)\,\omega + \ddot{g}(0)\frac{\omega^2}{2} + \cdots, \tag{5.2.6}$$

where, from Eq. (5.1.22),

$$g(0) = \theta/\pi. \tag{5.2.7}$$

From the Wiener-Khinchine relation [Eq. (3.3.12)], the first derivative of $g(\omega)$ can be expressed as follows:

$$\dot{g}(\omega) = \frac{2}{\pi}\int_0^\infty \rho(\tau)\frac{d\cos\omega\tau}{d\omega}d\tau = -\frac{2}{\pi}\int_0^\infty \tau\rho(\tau)\sin\omega\tau\,d\tau. \tag{5.2.8}$$

For $\omega \to 0$, since $\sin\omega\tau \to \omega\tau$, the limiting form of $\dot{g}(\omega)$ near $\omega = 0$ is

$$\dot{g}(\omega) = -\frac{2}{\pi}\,\omega\int_0^\infty \tau^2\rho(\tau)\,d\tau = -\frac{\omega}{\pi}\,\theta_2, \tag{5.2.9}$$

where θ_2 is the second moment of $\rho(\tau)$, defined by Eq. (5.1.20). Evidently, if θ_2 is finite, the slope of $g(\omega)$ at $\omega = 0$ will be zero:

$$\dot{g}(0) = 0, \quad \text{if } |\theta_2| < \infty. \tag{5.2.10}$$

The second derivative of $g(\omega)$ at $\omega = 0$ is

$$\begin{aligned}\ddot{g}(0) &= \frac{2}{\pi}\int_0^\infty \rho(\tau)\left[\frac{d^2\cos\omega\tau}{d\omega^2}\right]d\tau\bigg|_{\tau=0}\\ &= -\frac{2}{\pi}\int_0^\infty \tau^2\rho(\tau)\,d\tau = -\frac{\theta_2}{\pi}.\end{aligned} \tag{5.2.11}$$

We conclude from the preceding analysis that the equivalent conditions $\dot{g}(0) = 0$ and $|\theta_2| < \infty$ serve, in effect, as sufficient conditions for the existence of θ. If $|\theta_2| < \infty$, $g(\omega)$ takes the following form near $\omega = 0$;

$$g(\omega) = \frac{\theta}{\pi} - \frac{\theta_2}{2\pi}\omega^2 + \cdots. \tag{5.2.12}$$

The condition $\dot{g}(0) = 0$ implies the existence of a *local extremum* at $\omega = 0$. If $\theta_2 > 0$, $g(0)$ will be a local maximum; if $\theta_2 < 0$, $g(0)$ will be a local minimum. To see just how fast $g(\omega)$ changes near the frequency origin, it is useful to introduce the frequency ω_2 as follows:

$$\frac{2\pi}{\omega_2} = \left(\frac{|\theta_2|}{\theta}\right)^{1/2}, \tag{5.2.13}$$

where $(\omega_2)^{-1}$ has the same dimension as the scale of fluctuation. The expression for $g(\omega)$ near $\omega = 0$ can now be stated as follows:

$$g(\omega) = \frac{\theta}{\pi}\left[1 \pm 2\pi^2\left(\frac{\omega}{\omega_2}\right)^2 + \cdots\right], \tag{5.2.14}$$

where the sign chosen must be the opposite of the sign of θ_2. For example, $\theta_2 > 0$ implies that $g(0)$ is a local maximum; hence the negative sign is appropriate. Some examples are offered below to illustrate, and provide further interpretation of, these findings.

Results for Two Families of Correlation Models

Case 1. The one-sided spectral density function of the family of autoregressive models considered in Sec. 3.7 has the form [see Eq. (3.7.26)]:

$$G(\omega) = \frac{G_0}{(\omega^2 + \alpha^2)^m}, \quad \alpha > 0,\ m = 0, 1, 2, \ldots, \tag{5.2.15}$$

where m indicates "memory length" of the autoregressive random process. The case $m = 0$ represents ideal white noise ($G(\omega) = G_0$), while $m = 1$ corresponds to the simple exponential correlation model. Integrating over positive frequencies yields the relationship between σ^2 and G_0:

$$G_0 = \frac{(2\alpha)^{2m-1}[(m-1)!]^2}{\pi(2m-2)!}\sigma^2, \quad m = 1, 2, \ldots. \tag{5.2.16}$$

The scale of fluctuation is

$$\theta = \pi g(0) = \frac{G_0}{\sigma^2\alpha^{2n}} = \frac{[(m-1)!]^2 2^m}{(2m-2)!}\frac{1}{\alpha}, \quad m = 1, 2, \ldots. \tag{5.2.17}$$

Since $m = 0$ represents white noise, Eq. (5.2.17) correctly predicts $\theta = \pi G_0/\sigma^2$, where $\theta \to 0$ as $\sigma^2 \to \infty$. This special case is discussed in more detail in Sec. 5.5. The first derivative of $G(\omega)$ is

$$\dot{G}(\omega) = \frac{dG(\omega)}{d\omega} = \frac{-2\omega m G_0}{(\omega^2 + \alpha^2)^{m+1}}, \quad m \geq 0. \tag{5.2.18}$$

Hence the basic condition,

$$\dot{G}(0) = \dot{g}(0) = 0 \tag{5.2.19}$$

is satisfied for any $m \geq 0$ and $\alpha > 0$. The second moment of the correlation function is

$$\theta_2 = -\pi\ddot{g}(0) = \frac{2\pi m G_0}{\sigma^2 \alpha^{2(m+1)}}, \quad m \geq 0. \tag{5.2.20}$$

Note that θ_2 is positive for $m > 0$, implying that $g(\omega)$ has a local maximum at $\omega = 0$. (For $m = 0$, $G(\omega)$ is constant and its derivatives are zero.) The parameter governing the behavior of $g(\omega)$ near $\omega = 0$ [see Eq. (5.2.13)] is

$$\omega_2 = 2\pi\left(\frac{\theta}{|\theta_2|}\right)^{1/2} = \pi\alpha\sqrt{2/m}, \quad m > 0. \tag{5.2.21}$$

Finally, combining Eqs. (5.2.14) and (5.2.21) yields the series expansion

$$g(\omega) = \frac{\theta}{\pi}\left[1 - m\left(\frac{\omega}{\alpha}\right)^2 + \cdots\right], \quad m > 0. \tag{5.2.22}$$

At frequencies ω much smaller than α the value of $g(\omega)$ will indeed be very close to $g(0) = \theta/\pi$, provided m is relatively small.

Case 2. Another important family of mathematical models for the correlation structure is described by

$$\rho(\tau) = e^{-a|\tau|}\cos\omega_0\tau, \quad a > 0,\ \omega_0 \geq 0, \tag{5.2.23}$$

or the corresponding s.d.f.

$$g(\omega) = \frac{2}{\pi}\frac{a(a^2 + \omega_0^2 + \omega^2)}{\{a^2 + (\omega_0 - \omega)^2\}\{a^2 + (\omega_0 + \omega)^2\}}. \tag{5.2.24}$$

The case $a \to 0$ is associated with a sinusoid with random phase angle, while the combination $\omega_0 = 0$ and $a \to \infty$ is indicative of white noise. When $\omega_0 = 0$, Eq. (5.2.23) reduces to the exponential correlation model (the same as when $m = 1$ is taken in Case 1 above). The scale of fluctuation, in the general case, is

$$\theta = \pi g(0) = \frac{2a}{a^2 + \omega_0^2}. \tag{5.2.25}$$

The second moment of the correlation function is finite (and hence $\dot{g}(0) = 0$) for all combinations of a and ω_0 except the degenerate case $a = \omega_0 = 0$ (when $\rho(\tau) = 1$ for all τ). The expression for θ_2 is

$$\theta_2 = \frac{4a(a^2 - 3\omega_0^2)}{(a^2 + \omega_0^2)^3}. \tag{5.2.26}$$

Note that θ_2 is positive if $a > \sqrt{3}\,\omega_0$ and negative if $a < \sqrt{3}\,\omega_0$. Recall that the parameter ω_2 [see Eq. (5.2.13)] governs the behavior of $g(\omega)$ near the frequency origin. We have, in this case,

$$\omega_2 = \pi\sqrt{2}\frac{a^2 + \omega_0^2}{|a^2 - 3\,\omega_0^2|^{1/2}}. \tag{5.2.27}$$

For $\omega_0 \to 0$, $\omega_2 \to \pi\sqrt{2}\,a$ and $g(0)$ is a local maximum, while for $a \to 0$, $\omega_2 \to \pi\sqrt{2/3}\,\omega_0$ and $g(0)$ is a local minimum.

Variance Functions for "$\theta = 0$" Narrow-Band Processes

Suppose $X(t)$ is a band-limited white noise with variance σ^2. In particular, its spectral density function is constant within a narrow frequency band $(\omega_0 - \Delta\omega,\ \omega_0 + \Delta\omega)$ and zero elsewhere, as shown in Fig. 5.4. The scale of fluctuation θ will be zero, since $g(0) = 0$. It is easy to verify that the limiting decay of the variance function of this process is inversely proportional to T^2, not to T. The variance of $X_T(t)$ can be obtained directly from Eq. (3.7.39):

$$\sigma_T^2 = \frac{\sigma^2}{2\Delta\omega}\int_{\omega_0-\Delta\omega}^{\omega_0+\Delta\omega}\left[\frac{\sin(\omega T/2)}{\omega T/2}\right]^2 d\omega \simeq \sigma^2\left[\frac{\sin(\omega_0 T/2)}{\omega_0 T/2}\right]^2. \tag{5.2.28}$$

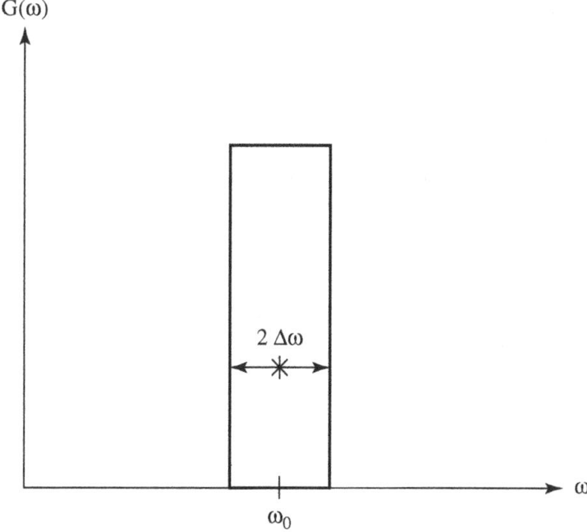

Fig. 5.4 Spectral density function of narrow-band-limited white noise (with $\theta = 0$).

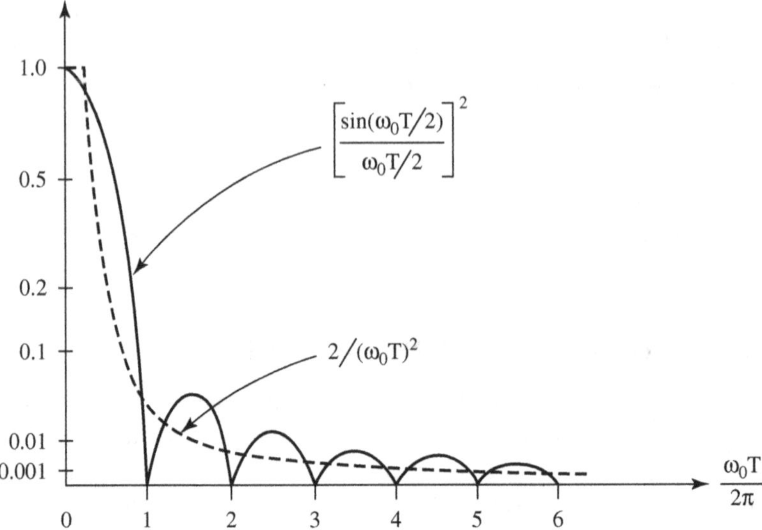

Fig. 5.5 Variance function of a $\theta = 0$ narrow-band stationary random process.

The function multiplying σ^2 is by definition the variance function $\gamma(T)$ of such a narrow-band random process $X(t)$ with central frequency ω_0:

$$\gamma(T) \simeq \left[\frac{\sin(\omega_0 T/2)}{\omega_0 T/2}\right]^2. \tag{5.2.29}$$

Fig. 5.5 shows that the function equals one when $T = 0$ and repeatedly drops to zero, at the values $T = 2n\pi/\omega_0$ $(n = 1, 2, \ldots)$. In between zero-values the function rises to smaller and smaller maxima as T increases. These local maxima occur near points where the sine function equals one. A useful "best estimate" of $\gamma(T)$ is obtained by replacing $[\sin(\omega_0 T/2)]^2$ by $1/2$, its average value over a typical half-period interval, while at very small values of T the variance function is very close to one. This suggests the following approximation:

$$\gamma(T) \simeq \begin{cases} 1, & T \leq \sqrt{2}/\omega_0, \\ \dfrac{2}{(\omega_0 T)^2}, & T \geq \sqrt{2}/\omega_0. \end{cases} \tag{5.2.30}$$

In terms of the "average period" $T_0 = 2\pi/\omega_0$, Eq. (5.2.30) indicates no variance reduction for values $T \leq T_0/(\sqrt{2}\pi) = 0.23\ T_0$. At $T = T_0$ the variance function is predicted to have dropped down to $2/(2\pi)^2 \simeq 0.05$. The variance of the derivative of $X_T(t)$ is obtained by multiplying $G_T(\omega)$ by ω^2 and integrating over all frequencies:

$$\sigma^2_{\dot{X}_T} = \frac{\sigma^2}{2\Delta\omega}\int_{\omega_0-\Delta\omega}^{\omega_0+\Delta\omega} \omega^2 \left[\frac{\sin(\omega T/2)}{(\omega T/2)}\right]^2 d\omega$$

$$\simeq \sigma^2 \left[\frac{2}{T}\sin\left(\frac{\omega_0 T}{2}\right)\right]^2. \tag{5.2.31}$$

The ratio of standard deviations $\sigma_{\dot{X}_T}/\sigma_T$ is exactly equal to ω_0 when the bandwidth vanishes, $\Delta\omega \to 0$. Also, when $T \to 0$, $\sin(\omega_0 T/2) \to \omega_0 T/2$; hence $\gamma(T) \to 1$ and $\sigma^2_{\dot{X}_T} \to \omega_0^2\sigma^2$.

5.3 Covariance of Local Integrals or Local Averages

Algebraic Relations Involving Local Integrals

Consider the random function $X(t)$ in relation to two line segments, T and T', arbitrarily located on the t-axis, as shown in Fig. 5.6. Interest focuses

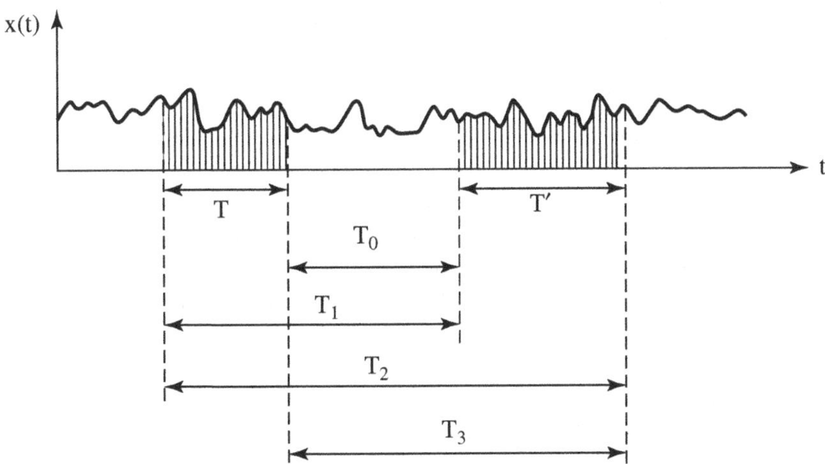

Fig. 5.6 Definition of the intervals T_0, T_1, T_2 and T_3.

on the *covariance* of the local integrals

$$I_T = I_T(t) = \int_{t-T/2}^{t+T/2} X(t)dt \tag{5.3.1}$$

and

$$I_{T'} = I_{T'}(t') = \int_{t'-T'/2}^{t'+T'/2} X(t)dt, \tag{5.3.2}$$

corresponding to the local averages

$$X_T = I_T/T$$

and

$$X_{T'} = I_{T'}/T'.$$

It is useful to characterize the intervals T and T' on the t-axis in terms of four distances T_0, T_1, T_2, and T_3 depicted in Fig. 5.6. For the sake of general applicability, these distances must be interpreted as follows:

$T_0 =$ distance from the end of the first interval to the beginning of the second interval,
$T_1 =$ distance from the beginning of the first interval to the beginning of the second interval,
$T_2 =$ distance from the beginning of the first interval to the end of the second interval,
$T_3 =$ distance from the end of the first interval to the end of the second interval.

Fig. 5.7 illustrates this interpretation for two cases: (a) the intervals T and T' overlap; and (b) interval T' lies completely inside interval T. The various local integrals are related by the following algebraic identity:

$$2I_T I_{T'} = I_{T_0}^2 - I_{T_1}^2 + I_{T_2}^2 - I_{T_3}^2 = \sum_{k=0}^{3}(-1)^k I_{T_k}^2, \tag{5.3.3}$$

where the quantities I_T, I_{T_0} etc., are the *local integrals* of $X(t)$ over the respective intervals, T, T_0, etc. Note from Fig. 5.6 that the definition of the intervals implies

$$
\begin{aligned}
I_{T_1} &= I_T + I_{T_0}, \\
I_{T_3} &= I_{T'} + I_{T_0}, \\
I_{T_2} &= I_T + I_{T'} + I_{T_0}.
\end{aligned}
\tag{5.3.4}
$$

In Fig. 5.7 the relationships are similar, except that the plus sign in front of I_{T_0} must be replaced by a minus sign. In either case, it suffices to insert these relations into Eq. (5.3.3) (and perform the necessary algebra) to prove the algebraic identity.

Correlation between Local Integrals or Local Averages

The algebraic identity, Eq. (5.3.3), leads to the following expression for the covariance of the local integrals I_T and $I_{T'}$ in terms of the (integral-based)

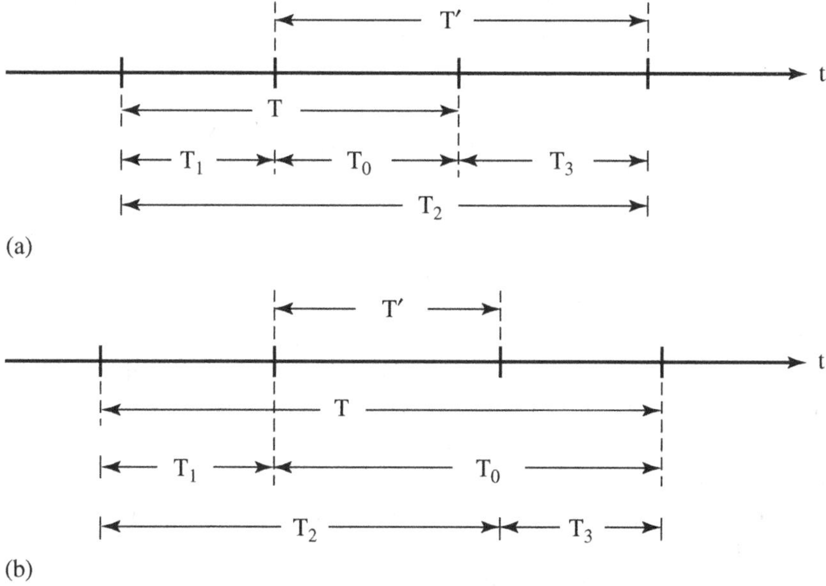

Fig. 5.7 Special cases: (a) the intervals T and T' overlap partially; (b) interval T contains interval T'.

variance function of $X(t)$:

$$\mathrm{Cov}[I_T, I_{T'}] \equiv T\,T'\mathrm{Cov}[X_T, X_{T'}] \tag{5.3.5}$$

$$= \frac{\sigma^2}{2}[\Delta(T_0) - \Delta(T_1) + \Delta(T_2) - \Delta(T_3)] = \frac{\sigma^2}{2}\sum_{k=0}^{3}(-1)^k\Delta(T_k)$$

To prove this equation, take the expectation on both sides of Eq. (5.3.3):

$$2E[I_T I_{T'}] = \sum_{k=0}^{3}(-1)^k E[I_T^2]. \tag{5.3.6}$$

The identity may also be expressed in terms of mean values, each of which is just proportional to the time interval involved (for example, $E[I_T] = m_X T$ and $E[I_{T'}] = m_X T'$). We may write:

$$2E[I_T]E[I_{T'}] = \sum_{k=0}^{3}(-1)^k E^2[I_{T_k}]. \tag{5.3.7}$$

Subtracting Eq. (5.3.7) from Eq. (5.3.6) and accounting for $\mathrm{Var}[I_{T_k}] = \sigma^2\Delta(T_k)$ completes the proof of Eq. (5.3.5). The coefficient of correlation between the local averages X_T and $X_{T'}$ (or local integrals I_T and $I_{T'}$) is

$$\rho_{X_T X_{T'}} \equiv \rho_{I_T I_{T'}} = \frac{\mathrm{Cov}[X_T, X_{T'}]}{\sigma_T \sigma_T'} = \frac{\sum_{k=0}^{3}(-1)^k\Delta(T_k)}{2[\Delta(T)\Delta(T')]^{1/2}}. \tag{5.3.8}$$

This result provides the basis for methodology to analyze a wide range of stochastic problems involving one-dimensional random variation. It enables one to generate the matrix of correlation coefficients between pairs of local averages (or local integrals) of some random property $X(t)$ associated with different segments (or "finite elements") along the t-axis. Only the variance function (in either of its forms, $\Delta(T)$ of $\gamma(T) \equiv \Delta(T)/T^2$) is needed; moreover, based on approximations for the variance function proposed in Sec. 5.1, knowledge of the scale θ usually suffices to obtain reasonable estimates for the correlation coefficients required in practical applications that involve single-scale random variation.

Consider the results obtained by using Eq. (5.3.8) for a few special cases. First, assume that the distance T_0 separating the intervals T and T' is much larger than θ. All the terms in the numerator then become essentially proportional to θ, leading to

$$\rho = \frac{\theta[T_0 - T_1 + T_2 - T_3]}{2[\Delta(T)\Delta(T')]^{1/2}} = 0. \tag{5.3.9}$$

Predictably, the coefficient of correlation between local averages over time intervals that are widely separated (relative to the scale θ) vanishes. Next, consider the case when the total time interval T_2 is much smaller than the correlation time θ so that the (integral-based) variance function $\Delta(u)$ may be replaced by u^2 in Eq. (5.3.8). The expected result – perfect unit correlation – can be confirmed by simple algebra. One obtains the same result in case the intervals T and T' entirely overlap. As indicated in Fig. 5.7, this implies $T = T' = T_0 = T_2$ and $T_1 = T_3 = 0$, and based on Eq. (5.3.8), $\rho = 2\Delta(T)/[2\Delta(T)] = 1$.

Covariance of Point Values and Local Averages

By defining the intervals T and $T' = 2\tau$ in the manner illustrated in Fig. 5.8, some very interesting and useful results can be obtained for the covariance between the local average $X_T(t)$ over an interval T centered at time t and a "point value" of the process, $X(t')$. In Case (a), the point t' lies inside the averaging interval T. The distances a and b identify the relative positions of the point and the segment. The covariance can be then obtained from

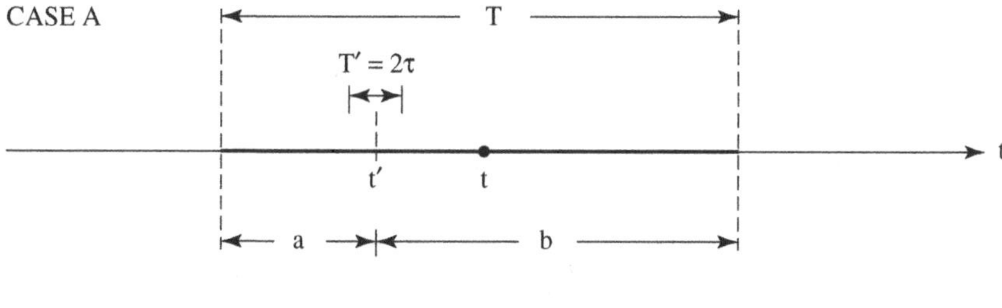

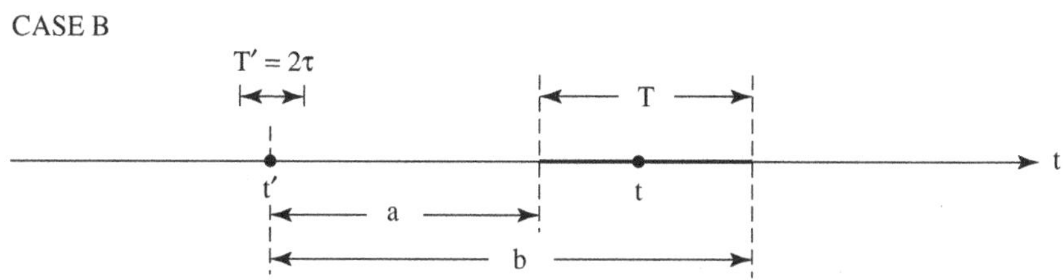

Fig. 5.8 (a) Point t' inside interval T centered at point t; (b) point t' outside interval T centered at t.

Eq. (5.3.5) by introducing the local average $X_{T'}(t')$, where $T' = 2\tau$, and then letting τ vanish. We may write:

$$\begin{aligned}&(2\tau T)\,\mathrm{Cov}[X_T(t), X(t')]\\&\quad = \frac{\sigma^2}{2}[\Delta(a+\tau) - \Delta(a-\tau) + \Delta(b+\tau) - \Delta(b-\tau)].\end{aligned} \tag{5.3.10}$$

Dividing both sides by $(2\tau T)$ and taking the limit $\tau \to 0$ yields

$$\mathrm{Cov}[X_T(t), X(t')] = \frac{\sigma^2}{2T}[\dot{\Delta}(a) + \dot{\Delta}(b)], \quad (t' \text{ inside interval } T) \tag{5.3.11}$$

where $\dot{\Delta}(u)$ denotes the first derivative of the variance function:

$$\dot{\Delta}(u) = \left.\frac{d\Delta(T)}{dT}\right|_{T=u}. \tag{5.3.12}$$

In Case B, the point t' lies outside the interval T, and the quantities a and b are again defined as the distances between the point t' and the respective ends of the interval T. We now have $T = b - a$, compared to $T = a + b$ for Case A. The result is

$$\mathrm{Cov}[X_T(t), X(t')] = \frac{\sigma^2}{2T}[\dot{\Delta}(b) - \dot{\Delta}(a)], \quad (t' \text{ outside interval } T). \tag{5.3.13}$$

A simple and useful approximation for the first derivative of $\Delta(T)$,

Table 5.1 Covariance between the point value of a random process and the local average of the process over an interval T.

	Point inside interval[a]	Point outside interval[b]
$a = 0$	$\frac{\sigma^2}{2T}\dot{\Delta}(b)$	$\frac{\sigma^2}{2T}\dot{\Delta}(b)$
$a, b \gg \theta$	$\frac{\sigma^2}{2T}2\theta = \sigma^2\frac{\theta}{T}$	$\frac{\sigma^2}{2T}(\theta - \theta) = 0$
$a, b \ll \theta$	$\frac{\sigma^2}{2T}(2a + 2b) = \sigma^2$	$\frac{\sigma^2}{2T}(2b - 2a) = \sigma^2$

[a]See Fig. 5.8, Case A.
[b]See Fig. 5.8, Case B.

compatible with Eq. (5.1.39) in which $k_1 = 1/3$, has the form:

$$\dot{\Delta}(u) = \begin{cases} 2u - u^2/\theta, & 0 \le u \le \theta, \\ \theta, & u \ge \theta. \end{cases} \tag{5.3.14}$$

Some specific results, obtained by using Eqs. (5.3.11) through (5.3.14), are summarized in Table 5.1.

Decomposition in Terms of Local Averages

The difference $[X(t') - X_T(t)]$, in case t' lies inside the interval $(t - T/2,\ t + T/2)$, represents the deviation from the (observed or predicted) local average $X_T(t)$. Using Eq. (5.3.11), its variance may be expressed as follows:

$$\begin{aligned} \mathrm{Var}[X(t') - X_T(t)] &= \sigma^2 + \sigma^2\gamma(T) - 2\mathrm{Cov}[X(t'), X_T(t)] \\ &= \sigma^2 \left\{ 1 + \gamma(T) - [\dot{\Delta}(a) + \dot{\Delta}(b)]/T \right\}, \end{aligned} \tag{5.3.15}$$

The distances a and $b = T - a$ unambiguously locate t' inside the interval.

In case $t' = t$, and hence $a = b = T/2$, a good approximation for the variance of $[X(t') - X_T(t)]$ is

$$\mathrm{Var}[X(t) - X_T(t)] \simeq \sigma^2[1 - \gamma(T)]. \tag{5.3.16}$$

Eq. (5.3.16) implies $\mathrm{Cov}[X(t) - X_T(t)] \simeq \sigma^2\gamma(T)$, and also that the local average $X_T(t)$ and the deviation $[X(t) - X_T(t)]$ are approximately uncorrelated. Taking $m = 0$ (for convenience), we may write

$$E[(X(t) - X_T(t))X_T(t)] = E[X(t)X_T(t)] - \sigma^2\gamma(T) \simeq 0. \tag{5.3.17}$$

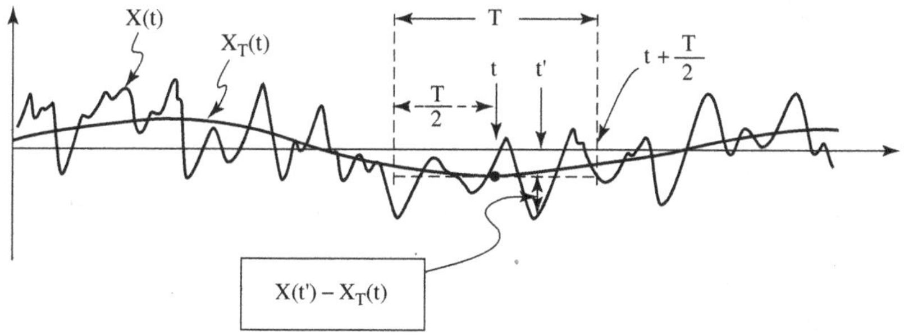

Fig. 5.9 Decomposition of $X(t')$ in terms of a local average process $X_T(t)$ and a local deviation $[X(t') - X_T(t)]$.

A useful interpretation of these results, illustrated in Fig. 5.9, is as follows. The random process $X(t)$ may be expressed as the sum of two "effectively uncorrelated" processes, the local average $X_T(t)$ and the (local) deviation $[X(t) - X_T(t)]$:

$$X(t) = X_T(t) + [X(t) - X_T(t)]. \tag{5.3.18}$$

The variances of the component processes are $\sigma^2\gamma(T)$ and $\sigma^2[1 - \gamma(T)]$, respectively. The conditional statistics of $X(t)$ *given* $X_T(t)$ become

$$E[X(t)|X_T(t)] = X_T(t), \tag{5.3.19}$$

$$\begin{aligned} \text{Var}[X(t)|X_T(t)] &= \text{Var}[X(t) - X_T(t)|X_T(t)] \\ &= \text{Var}[X(t) - X_T(t)] \simeq \sigma^2[1 - \gamma(T)]. \end{aligned} \tag{5.3.20}$$

The theory of conditional expectation is closely related to these results and offers further operational uses for the variance function $\gamma(T)$. A priori, the conditional mean of $X(t)$ equals the mean of $X_T(t)$, namely, $m = E[X]$. If the local average $X_T(t)$ is observed or predicted, then the conditional (posterior) mean of $X(t)$ becomes equal to the observation or prediction. The conditional variance of $X(t)$ given $X_T(t)$, normalized with respect to σ^2, equals $[1 - \gamma(T)]$; it can be read directly from the plot of the variance function $\gamma(T)$. These concepts provide underpinning for a "local average subdivision" (LAS) approach to analyzing random processes (Fenton and Vanmarcke [60]). The random function $X(t)$, according to this approach, may be expressed as a sum of (approximately uncorrelated) components, each representing the deviation of fine-scale (high-resolution) local averages from coarser-scale (lower-resolution) local averages, so that each component accounts for a fraction of the total variance, and hence for a distinct increment of the variance function, of $X(t)$.

Covariance Function of Local Average Processes

Eq. (5.3.5) may be used to derive not only covariances but also covariance functions of moving averages of a stationary random function. Specifically, by taking $T = T'$, $T_1 = T_3 = \tau$, $T_0 = T - \tau$, and $T_2 = T + \tau$ (see Fig. 5.10), Eq. (5.3.5) generates the covariance function of the local average process $X_T(t)$, namely:

$$B_T(\tau) = \frac{\sigma^2}{2T^2}[\Delta(T + \tau) - 2\Delta(\tau) + \Delta(T - \tau)]. \tag{5.3.21}$$

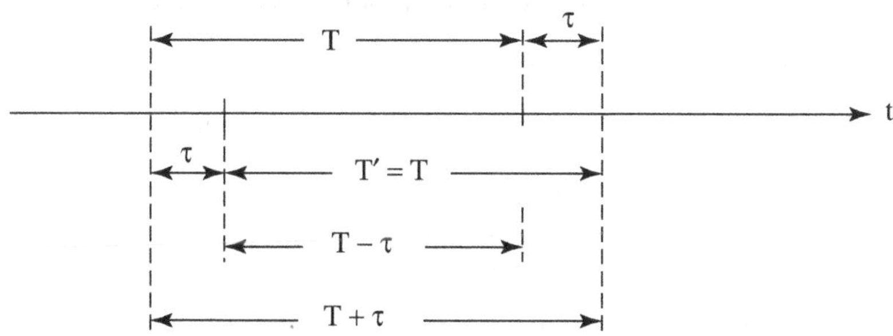

Fig. 5.10 Definition of intervals needed to evaluate the covariance function of $X_T(t)$.

Consider its behavior in the limiting case when both T and $\tau \to 0$:

$$B_T(\tau) = \frac{\sigma^2}{2T^2}[(T+\tau)^2 - 2\tau^2 + (T-\tau)^2] \longrightarrow \sigma^2, \tag{5.3.22}$$

where $\sigma \equiv \sigma_X$. For $T \gg \theta$ and $\tau \to 0$, Eq. (5.3.21) yields

$$B_T(\tau) = \frac{\sigma^2}{2\,T^2}[(T+\tau)\,\theta - 2\tau^2 + (T-\tau)\,\theta] \longrightarrow \sigma^2\theta/T. \tag{5.3.23}$$

The expression [Eq. (5.3.21)] may be used to derive the "inverse" of the relationship between $\gamma(T)$ and $\rho(\tau)$ [Eq. (5.1.6)]. When $T \to 0$, Eq. (5.3.21) can be arranged in the form of a second-order derivative:

$$\begin{aligned} B(\tau) &= \lim_{T\to 0}\frac{\sigma^2}{2T}\left[\left(\frac{\Delta(\tau+T)-\Delta(\tau)}{T}\right) - \left(\frac{\Delta(\tau)-\Delta(T-\tau)}{T}\right)\right] \\ &= \frac{\sigma^2}{2}\left[\frac{d^2\Delta(T)}{dT^2}\right]_{T=|\tau|} = \frac{\sigma^2}{2}\left[\frac{d^2\Delta(\tau)}{d\tau^2}\right]. \end{aligned} \tag{5.3.24}$$

Dividing by σ^2 and replacing $\Delta(\tau)$ by $\tau^2\gamma(\tau)$ leads to

$$\rho(\tau) = \frac{1}{2}\frac{d^2}{d\tau^2}[\tau^2\gamma(\tau)] = \gamma(\tau) + 2\tau\frac{d\gamma(\tau)}{d\tau} + \frac{\tau^2}{2}\frac{d^2\gamma(\tau)}{d\tau^2}. \tag{5.3.25}$$

Eqs. (5.3.25) and (5.1.6) constitute a pair of relations linking the functions $\gamma(T)$ and $\rho(\tau)$.

The expression for the covariance function of $X_T(t)$ [Eq. (5.3.21)] provides the starting point for the derivation of important results for the mean square derivative and related statistical properties of $X_T(t)$.

5.4 Mean Square Derivative and Spectral Moments

Existence of the Mean Square Derivative

We show in the first part of this section that even a minimal amount of local averaging renders a random process $X(t)$ for which the mean square derivative $\sigma_{\dot{X}}^2 = E[\dot{X}^2]$ does not exist "mean square differentiable." Recall that $\dot{B}(0) = 0$ is the necessary and sufficient condition for the existence of the mean square derivative. Likewise, for the local average process $X_T(t)$, mean square differentiability hinges on whether $\dot{B}_T(\tau)$ vanishes at $\tau = 0$. The derivative of the covariance function of $X_T(t)$ may be obtained from Eq. (5.3.21):

$$\dot{B}_T(\tau) = \frac{\sigma^2}{2T^2}\left[\frac{d}{d\tau}\Delta(T+\tau) - 2\frac{d}{d\tau}\Delta(\tau) + \frac{d}{d\tau}\Delta(T-\tau)\right]$$

$$= \frac{\sigma^2}{2T^2}[\dot{\Delta}(T+\tau) - 2\dot{\Delta}(\tau) - \dot{\Delta}(T-\tau)], \tag{5.4.1}$$

where $\dot{\Delta}(u)$ denotes the derivative of the variance function evaluated at u.

In general, we can express the covariance function of $X(t)$ as a polynomial near $\tau = 0$, for $\tau \geq 0$, as follows:

$$B(\tau) = \sigma^2\left[1 + b\tau + \frac{c}{2}\tau^2 + \cdots\right], \tag{5.4.2}$$

in which value of the coefficient

$$b = \dot{\rho}(0) = \dot{B}(0)/\sigma^2 \tag{5.4.3}$$

determines whether $X(t)$ is mean square differentiable: b must be zero for $\sigma_{\dot{X}}^2$ to be finite. The series expansion for $\gamma(T)$ is obtained by combining Eqs. (5.4.2) and (5.1.6):

$$\gamma(T) = \frac{2}{T}\int_0^T \left(1 - \frac{\tau}{T}\right)\left[1 + b\tau + \frac{c}{2}\tau^2 + \cdots\right] d\tau$$

$$= 1 + \frac{b}{3}T + \frac{c}{12}T^2 + \cdots. \tag{5.4.4}$$

Furthermore, since $\Delta(T) = T^2\gamma(T)$,

$$\Delta(T) = T^2 + \frac{b}{3}T^3 + \frac{c}{12}T^4 + \cdots. \tag{5.4.5}$$

The corresponding series for the derivative of $\Delta(T)$ is needed to permit evaluation of $\dot{B}_T(\tau)$ based on Eq. (5.4.1). From Eq. (5.4.5) we obtain

$$\dot{\Delta}(T) = \frac{d\Delta(T)}{dT} = 2T + bT^2 + \frac{c}{3}T^3 + \cdots. \tag{5.4.6}$$

Combining Eqs. (5.4.1) and (5.4.6) yields the series expansion for $\dot{B}_T(\tau)$:

$$\dot{B}_T(\tau) = \frac{b\sigma^2}{T^2}(2\tau T - \tau^2) + c\sigma^2\tau + \cdots, \tag{5.4.7}$$

from which the desired result follows directly, namely:

$$\dot{B}_T(0) = 0, \quad \text{for } T > 0, \tag{5.4.8}$$

regardless of the value of b.

Mean Square Derivative of Local Average Processes

One concludes from the preceding analysis that the local average process $X_T(t)$ and the local integral process $I_T(t) = TX_T(t)$ will be mean square differentiable, for $T > 0$, even if the mean square derivative of the original random process $X(t)$ does not exist. Our next task is to evaluate the mean square derivative $\sigma^2_{\dot{X}_T} = E[\dot{X}^2_T]$ in terms of the second-order properties of $X(t)$ and the averaging interval T. The analysis is carried out in the time domain, and the predicted "asymptotic" behavior of $\sigma^2_{\dot{X}_T}$ is then confirmed by a complementary analysis in the frequency domain.

In the time domain the basic formula for the mean square derivative, applied to the local average processes $X_T(t)$, is as follows:

$$E[(\dot{X}_T(t))^2] = \sigma^2_{\dot{X}_T} = B_{\dot{X}_T}(0) = -\left.\frac{d^2B_T(\tau)}{d\tau^2}\right|_{\tau=0}. \tag{5.4.9}$$

The second derivative of $B_T(\tau)$ can be evaluated by combining Eqs. (5.3.21) and (5.3.24):

$$\begin{aligned} B_{\dot{X}_T}(\tau) &= -\frac{d^2B_T(\tau)}{d\tau^2} = \frac{\sigma^2}{2T^2}\left\{\frac{d^2}{d\tau^2}[2\Delta(\tau) - \Delta(T+\tau) - \Delta(T-\tau)]\right\} \\ &= \frac{\sigma^2}{2T^2}\left[2\frac{d^2}{d\tau^2}\Delta(\tau) - \frac{d^2}{d\tau^2}\Delta(T+\tau) - \frac{d^2}{d\tau^2}\Delta(T-\tau)\right] \\ &= \frac{1}{T^2}[2B(\tau) - B(T+\tau) - B(T-\tau)]. \end{aligned} \tag{5.4.10}$$

Taking $\tau = 0$ yields the mean square derivative of $\dot{X}_T(t)$,

$$\sigma^2_{\dot{X}_T} = \frac{2}{T^2}[B(0) - B(T)] = \frac{2\sigma^2}{T^2}[1 - \rho(T)]. \tag{5.4.11}$$

Using $\sigma^2_T = \sigma^2\Delta(T)/T^2$, Eq. (5.4.11) can be restated as follows:

$$\sigma^2_{\dot{X}_T} = 2\sigma^2_T\left(\frac{1 - \rho(T)}{\Delta(T)}\right), \tag{5.4.12}$$

where $\sigma^2_T \equiv \sigma^2_{X_T}$ is the variance of $X_T(t)$; $\sigma^2_{\dot{X}_T}$ is the variance of $\dot{X}_T(t)$; and $\rho(\cdot)$ and $\Delta(\cdot)$ are, respectively, the correlation function and the integral-based variance function of $X(t)$. This important result leads to much new information about (and stable estimates for the statistics of) level crossings and extremes of the derived random processes $X_T(t)$ and $I_T(t)$.

Asymptotic Behavior

At relatively large values of the ratio T/θ, $\rho(T) \to 0$ and the mean square derivative of $X_T(t)$ obeys the asymptotic expression:

$$\sigma^2_{\dot{X}_T} = 2\sigma^2/T^2, \quad T \gg \theta. \tag{5.4.13}$$

For $T \to 0$, we expect $\sigma^2_{\dot{X}_T}$ to converge to $\sigma^2_{\dot{X}}$, provided the latter exists. Taking $T = 0$ in Eq. (5.4.11) leads to the indeterminate result $0/0$. Evidently, $\sigma^2_{\dot{X}_T}$ will be properly defined only if

$$[1 - \rho(T)] \propto T^2, \quad \text{when } T \to 0. \tag{5.4.14}$$

Using l'Hospital's rule, we twice differentiate both the numerator and denominator in Eq. (5.4.11) and set $T = 0$. As expected, the result agrees with Eq. (3.6.7):

$$\sigma^2_{\dot{X}} = \lim_{T\to 0} \sigma^2_{\dot{X}_T} = -\left[\frac{d^2}{dT^2}B(T)\right]_{T=0}. \tag{5.4.15}$$

Complementary Analysis in the Frequency Domain

An even clearer picture emerges in the frequency domain. Replacing $\cos\omega T$ in the Wiener-Khinchine relation [Eq. (3.3.12)],

$$\rho(T) = \int_0^\infty g(\omega)\cos\omega T\, d\omega, \tag{5.4.16}$$

by its series expansion,

$$\cos\omega T = 1 - \frac{\omega^2 T^2}{2} + \cdots, \tag{5.4.17}$$

and then inserting Eq. (5.4.16) into Eq. (5.4.11), one obtains:

$$\begin{aligned}\sigma_{\dot{X}}^2 = \lim_{T\to 0} \sigma_{\dot{X}_T}^2 &= \frac{2\sigma^2}{T^2}\left\{1 - \int_0^\infty g(\omega)\left[1 - \frac{\omega^2 T^2}{2}\right] d\omega\right\} \\ &= \frac{2\sigma^2}{T^2}\frac{T^2}{2}\left\{\int_0^\infty \omega^2 g(\omega)\, d\omega\right\} = \sigma^2\Omega_2^2,\end{aligned} \tag{5.4.18}$$

provided the second spectral moment of $g(\omega)$ exists.

Recall that $\sigma_{\dot{X}_T}^2$ equals $\lambda_{2,T}$, the second spectral moment of $X_T(t)$. The characteristic frequency related to the second spectral moment is

$$\Omega_{2,T} = \frac{\sigma_{\dot{X}_T}}{\sigma_T} = \frac{\sqrt{2}}{T}\left[\frac{1-\rho(T)}{\gamma(T)}\right]^{1/2}. \tag{5.4.19}$$

For $T \gg \theta$, since $\rho(T) \to 0$ and $\gamma(T) \to \theta/T$, Eq. (5.4.19) leads to

$$\Omega_{2,T} \to \sqrt{\frac{2}{T\theta}}, \quad T \gg \theta. \tag{5.4.20}$$

When $T \to 0$, $\Omega_{2,T} \to \Omega_2$, provided the latter exists.

Spectral Moments and Spectral Bandwidth Measures

We already analyzed the spectral moments $\lambda_{0,T} = \sigma_T^2$ and $\lambda_{2,T} = \sigma_{\dot{X}_T}^2$ of the local average process $X_T(t)$ for the full range of averaging intervals T. To evaluate the spectral bandwidth factor δ_T of $X_T(t)$, we also need the first spectral moment,

$$\lambda_{1,T} = \int_0^\infty \omega\, G_T(\omega)\, d\omega = \sigma^2 \int_0^\infty \omega g(\omega)\left[\frac{\sin(\omega T/2)}{\omega T/2}\right]^2. \tag{5.4.21}$$

A somewhat tedious derivation (see RF1 [137]) yields the following approximation for $\lambda_{1,T}$ for relatively large values of the ratio T/θ:

$$\lambda_{1,T} \simeq \frac{2\sigma^2\theta}{\pi T^2}\ln\left(5.26\,\frac{T}{\theta}\right), \qquad T \gg \theta. \tag{5.4.22}$$

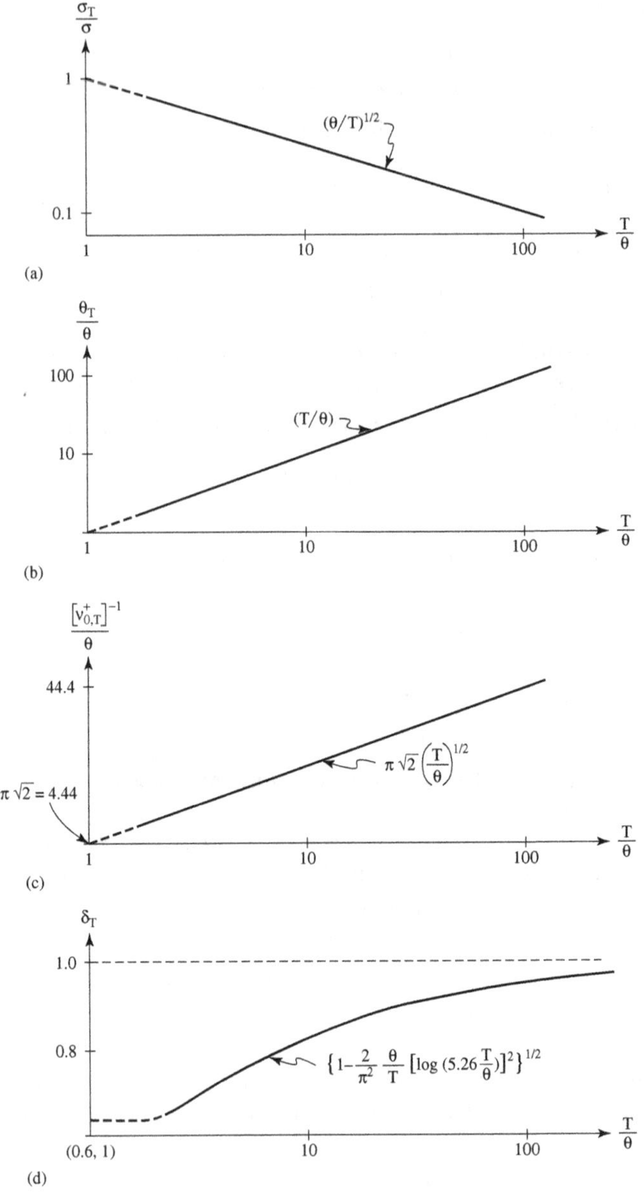

Fig. 5.11 Statistics of the moving average process $X_T(t)$ plotted against the normalized averaging interval T/θ: (a) ratio of standard deviations σ_T/σ; (b) ratio of scales of fluctuation θ_T/θ; (c) inverse of the average rate of upcrossing of the mean, normalized with respect to the scale θ (assuming $X_T(t)$ is Gaussian); (d) the spectral bandwidth measure δ_T.

Inserting the above expression, along with $\lambda_{0,T} = \sigma^2\theta/T$ and $\lambda_{2,T} = 2\sigma^2/T^2$, into the definition [Eq. (4.1.5)] of the spectral bandwidth measure, yields an approximate asymptotic expression for the square of δ_T:

$$\delta_T^2 \simeq 1 - \frac{2}{\pi^2}\frac{\theta}{T}\left[\ln\left(5.26\frac{T}{\theta}\right)\right]^2, \quad T \gg \theta. \tag{5.4.23}$$

The corresponding result for δ_T is plotted versus T/θ in Fig. 5.11(d). As T increases, δ_T slowly converges toward its unit upper bound. This information about δ_T and $\sigma_{\dot{X}_T}$ enables quantitative treatment of the fluctuations of the envelope of $X_T(t)$ [see Sec. 5.5].

For completeness, consider also the fourth spectral moment of $X_T(t)$,

$$\begin{aligned}\lambda_{4,T} &= \int_0^\infty \omega^4 G_T(\omega)\,d\omega = \sigma^2\int_0^\infty \omega^4 g(\omega)\left[\frac{\sin(\omega T/2)}{\omega T/2}\right]^2 d\omega \\ &= \frac{4\sigma^2}{T^2}\int_0^\infty \omega^2 g(\omega)\sin^2\left(\frac{\omega T}{2}\right)d\omega.\end{aligned} \tag{5.4.24}$$

The effect of the function $\sin^2(\omega T/2)$, when $T \to \infty$, is to reduce all contributions to the integral of $\omega^2 g(\omega)\,d\omega$ by the factor 2. Provided the second spectral moment of $X(t)$ exists, the asymptotic expression for $\lambda_{4,T}$ is

$$\lambda_{4,T} = 2\sigma^2\Omega_2^2/T^2, \quad T \gg \theta. \tag{5.4.25}$$

Combining this result with $\lambda_{0,T} = \sigma^2\theta/T$ and $\lambda_{2,T} = 2\sigma^2/T^2$, the limiting expression for the square of the spectral bandwidth factor ε_T [see Eq. (4.1.7)] becomes:

$$\varepsilon_T^2 = 1 - \frac{\lambda_{2,T}^2}{\lambda_{0,T}\lambda_{4,T}} \simeq 1 - \frac{2}{T\theta\,\Omega_2^2}, \quad T \gg \theta. \tag{5.4.26}$$

Note that the existence of $\lambda_{4,T}$ and ε_T requires $X(t)$ to be mean square differentiable, that is, Ω_2 must be finite. The bandwidth measure ε_T, like δ_T, approaches its unit upper bound as $T \to \infty$.

5.5 Level-Crossing and Extreme-Value Statistics

Level-crossing and extreme-value statistics of a stationary random process $X(t)$ depend on quantities such as $\sigma_{\dot{X}_T}$ and δ, which are in turn related to the second spectral moment of $X(t)$ or the second derivative of $\rho(\tau)$ evaluated at $\tau = 0$. These quantities, by implication, will be sensitive to details of the correlation structure of $X(t)$ about which information is usually lacking, namely the behavior at very high frequencies ω or near-zero time lags τ. This basic problem of high sensitivity or instability of key statistics motivates the introduction of the local average process $X_T(t)$ that converges to $X(t)$ in the limit when $T \to 0$. Actually, taking the limit $(T \to 0)$ is often unnecessary (considering the purpose of the stochastic model) or unjustified (in light of the limited resolution of data validating the model). Generally, it suffices to make the averaging window T suitably small but nonzero. This way, analysis results will not be unduly sensitive to hard-to-verify assumptions about the fluctuations on the microscale.

The results of Sec. 4.4, all stated in terms of $X(t)$, are also applicable to $X_T(t)$. If $X(t)$ is Gaussian, $X_T(t)$ will of course also be Gaussian. But even if $X(t)$ is non-Gaussian, for example if it is a discrete-state process, $X_T(t)$ will tend toward the Gaussian process as the ratio T/θ increases, by virtue of the Central Limit Theorem. There is also evidence that, as T increases, the p.d.f. of the derivative process $\dot{X}_T$ approaches the Gaussian distribution (much) more rapidly than the process X_T itself. For instance, consider a continuous-parameter Markov process $X(t)$ that makes sudden, random jumps from one state to another. At the points of discontinuity the derivative $\dot{X}(t)$ is infinite (with equal chance, $+\infty$ or $-\infty$). A minimal amount of local averaging eliminates these discontinuities, and dramatically affects $f_{\dot{X}}(\dot{x})$ without necessarily changing $f_X(x)$, the p.d.f. of $X(t)$, by much. Consider also that a derivative process may be Gaussian even if the original process is not. The envelope of a Gaussian process provides a good example: $R(t)$ is Rayleigh distributed but $\dot{R}(t)$ is Gaussian with mean zero. Symmetry about a zero mean is the rule for derivatives of stationary random processes.

Mean Upcrossing Rate

If $X_T(t)$ is Gaussian with mean zero, the mean rate of upcrossings of the threshold level b by $X_T(t)$ is [see Eqs. (4.4.3) and (4.4.5)]:

$$\nu^+_{b,T} = \frac{1}{2}E[|\dot{X}_T|]f_T(b) = \nu^+_{0,T}\exp\left\{\frac{-b^2}{2\sigma^2_T}\right\}, \tag{5.5.1}$$

where $f_T(b)$, σ_T, and $\nu^+_{0,T}$ denote the p.d.f., the standard deviation, and the mean zero-upcrossing rate of $X_T(t)$, respectively. From Eq. (5.1.2),

$\sigma_T^2 = \sigma^2\gamma(T)$, and based on Eq. (5.4.19) we may write:

$$\nu_{0,T}^+ = \frac{1}{2\pi}\frac{\sigma_{\dot{X}_T}}{\sigma_T} = \frac{1}{\sqrt{2}\pi T}\left[\frac{1-\rho(T)}{\gamma(T)}\right]^{1/2}. \tag{5.5.2}$$

For $T \gg \theta$, the mean upcrossing rate for $X_T(t)$ of level b becomes

$$\nu_{b,T}^+ = \frac{1}{\pi\sqrt{2\theta T}}\exp\left\{-\frac{b^2 T}{2\sigma^2\theta}\right\}, \quad T \gg \theta. \tag{5.5.3}$$

Taking $b = 0$ yields the asymptotic expression for the mean rate of zero-upcrossings, $\nu_{0,T}^+ = (\pi\sqrt{2\theta T})^{-1}$, for $T \gg \theta$.

Extreme Value Distribution

We can now express the probability distribution of the maximum value of the local-average random process $X_T(t)$, based on the Poisson-process approximation for arrivals of rare events. The probability that $X_T(t)$ (with mean $m = m_T = 0$, for convenience) will not exceed a high barrier level b within the time interval $(0, t_0)$ is

$$\begin{aligned} L_b(t_0) &= P\Big[\underset{t_0}{\text{Max}}\, X_T(t) \le b\Big] = \exp\{-\nu_{b,T}^+ t_0\} \\ &= \exp\left\{-\frac{t_0}{\pi\sqrt{2\theta T}}\exp\left(-\frac{b^2 T}{2\sigma^2\theta}\right)\right\}, \quad t_0 \ge 0,\ T \gg \theta. \end{aligned} \tag{5.5.4}$$

If $m \neq 0$, it suffices to replace b^2 by $(b-m)^2$. Eq. (5.5.4) gives the exact limiting distribution of extreme values of a large class of stationary random processes (non-Gaussian, discrete state, and so on); under sufficient local averaging, these derived random processes enter into the "Gaussian sphere of attraction" by virtue of the Central Limit Theorem, rendering Eq. (5.5.4) asymptotically exact as $b \to \infty$ and $T \to \infty$. In practical applications, it is expected to give good results when $T > \theta$ and $b \ge 2\sigma_T$.

Various improvements to, and extensions of, Eq. (5.5.4) are possible, in particular by making use of Eq. (5.1.39) to quantify the variance function (for any value of T/θ) and by considering the refinements that account for lower barrier levels and for the effect of clustering of level crossings.

Other Level-Crossing Statistics

The mean duration of an excursion by $X_T(t)$ above the threshold b [from Eq. (4.4.14)] is:

$$E[\mathscr{T}_{b,T}] = \frac{F^c_{T}(b)}{\nu^+_{b,T}} = \frac{2}{E[|\dot{X}_T|]}\frac{F^c_{T}(b)}{f_T(b)}, \tag{5.5.5}$$

where $F^c_T(b)$ is the c.d.f. of $X_T \equiv X_T(t)$. If $\dot{X}_T(t)$ is Gaussian, the mean of the absolute value of $\dot{X}_T$ is given by

$$E[|\dot{X}_T|] = \sqrt{\frac{2}{\pi}}\sigma_{\dot{X}_T} = \frac{2}{\sqrt{\pi}T}[1-\rho(T)]^{1/2}. \tag{5.5.6}$$

This equation is valid over the entire range of values of the averaging interval T, from 0 to ∞. For $T \gg \theta$, assuming $X_T(t)$ is Gaussian, the mean excursion time approaches the following analytical form [see Eq. (4.4.16)]:

$$E[\mathscr{T}_{b,T}] \to \frac{\sigma_T}{\sqrt{2\pi}\,b\,\nu^+_{0,T}} = \sqrt{\pi}\,\theta\,\frac{\sigma}{b}, \quad T \gg \theta,\ b \gg \sigma_T. \tag{5.5.7}$$

Note the lack of dependence on T. The corresponding expression for $E[\mathscr{T}_b]$, the mean excursion length for the original process $X(t)$, is

$$E[\mathscr{T}_b] \to \frac{\sigma}{\sqrt{2\pi}\,b\,\nu^+_0}, \quad b \gg \sigma, \tag{5.5.8}$$

provided ν^+_0 exists. The results, presented in Sec. 4.4, for the envelope excursion statistics, the mean clump size, and the improved reliability estimates can now all be applied to the derived random process $X_T(t)$.

5.6 Invariant and Regenerative Properties

Invariance under Linear Transformation

Assume that a stationary random process $X(t)$ with mean m_X, variance σ^2_X, and scale of fluctuation θ_X excites a time-invariant linear system characterized by the transfer function $H(\omega)$ and impulse response function $h(t)$. The corresponding statistics of the response process $Y(t)$ are m_Y, σ^2_Y and

θ_Y. The following relations for m_Y and σ_Y^2 are well known:

$$m_Y = H(0)m_X = m_X \int_{-\infty}^{+\infty} h(t)\,dt, \tag{5.6.1}$$

$$\sigma_Y^2 = \int_0^{\infty} |H(\omega)|^2 G_X(\omega)\,d\omega = \sigma_X^2 \int_0^{\infty} |H(\omega)|^2 g_X(\omega)\,d\omega. \tag{5.6.2}$$

The scale of fluctuation of the response process is

$$\theta_Y = \pi g_Y(0) = \frac{\pi G_Y(0)}{\sigma_Y^2} = \frac{\pi}{\sigma_Y^2} G_X(0)|H(0)|^2 = \frac{\sigma_X^2\,\theta_X}{\sigma_Y^2}|H(0)|^2. \tag{5.6.3}$$

Combining Eqs. (5.6.1) and (5.6.3) yields, for $m_Y \neq 0$,

$$\theta_Y \left(\frac{\sigma_Y}{m_Y}\right)^2 = \theta_X \left(\frac{\sigma_X}{m_X}\right)^2, \tag{5.6.4}$$

where (σ_Y/m_Y) and (σ_X/m_X) are the *coefficients of variation* of $Y(t)$ and $X(t)$, respectively. Clearly, the scale of fluctuation θ_Y of the response of a linear system to stationary random input can be determined without additional effort once the response mean and variance are known.

Invariance under Local Averaging

If the linear system is a "moving average" operation, the integral of the "weighting function" $h(t)$ is, by definition, equal to one:

$$H(0) = \int_{-\infty}^{+\infty} h(t)\,dt = 1. \tag{5.6.5}$$

Based on Eq. (5.6.3), for any linear transformation in which the mean is preserved ($m_X = m_Y$), the following relationship holds:

$$\sigma_Y^2\,\theta_Y = \sigma_X^2\,\theta_X. \tag{5.6.6}$$

If the system function $h(t)$ is a rectangular box having unit area and width T, then the response process $Y(t)$ is identical to the moving average process $X_T(t)$, whose variance is

$$\sigma_T^2 \equiv \sigma_{X_T}^2 = \sigma^2\gamma(T). \tag{5.6.7}$$

From Eqs. (5.6.6) and (5.6.7), the scale of fluctuation of $X_T(t)$ is:

$$\theta_{X_T} \equiv \theta_T = \theta/\gamma(T). \tag{5.6.8}$$

When T is very small, $\theta_T \simeq \theta$. In case T/θ is much greater than one, $\gamma(T) \to \theta/T$, and the scale θ_T approaches T itself, as displayed in Fig. 5.11(b).

Two "Laws" of Stochastic Behavior under Linear Transformation

Local averaging of a random function reduces its point-to-point variability and increases its scale of fluctuation, in such a way as to keep their product invariant. Eqs. (5.6.4) and (5.6.6) may be said to quantify a principle of "conservation of uncertainty" under linear transformation: as the variability, represented by the coefficient of variation (σ/m), is reduced, the degree of (temporal) correlation grows in such a way as to keep $\theta(\sigma/m)^2$ invariant. In other words, there exists a basic complementarity of the intensity of variability (σ^2/m^2) and the measure of persistence (θ) of random functions.

When a zero-mean random process undergoes sustained local averaging (as occurs in diffusion), the variance σ^2 decreases monotonically, while the scale of fluctuation (θ) simultaneously grows, keeping their product constant. Simultaneously, the spectral bandwidth factors δ and ε keep growing [see Eqs. (5.4.23) and (5.4.26)], tending toward their upper bounds $\delta = \varepsilon = 1$. The quantities δ and ε are indicative of the degree of disorder (roughness, chaos) of a random process; recall that small values of δ and ε typify narrow-band fluctuations. The connection to the second law of thermodynamics is notable, especially since the principle of "conservation of uncertainty", mentioned above, evokes the first law of thermodynamics.

In conclusion, local averaging reduces the point variance σ^2, preserves the "density of uncertainty" $\sigma^2\theta$, and moves the bandwidth measures δ and ε closer to their upper limit, indicative of a greater degree of disorder.

Regenerative Property

It is advantageous to model random phenomena at a level of aggregation compatible with the way information (about the phenomenon) is acquired or processed. The act of measuring often involves some degree of local averaging. In modeling the variation of attributes of multiphase materials or "impulse" processes, it may be useful to define a minimum distance T^* below which the behavior of the variance function is unknown or holds no practical interest. Information about the process $X(t)$ may take the form of observations of some local average process $X_T(t)$, where $T \geq T^*$.

An advantage of the analysis format based on variance functions is that the "reference distance" T^* can be chosen flexibly. To see this, note that the variance of $X_T(t)$, for $T \geq T^*$, can be expressed in two ways, as follows:

$$\sigma_T^2 = \sigma^2\gamma(T) = \sigma_{T^*}^2\gamma_{T^*}(T), \quad T \geq T^*, \tag{5.6.9}$$

where the last expression is the product of the *variance* and the *variance function* of $X_{T^*}(t)$. Since $\sigma_{T^*}^2 = \sigma^2\gamma(T^*)$, we must have

$$\gamma(T) = \gamma(T^*)\gamma_{T^*}(T), \quad T \geq T^*, \tag{5.6.10}$$

or

$$\gamma_{T^*}(T) = \frac{\gamma(T)}{\gamma(T^*)}, \quad T \geq T^*. \tag{5.6.11}$$

In words, the variance function of $X_{T^*}(t)$ is proportional to the variance function of $X(t)$ for $T \geq T^*$, the proportionality factor being the variance ratio $\sigma^2/\sigma_{T^*}^2 = [\gamma(T^*)]^{-1}$. Clearly, to obtain $\gamma_{T^*}(T)$, it suffices to renormalize $\gamma(T)$ so that $\gamma(T^*) = 1$. The new variance function can now describe, presumably without loss of useful information, the correlation structure of local averages of the original process.

Related Results for White Noise and the Poisson Process

Ideal white noise has a constant spectral density $G(\omega) = G_0$ at all non-negative frequencies. It may be viewed as the limiting case, for $\omega_0 \to \infty$, of *band-limited* (low-pass) white noise whose spectral density equals G_0 for $0 \leq \omega \leq \omega_0$ and is zero elsewhere, and whose variance is

$$\sigma^2 = \omega_0 G_0. \tag{5.6.12}$$

The correlation function is

$$\rho(\tau) = \frac{\sin \omega_0\tau}{\omega_0\tau}. \tag{5.6.13}$$

The variance function, obtained from Eq. (5.1.6), equals

$$\gamma(T) = \frac{2}{T}\int_0^T \left(1 - \frac{\tau}{T}\right)\frac{\sin \omega_0\tau}{\omega_0\tau}d\tau = \frac{\pi}{\omega_0 T} + \frac{2}{\omega_0^2T^2}\cos\omega_0 T. \tag{5.6.14}$$

Its limiting expression, for large values of T, is

$$\gamma(T) = \frac{\pi}{\omega_0 T}, \tag{5.6.15}$$

and hence

$$\theta = \pi/\omega_0. \tag{5.6.16}$$

The same result is obtained by integrating $\rho(\tau)$, or by evaluating $\pi g(0)$.

In the ideal white noise case, when $\omega_0 \to \infty$, note that $\sigma^2 \to \infty$ and $\theta \to 0$ in such a way that the product $\sigma^2\theta$ remains constant:

$$\sigma^2\theta = \pi G_0. \tag{5.6.17}$$

Of course, ideal white noise is physically unrealizable. In any practical situation the variance is finite because there is, in effect, an upper limit on the frequencies of contributing components. In such cases, one can estimate the scale of fluctuation from Eq. (5.6.17).

We have seen in Sec. 2.7 that ideal white noise is closely related to the Poisson process. Both random processes are characterized by the fact that values in nonoverlapping time intervals are statistically independent; hence, both give rise to (integral) processes with independent increments. This implies that the variance of the local integral process $I_T(t)$ [defined by Eq. (5.1.27)] is proportional to T. We know that, in general, $\sigma^2_{I_T}$ becomes proportional to T when $T \gg \theta$,

$$\sigma^2_{I_T} = \sigma^2\theta T, \tag{5.6.18}$$

for a stationary random process $X(t)$ for which θ exists. If $X(t)$ is a Poisson process with mean arrival rate λ, then $\sigma^2_{I_T} = \lambda T$, implying

$$\sigma^2\theta = \lambda. \tag{5.6.19}$$

Note that a stationary random process derived from white noise or from a Poisson process by linear transformation will have a well-defined scale of fluctuation, since θ_Y in Eq. (5.6.3) depends only on the product $\sigma^2\theta$.

5.7 Parallel Results for Random Series and Point Processes

Random Series: Variance Function

Many of the results presented in this chapter can be restated, with minor modifications, in case the one-dimensional process at hand is a stationary random series, namely a series of random variables $X(t)$ associated with equidistant points on the coordinate axis. Assume that the observations are made at integer values $t = \ldots, -1, 0, 1, 2, \ldots$. (The random series may or not be sampled from an underlying continuous-parameter random process.) The derived time series of interest is the "moving sum" process

$$I_T(t) = \sum_{k=t+1}^{t+T} X(k), \quad t = \ldots, -1, 0, 1, 2, \ldots ; \; T = 1, 2, \ldots, \tag{5.7.1}$$

or the "moving average" process

$$X_T(t) = I_T(t)/T. \tag{5.7.2}$$

The associated variance functions are, respectively,

$$\gamma(T) = \frac{\sigma^2_{X_T}}{\sigma^2}, \quad T = 1, 2, \ldots, \tag{5.7.3}$$

and

$$\Delta(T) = \frac{\sigma^2_{I_T}}{\sigma^2} = T^2\gamma(T), \quad T = 1, 2, \ldots. \tag{5.7.4}$$

If the correlation function of $X(t)$ is known, the variance function $\gamma(T)$ may be obtained from the relation that parallels Eq. (5.1.6), namely:

$$\gamma(T) = \frac{1}{T} + \frac{2}{T}\sum_{\tau=1}^{T}\left(1 - \frac{\tau}{T}\right)\rho(\tau). \tag{5.7.5}$$

Example 1. $X(t)$ is an uncorrelated random series for which

$$\rho(\tau) = \delta(\tau) = \begin{cases} 1, & |\tau| = 0, \\ 0, & |\tau| = 1, 2, \ldots. \end{cases} \tag{5.7.6}$$

Its variance function is

$$\gamma(T) = 1/T, \quad T = 1, 2, \ldots. \tag{5.7.7}$$

Example 2. $X(t)$ is a discrete first-order autoregressive with a geometrically decreasing correlation function:

$$\rho(\tau) = r^{|\tau|}, \quad 0 \le r < 1, \; |\tau| = 0, 1, 2, \ldots. \tag{5.7.8}$$

Its variance function, obtainable from Eq. (5.7.5), is

$$\gamma(T) = \frac{1}{T}\left(\frac{1+r}{1-r}\right) - \frac{2}{T^2}\frac{r(1-r^T)}{(1-r)^2}, \quad T = 1, 2, \ldots. \tag{5.7.9}$$

For $T = 1$ and $T = 2$, Eq. (5.7.9) yields, respectively:

$$\gamma(1) = 1, \quad \text{and} \quad \gamma(2) = (1 + r)/2, \tag{5.7.10}$$

and for large values of T, since $r^T \to 0$, we have

$$\gamma(T) \to \left(\frac{1+r}{1-r}\right)\frac{1}{T}. \tag{5.7.11}$$

Scale of Fluctuation

In parallel with Eq. (5.1.25), the expressions for the scale of fluctuation θ of a stationary random series are

$$\begin{cases} \theta = \lim_{T\to\infty} T\gamma(T), \\ \theta = \sum_{T=-\infty}^{\infty} \rho(\tau) = 1 + 2\sum_{i=1}^{\infty} \rho(\tau), \\ \theta = \pi g(0). \end{cases} \tag{5.7.12}$$

For an uncorrelated random series, the "scale" equals the sampling interval:

$$\theta = 1. \tag{5.7.13}$$

This agrees with Eq. (5.7.12c) since

$$g(\omega) = \begin{cases} 1/\pi, & 0 \le \omega \le \pi, \\ 0, & \text{elsewhere}, \end{cases} \tag{5.7.14}$$

for an uncorrelated random series. The scale of fluctuation of the first-order autoregressive process [Eq. (5.7.8)] is

$$\theta = \frac{1+r}{1-r}. \tag{5.7.15}$$

If the one-step correlation coefficient r is close to one, θ will be much larger than one. When $r \to 0$, $\theta \to 1$. When $r \to 1$, by taking $1 - r = 1/b$, the scale θ is seen approaching the value $2b$ that characterizes the exponential correlation function [see Eq. (5.1.15)].

Covariance of Moving Averages

Consider two random variables I_T and $I_{T'}$, both defined as sums of consecutive observations of the random series $X(t)$:

$$I_T = \sum_{t+1}^{t+T} X(t) \quad \text{and} \quad I_{T'} = \sum_{t'+1}^{t'+T'} X(t), \tag{5.7.16}$$

where t and t' are arbitrarily located on the t-axis. As in the continuous-parameter case we define the distances (now integers) T_0, T_1, T_2, and T_3, similar to those shown in Fig. 5.6. The covariance of I_T and $I_{T'}$ is related to the variance function $\Delta(T)$ in the same way as in the continuous-parameter case [Eq. (5.3.5)]:

$$\begin{aligned} \text{Cov}[I_T, I_{T'}] &= \frac{\sigma^2}{2}[\Delta(T_0) - \Delta(T_1) + \Delta(T_2) - \Delta(T_3)] \\ &= \frac{\sigma^2}{2}\sum_{k=0}^{3}(-1)^k \Delta(T_k). \end{aligned} \tag{5.7.17}$$

The covariance of the process $X(t)$ is obtained by taking $T = T' = 1$ and $T_0 = \tau - 1$:

$$B(\tau) = \frac{\sigma^2}{2}[\Delta(\tau + 1) - 2\Delta(\tau) + \Delta(\tau - 1)]. \tag{5.7.18}$$

This equation may be interpreted as the inverse of Eq. (5.7.5), and as the random series equivalent of Eq. (5.3.24).

Closely related to the sums I_T and $I_{T'}$ are the moving averages

$$X_T = I_T/T \quad \text{and} \quad X_{T'} = I_{T'}/T', \tag{5.7.19}$$

whose covariance differs from Eq. (5.7.17) by the factor TT'. The covariance function of the random series of moving averages

$$X_T(t) = \frac{1}{T}\sum_{k=t+1}^{t+T} X(k), \quad T = 1, 2, \ldots, \tag{5.7.20}$$

may be obtained by taking $T = T' = T$, $T_1 = T_2 = \tau$, etc., in Eq. (5.7.17). The result is

$$B_T(\tau) \equiv B_{X_T}(\tau) = \frac{\sigma^2}{2T^2}[\Delta(\tau + T) - 2\Delta(\tau) + \Delta(\tau - T)], \tag{5.7.21}$$

which for $T = 1$ reduces to Eq. (5.7.18), as expected. It is the random series equivalent of (and is identical in form to) Eq. (5.3.21).

5.8 Role of the Scale of Fluctuation in Optimal Sampling

An important practical question is where to sample in the parameter space of a random field in order to gain the most information for a given level of expenditure. This section focuses on the role of the scale of fluctuation in residual error assessment and sampling plan design for *single-scale* random processes. Consider a decision situation in which one must choose the rate at which a random process $X(t)$ is to be sampled for the purpose of reconstructing the time history with minimum error.

In the classical formulation of the optimal linear interpolation problem, reviewed in Sec. 2.6, error-free observations of a random process $X(t)$ are made at equal time intervals Δt, thus yielding the sequence $\ldots x(-2\Delta t)$, $x(-\Delta t)$, $x(0)$, $x(\Delta t)$, $x(2\Delta t)$, $\ldots$, and the expectation (or best estimate) of $X(t)$ at any time t is expressed as a linear combination of the observations

$$E[X(t)] = \sum_{k=-\infty}^{+\infty} a_k x(k\Delta t). \tag{5.8.1}$$

The objective is to minimize the mean square error of this linear prediction, i.e., to minimize the updated or posterior variance of $X(t)$. The theory (presented in Sec. 2.6) provides the optimal coefficients a_k as well as the expression for the mean square interpolation error.

It is also possible to find the conditions under which a random process can be approximated arbitrarily closely by linearly combining the observations $x(k\Delta t)$ (Yaglom [158]). Specifically, if the spectral density function $G(\omega)$ of the random process $X(t)$ vanishes at frequencies exceeding Ω, perfect reconstruction (with zero interpolation error) is possible by taking as the interval between observations:

$$\Delta t = \pi/\Omega, \tag{5.8.2}$$

The corresponding optimal coefficients are

$$a_k = \frac{\sin(\Omega t - k\pi)}{\Omega t - k\pi}. \tag{5.8.3}$$

In combination, the above three equations constitute Shannon's sampling theorem, widely used in communication engineering (see Middleton [96]).

In case $G(\omega)$ does not vanish at values $\omega > \Omega$, the size of the interpolation error (presuming observations are spaced at intervals Δt) is approximately equal to the area of $G(\omega)$ associated with frequencies larger than $\Omega = \pi/\Delta t$ [see Yaglom [158]). For single-scale wide-band processes the decay of $G(\omega)$ at high frequencies is strongly dependent on the scale θ. This stems from the fact that $g(0) = \theta/\pi$ and $\dot{g}(0) = 0$, while the area under $g(\omega)$ must equal one. Recall, in particular, that the s.d.f. of one-dimensional random variation associated with a spatial (3-D) isotropic random field must decrease monotonically with frequency: $\dot{g}(\omega) \leq 0$ for all ω (see Sec. 3.4). Based on simple constraints on the shape of $g(\omega)$ implied by this information, one concludes that only a small fraction of the area under the spectral density function can come from frequencies above, say, $\Omega = 2\pi/\theta$, corresponding to the sampling interval

$$\Delta t = \pi/\Omega = \theta/2. \tag{5.8.4}$$

Hence, if a random process $X(t)$ is sampled at intervals $\Delta t = \theta/2$, nearly error-free interpolation can be achieved by using the linear combination, Eq. (5.8.1), of the observations $x(k\Delta t)$, with Ω in Eq. (5.8.3) set at $2\pi/\theta$.

Sampling a single-scale random process at much smaller intervals would be wasteful, since very little accuracy would be gained while the cost of sampling would increase. Larger sampling intervals may of course be justifiable on a cost-benefit basis. The resulting interpolation errors can be approximated (as a function of the ratio $\Delta t/\theta$) by adopting an approximate analytical form for $g(\omega)$ or $\rho(\tau)$. In the special case when the random process is a low-pass white noise with s.d.f. $G(\omega) = G_0$, $0 \leq \omega \leq \omega_0$, the sampling interval yielding zero interpolation error is $\Delta t = \pi/\omega_0 = \theta$.

The main conclusion of the preceding analysis is that one may expect optimal sampling intervals to be proportional to the scale of fluctuation. When the scale increases (for example, due to local averaging), optimal sampling intervals will increase proportionately. In general, the optimal value of the ratio $\Delta t/\theta$ will depend on economic factors (including the cost of sampling and the monetary consequences of residual errors).

5.9 Composite Random Processes and the Scale Spectrum

Basic Relations

The wide-band correlation functions encountered in preceding sections all have a simple functional form; they generally decrease (from one to zero) in an uncomplicated manner. The pattern of decay is usually governed by a single parameter, which is in turn related to the scale of fluctuation θ. However, correlation functions associated with real phenomena are often more complex. As Fig. 5.12 illustrates, the random variation may be composed of contributions from fundamentally different sources, with different correlation structure. A useful model in many applications is a *composite* random process $X(t)$ defined as the sum of *independent* component processes $X_i(t)$:

$$X(t) = X_1(t) + X_2(t) + \cdots = \sum_i X_i(t). \tag{5.9.1}$$

For example, one of the components may represent a random measurement error (with $\theta_i \to 0$) and another a strongly correlated (systematic) error source (with $\theta_i \to \infty$). Owing to the assumed statistical independence, the first- and second-order statistics of the composite process are related quite simply to those of the component processes. The mean of $X(t)$ is

$$m = m_1 + m_2 + \cdots = \sum_i m_i, \tag{5.9.2}$$

and its variance is

$$\sigma^2 = \sigma_1^2 + \sigma_2^2 + \cdots = \sum_i \sigma_i^2. \tag{5.9.3}$$

In terms of the fractional contributions of the component processes $X_i(t)$, $i = 1, 2, \ldots$, to the total variance,

$$q_i = \frac{\sigma_i^2}{\sigma^2} = \frac{\sigma_i^2}{\sum_i \sigma_i^2}, \tag{5.9.4}$$

satisfying

$$q_i \geq 0 \quad \text{and} \quad \sum_i q_i = 1, \tag{5.9.5}$$

the composite unit-area spectral density function may be expressed as a weighted combination of the component functions, as follows:

$$g(\omega) = \sum_i q_i\, g_i(\omega). \tag{5.9.6}$$

The correlation functions are likewise linked:

$$\rho(\tau) = \sum_i q_i\, \rho_i(\tau). \tag{5.9.7}$$

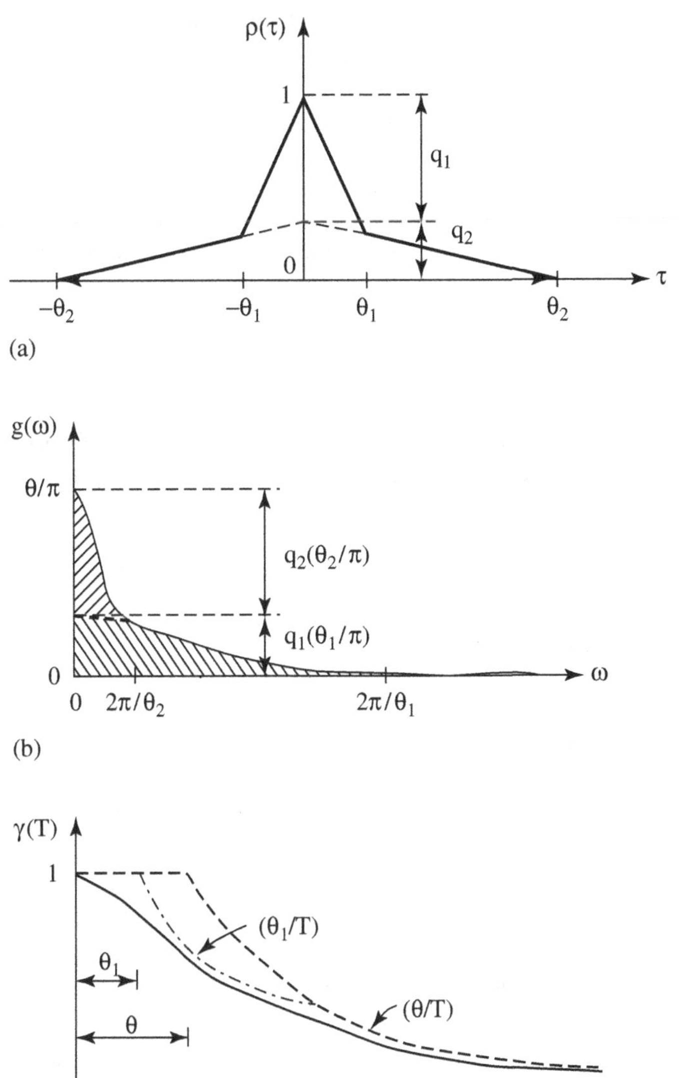

Fig. 5.12 Second-order properties of a composite random process whose two components have triangular correlation functions: (a) composite correlation function; (b) composite unit-area spectral density function; (c) composite variance function.

The relationship between the scales of fluctuation is obtained by taking $\omega = 0$ in Eq. (5.9.6) and multiplying both sides by π, or by integrating both sides of Eq. (5.9.7) over all values of τ, yielding

$$\theta = \sum_i q_i\,\theta_i. \tag{5.9.8}$$

The local average of $X(t)$ also equals the sum of local averages of the component processes $X_i(t)$:

$$X_T(t) = \sum_i X_{T,i}(t), \tag{5.9.9}$$

so that the variance function of $X(t)$, in its two alternate forms, $\gamma(T)$ or $\Delta(t)$, can also be expressed as a weighted average of component variance functions:

$$\begin{aligned}\gamma(T) &= \sum_i q_i \gamma_i(T),\\ \Delta(T) &= \sum_i q_i \Delta_i(T).\end{aligned} \tag{5.9.10}$$

The slope variance of $X_T(t)$,

$$\sigma^2_{\dot{X}_T} = \frac{2\sigma^2}{T^2}\left[1 - \sum_i q_i\,\rho_i(T)\right] = 2\sigma^2_{T}\left[\frac{1-\sum_i q_i\,\rho_i(T)}{\sum_i q_i \Delta_i(T)}\right], \tag{5.9.11}$$

can be used to predict threshold-crossing rates and extreme value statistics of sums or linear combinations of independent stationary random processes.

Finally, a relationship similar to that linking the scales of fluctuation [Eq. (5.9.8)] exists for the spectral moments of X_T and X_T, enabling the composite-process bandwidth measures (δ and ε) to be calculated.

Consider the case of a composite random process $X(t)$ consisting of two independent components with significantly different scales of fluctuation, the small scale labeled θ_1 and the large scale θ_2. In the example shown in Fig. 5.12, each component has a triangular correlation function with base $2\theta_i$ (and therefore area equal to θ_i), $i = 1, 2$. The correlation function of $X(t)$ is a composite of these triangular shapes. Note that the variance fractions q_1 and q_2 sum to one. The figure also shows the composite spectral density and variance functions.

Generalization of the Composite Random Process Model

The composite random process model [Eq. (5.9.1)] can accommodate both wide-band and "$\theta = 0$" narrow-band components; and any "$\theta = 0$" component with a relatively large bandwidth can itself be decomposed into two or more independent narrow-band components, each characterized by its mid-band frequency and its fractional contribution to the total variance.

In this context, we now propose a generalization of the concept of composite random processes. Any stationary stochastic process $X(t)$ with a given spectral density function can be represented as a sum of two types of component processes $X_i(t)$. Processes of the *first* type are wide-band (δ_i large) and characterized by a nonzero scale of fluctuation θ_i. Processes of the *second* type are narrow-band (δ_i small) and characterized by their mid-band frequency and $\theta_i = 0$.

As Fig. 5.13 indicates, the zero-frequency ordinate of the spectral density function $G(\omega)$ may be partitioned (by horizontal or slowly downward-sloping dividing lines) to separate the spectra of components of the first type. The remainder of the spectral mass may then be partitioned by vertical dividing lines, thereby conceptually producing components of the second type. The component spectral density functions fit together like pieces of a puzzle to form the composite s.d.f. $G(\omega)$.

Relationship to Fractional Noise

Consider again the composite variance function shown in Fig. 5.12(c). It tends toward inverse proportionality to the interval T when the latter exceeds θ_2. For $T \leq \theta_2$ the variance function will appear to be proportional to

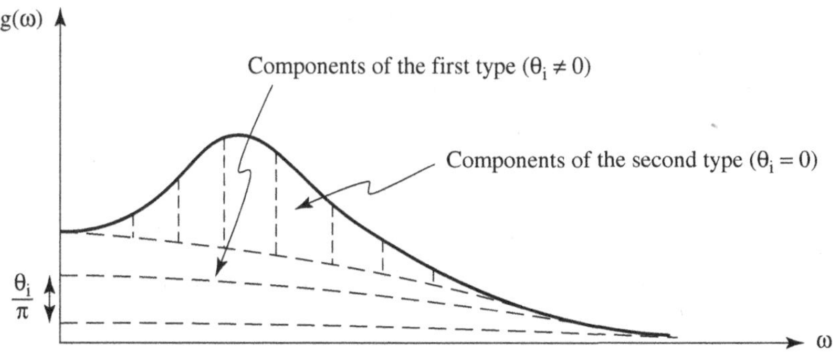

Fig. 5.13 Decomposition of the unit-area spectral density function.

T^{-b}, where $b < 1$. In other words, the presence of a slowly varying "random trend" makes it appear (to an impatient, data-starved observer) that the process is not following the expected $b = 1$ decay law for the variance function. Many natural phenomena exhibit this kind of behavior. Mandelbrot [88; 89] suggested that some of these can be modeled as *fractional noise*, which has a spectral density function proportional to ω^{-1} (see Fig. 5.14). Mandelbrot calls fractional noise "self-similar" because patterns of fluctuation appear similar regardless of the (time or distance) scale on which they are observed. For example, a wind speed record at a fixed location might appear similar when observed on a scale of seconds, minutes, hours, or years.

Ideal fractional noise, having an infinite scale of fluctuation (since $g(0) \to \infty$) as well as $\sigma^2 = \infty$ and $\delta = 1$ (see Eq. (4.7.11)), is physically unrealizable. Of course, there are limits to real observation times (or distances) as well as to the time (or distance) interval between observations. This implies that there exists a frequency range below and above which an ideal fractional noise model is unsupported by real data. The point is that phenomena expected to obey a fractional noise model can also be represented by a composite random process, comprised of (wide-band) components of the first type, as illustrated in Fig. 5.14. Such a model provides a powerful yet simple tool for stochastic modeling of multi-scale phenomena. Moreover, the concepts and techniques of composite modeling can be readily extended to multi-dimensional situations.

Scale Spectrum

For a composite random process $X(t) = \sum X_i(t)$, as defined in Eq. (5.9.1), it is useful to introduce the *scale spectrum* $q(\theta)$ which indicates how the variance of $X(t)$ is distributed among the components $X_i(t)$ having different scales of fluctuation θ_i. We may write:

$$q(\theta) = \sum_i q_i \, \delta_i(\theta - \theta_i), \tag{5.9.12}$$

where $\delta_i(\theta - \theta_i)$ is a "delta function" centered at $\theta = \theta_i$. Note that the composite scale θ, expressed by Eq. (5.9.8), corresponds to the "mean value" of the scale of fluctuation,

$$\theta = \overline{\theta} = \sum_i \theta_i \, q(\theta). \tag{5.9.13}$$

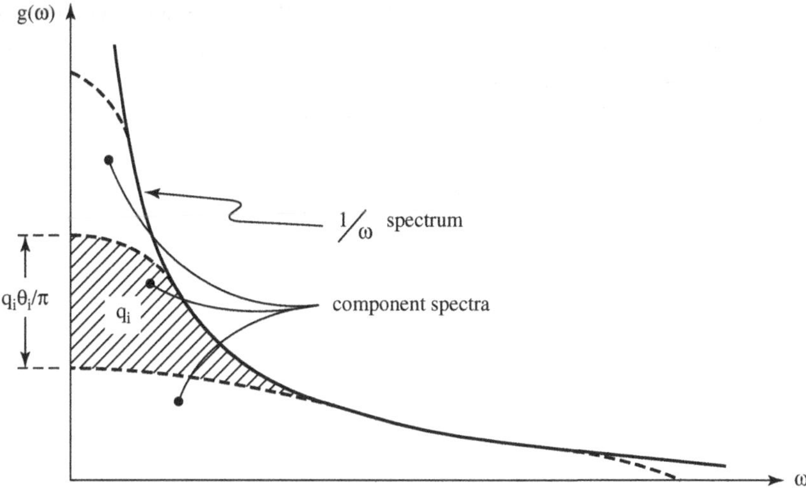

Fig. 5.14 Fractional noise approximated by a sum of uncorrelated component processes.

The spread or dispersion of the scale spectrum can be measured by the "variance" of the scale of fluctuation,

$$\mathrm{Var}[\theta] = \sum_i (\theta_i - \overline{\theta})^2 \, q(\theta), \tag{5.9.14}$$

and the corresponding "coefficient of variation" is

$$\text{c.o.v. of } \theta = \sqrt{\mathrm{Var}[\theta]}/\overline{\theta}. \tag{5.9.15}$$

This quantity is zero for *single-scale* wide-band random functions; in this case the characterization of $X(t)$ in terms of its statistics m, σ^2 and θ is clear and most useful. By contrast, for multi-scale random processes consisting of many components with widely varying scales θ_i, the width and shape of the "scale spectrum" $q(\theta)$, rather that the individual scales θ_i, will dominate the behavior (i.e., patterns of fluctuation) of the composite random process. Of particular interest is the case when the composite variance is equally distributed among components $X_i(t)$, $i = 1, 2, \ldots, M$, with steadily increasing scales θ_i (obeying, for instance, either an arithmetic- or geometric-growth law: $\theta_{i+1} = \theta_i + \Delta\theta$ or $\theta_{i+1} = \theta_i \exp\{\Delta\theta\}$). The key parameters of the scale spectrum, in this case, will depend on the limits (θ_1 and θ_M) of the range of scales, and on the increment $\Delta\theta$ or the (continuous-equivalent) scale-spectrum density. The writer applied the concept of the scale spectrum to model (aspects of) the energy density fluctuations arising during the inflation phase of the Big Bang [140]. The scale spectrum, in this case, is constant over a vast range of scales; the component scales follow the geometric-growth law; and the corresponding spectral density function is that of band-limited fractional ("$1/f$") noise.

Chapter 6

Two-Dimensional Local Average Processes

One of the brightest gems in the New England weather is the dazzling uncertainty of it.

Mark Twain

6.1 Variance Function and Measure of Correlation

Consider the random field $I_A(t_1, t_2)$ derived from a homogeneous two-dimensional random field $X(t_1, t_2)$ by local integration over a rectangular area $A = T_1T_2$ (see Fig. 6.1):

$$I_A(t_1, t_2) \equiv I_{T_1T_2}(t_1, t_2) = \int_{t_1-T_1/2}^{t_1+T_1/2} \int_{t_2-T_2/2}^{t_2+T_2/2} X(t_1, t_2)\, dt_1\, dt_2. \qquad (6.1.1)$$

The rectangular window is centered at (t_1, t_2), and its sides (having length T_1 and T_2) remain parallel to the respective coordinate axes. Dividing $I_A(t_1, t_2)$ by the area A yields the 2-D random process of local averages:

$$X_A(t_1, t_2) \equiv X_{T_1T_2}(t_1, t_2) = \frac{1}{A} I_A(t_1, t_2). \qquad (6.1.2)$$

To simplify the notation, we will generally omit reference to the parameters t_1 and t_2. The ratio of the variances of X_A and X is by definition $\gamma(U_1, U_2)$, the variance function of $X(t_1, t_2)$. We may write

$$\text{Var}\,[X_A] \equiv \sigma_A^2 \equiv \sigma_{T_1T_2}^2 = \sigma^2\gamma(T_1, T_2). \qquad (6.1.3)$$

Similarly, $\Delta(T_1, T_2)$ is the variance function associated with the local integral process, defined by the relation:

$$\text{Var}\,[I_{T_1T_2}] = \sigma^2\Delta(T_1, T_2), \qquad (6.1.4)$$

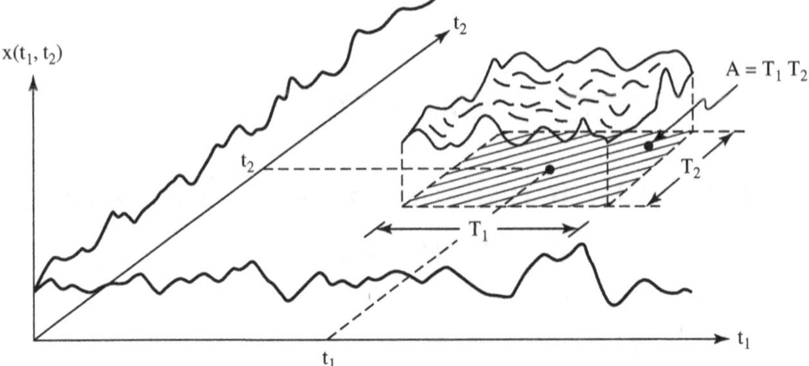

Fig. 6.1 Local averaging over a rectangular area $A = T_1T_2$ in the two-dimensional parameter space (t_1, t_2).

so the two variance functions differ only by the factor $A^2 = (T_1T_2)^2$:

$$\Delta(T_1, T_2) = (T_1T_2)^2 \gamma(T_1, T_2). \tag{6.1.5}$$

The relationship between $\rho(\tau_1, \tau_2)$ and $\gamma(T_1, T_2)$ follows from Eq. (3.6.47):

$$\gamma(T_1, T_2) = \frac{1}{T_1T_2} \int_{-T_1}^{+T_1} \int_{-T_2}^{+T_2} \left(1 - \frac{|\tau_1|}{T_1}\right)\left(1 - \frac{|\tau_2|}{T_2}\right) \rho(\tau_1, \tau_2)\, d\tau_1\, d\tau_2. \tag{6.1.6}$$

By expanding the product in the above integrand, $\gamma(T_1, T_2)$ can be expressed as a sum of four terms, one of which becomes dominant when T_1 and T_2 increase, *provided* the correlation function $\rho(\tau_1, \tau_2)$ decays sufficiently rapidly. (The exact conditions are examined in the next section.) When both T_1 and T_2 grow large, the variance function tends toward the asymptotic expression

$$\gamma(T_1, T_2) \to \frac{\alpha}{T_1T_2} = \frac{\alpha}{A}, \quad T_1,\ T_2 \text{ large}, \tag{6.1.7}$$

where the proportionality constant α is a "characteristic area" equal to the integral of $\rho(\tau_1, \tau_2)$:

$$\alpha = \int_{-\infty}^{+\infty} \int_{-\infty}^{+\infty} \rho(\tau_1, \tau_2)\, d\tau_1\, d\tau_2. \tag{6.1.8}$$

Table 6.1 Parallel interpretations of the correlation parameters θ and α.

Correlation distance θ	Correlation area α
$\lim_{T\to\infty} T\gamma(T)$	$\lim_{T_1,T_2\to\infty} T_1T_2\gamma(T_1,T_2)$
$\int_{-\infty}^{+\infty}\rho(\tau)\,d\tau$	$\int_{-\infty}^{+\infty}\int_{-\infty}^{+\infty}\rho(\tau_1,\tau_2)\,d\tau_1\,d\tau_2$
$2\pi s(0)$	$4\pi^2 s(0,0)$

The quantity α also has a simple interpretation in the frequency domain. Taking $\omega_1 = \omega_2 = 0$ in the Wiener-Khinchine relation , Eq. (3.3.9), yields

$$s(0,0) = \frac{1}{(2\pi)^2}\int_{-\infty}^{+\infty}\int_{-\infty}^{+\infty}\rho(\tau_1,\tau_2)\,d\tau_1\,d\tau_2. \tag{6.1.9}$$

Comparing Eqs. (6.1.8) and (6.1.9) leads to

$$\alpha = 4\pi^2 s(0,0). \tag{6.1.10}$$

Table 6.1 shows the alternate expressions for the characteristic area α in terms of the variance function [Eq. (6.1.7)], the correlation function [Eq. (6.1.8)] and the normalized spectral density function [Eq. (6.1.9)], parallelling those for the scale of fluctuation θ. Note that α cannot be negative, since it is proportional to the s.d.f. (evaluated at $\omega_1 = \omega_2 = 0$).

Behavior of $s(\omega_1, \omega_2)$ Near the Frequency Origin

The validity of the asymptotic expression for the variance function [Eq. (6.1.7)],

$$\gamma(T_1,T_2) \to \frac{\alpha}{T_1T_2}, \quad \text{as} \quad T_1,\ T_2 \to \infty,$$

is subject to certain conditions on the "moments" of the correlation function. As in the one-dimensional case, analysis in the frequency domain permits these conditions to be identified. First, consider how Eq. (6.1.7) can be derived by frequency domain analysis. From Eq. (3.6.43), the relationship between $s(\omega_1, \omega_2)$ and the variance function $\gamma(T_1, T_2)$ is

$$\gamma(T_1,T_2) = \int_{-\infty}^{+\infty}\int_{-\infty}^{+\infty}\left[\frac{\sin(\omega_1 T_1/2)}{\omega_1 T_1/2}\right]^2\left[\frac{\sin(\omega_2 T_2/2)}{\omega_2 T_2/2}\right]^2 s(\omega_1,\omega_2)\,d\omega_1\,d\omega_2. \tag{6.1.11}$$

The function $[(\sin z)/z]^2$ converges toward one when $z \to 0$. Therefore, Eq. (6.1.11) predicts, as expected, that there is no variance reduction when $T_i \to 0$ $(i = 1, 2)$:

$$\gamma(0,0) = \int_{-\infty}^{+\infty}\int_{-\infty}^{+\infty} s(\omega_1,\omega_2)\,d\omega_1\,d\omega_2 = 1. \tag{6.1.12}$$

At the opposite extreme when $T_i \to \infty$ $(i = 1,\ 2)$ in Eq. (6.1.11), assuming the function $s(\omega_1, \omega_2)$ varies smoothly near $\omega_1 = \omega_2 = 0$, it may be replaced by $s(0,0)$. By introducing a change of variables, $v_i = \omega_i T_i/(2\pi)$ $(i = 1, 2)$, Eq. (6.1.11) becomes

$$\gamma(T_1,T_2) = s(0,0)\,\frac{2\pi}{T_1}\frac{2\pi}{T_2}\int_{-\infty}^{+\infty}\left[\frac{\sin \pi v_1}{\pi v_1}\right]^2 dv_1 \int_{-\infty}^{+\infty}\left[\frac{\sin \pi v_2}{\pi v_2}\right]^2 dv_2. \tag{6.1.13}$$

Each integral on the right side equals one. Hence, using Eq. (6.1.10), we obtain

$$\gamma(T_1,T_2) = \frac{4\pi^2 s(0,0)}{T_1 T_2} = \frac{\alpha}{T_1 T_2} = \frac{\alpha}{A}, \quad T_1,\ T_2 \to \infty. \tag{6.1.14}$$

Implicit in the derivation is the assumption that $s(\omega_1, \omega_2)$ varies smoothly near $\omega_1 = \omega_2 = 0$. To examine this, consider the series expansion of $s(\omega_1, \omega_2)$ near the frequency origin:

$$s(\omega_1,\omega_2) = \frac{\alpha}{4\pi^2} + b_1\omega_1 + b_2\,\omega_2 + \frac{d_1}{2}\omega_1^2 + \frac{d_2}{2}\omega_2^2 + d_{12}\,\omega_1\omega_2 + \cdots, \tag{6.1.15}$$

in which the first term of the series agrees with Eq. (6.1.10). From Eq. (6.1.15), the coefficient b equals the first partial derivative of $s(\omega_1, \omega_2)$ with respect to ω_1, evaluated at the frequency-plane origin,

$$b_1 = \left.\frac{\partial s(\omega_1,\omega_2)}{\partial \omega_1}\right|_{\omega_1=\omega_2=0}. \tag{6.1.16}$$

Expressing $s(\omega_1, \omega_2)$ in terms of $\rho(\tau_1, \tau_2)$ by means of the Wiener-Khinchine relation [Eq. (3.3.7)], taking the partial derivative with respect to ω_1, and letting $\omega_i \to 0$ $(i = 1, 2)$ yields:

$$b_1 = -\frac{1}{(2\pi)^2}\left[\omega_1 \int_{-\infty}^{+\infty}\int_{-\infty}^{+\infty} \tau_1^2 \rho(\tau_1, \tau_2)\, d\tau_1\, d\tau_2 \right.$$
$$\left. + \omega_2 \int_{-\infty}^{+\infty}\int_{-\infty}^{+\infty} \tau_1\tau_2 \rho(\tau_1, \tau_2)\, d\tau_1\, d\tau_2\right]$$
$$= -\frac{1}{(2\pi)^2}[\omega_1 \alpha_2^{(1)} + \omega_2 \alpha_{11}], \tag{6.1.17}$$

in which, in general, the quantities $\alpha_k^{(i)}$ and α_{kl} refer to various *moments* of the 2-D correlation function:

$$\alpha_k^{(i)} = \int_{-\infty}^{+\infty}\int_{-\infty}^{+\infty} \tau_i^k \rho(\tau_1, \tau_2)\, d\tau_1\, d\tau_2, \qquad i = 1, 2,\ k = 1, 2, \ldots,$$
$$\alpha_{kl} = \int_{-\infty}^{+\infty}\int_{-\infty}^{+\infty} \tau_1^k \tau_2^l \rho(\tau_1, \tau_2)\, d\tau_1\, d\tau_2, \quad k,\ l = 1, 2, \ldots. \tag{6.1.18}$$

It follows from Eq. (6.1.17) that $b_1 = 0$, *provided* $|\alpha_2^{(1)}|$ and $|\alpha_{11}|$ are finite. Likewise, we have $b_2 = 0$, *provided* $|\alpha_2^{(2)}| < \infty$ and $|\alpha_{11}| < \infty$.

A similar analysis leads to the following results for the other coefficients in Eq. (6.1.15):

$$d_i = \left.\frac{\partial^2 s(\omega_1, \omega_2)}{\partial \omega_i^2}\right|_{\omega_1=\omega_2=0} = -\frac{\alpha_2^{(i)}}{4\pi^2}, \quad i = 1, 2,$$
$$d_{11} = \left.\frac{\partial^2 s(\omega_1, \omega_2)}{\partial \omega_1 \partial \omega_2}\right|_{\omega_1=\omega_2=0} = -\frac{\alpha_{11}}{4\pi^2}. \tag{6.1.19}$$

In conclusion, *provided the moments of second order of $\rho(\tau_1, \tau_2)$ are finite,* $s(\omega_1, \omega_2)$ will obey the following series expansion near $\omega_1 = \omega_2 = 0$:

$$s(\omega_1, \omega_2) = \frac{\alpha}{4\pi^2} - \frac{1}{8\pi^2}(\alpha_2^{(1)}\omega_1^2 + \alpha_2^{(2)}\omega_2^2 - 2\alpha_{11}\omega_1\omega_2) + \cdots. \tag{6.1.20}$$

An equivalent form of the (necessary and sufficient) conditions for the existence of α in the sense implied by Eq. (6.1.7), is that *the first-order partial derivatives of $s(\omega_1, \omega_2)$ must vanish at the frequency-plane origin.* Eq. (6.1.20) implies that $s(\omega_1, \omega_2)$ has a local extremum at $\omega_1 = \omega_2 = 0$. Moreover, if the second-order moments of $\rho(\tau_1, \tau_2)$ are positive, $s(0, 0)$ will be a local *maximum.*

Unidirectional Random Variation

The two-dimensional random field $X(t_1, t_2)$ generates many direction-dependent one-dimensional random functions, like those describing the variation along lines parallel to one of the coordinate axes, such as $X(t_1)$. Using the subscript $i = 1$ on t, τ, ω, or T to indicate direction, the second-order properties of $X(t_1)$ may be characterized by:

1. The correlation function

$$\rho(\tau_1) = \rho(\tau_1, 0). \tag{6.1.21}$$

2. The unit-area spectral density function

$$s(\omega_1) = \int_0^{\infty} s(\omega_1, \omega_2)\, d\omega_2. \tag{6.1.22}$$

3. The variance function

$$\gamma(T_1) = \gamma(T_1, 0). \tag{6.1.23}$$

Of course, similar expressions can be stated for $X(t_2)$. Based on the one-dimensional theory developed in Chap. 5, the *scale of fluctuation* of $X(t_i)$ can be expressed in one of the following ways:

$$\theta^{(i)} = \begin{cases} \displaystyle\int_{-\infty}^{+\infty} \rho(\tau_i)\, d\tau_i, \\ \pi g(\omega_i)|_{\omega_i=0}, & i = 1, 2. \\ \displaystyle\lim_{T_i \to \infty} T_i \gamma(T_i), \end{cases} \tag{6.1.24}$$

Recall that the basic condition for the validity of the one-dimensional theory is that the absolute value of the second moment of $\rho(\tau_i)$ be finite:

$$|\theta_2^{(i)}| < \infty, \qquad i = 1, 2, \tag{6.1.25}$$

where for $k = 1, 2, \ldots,$

$$\theta_k^{(i)} = \int_{-\infty}^{+\infty} \tau_i^k \rho(\tau_i)\, d\tau_i, \qquad i = 1, 2. \tag{6.1.26}$$

The equivalent condition in the frequency domain is that the first-order derivative of $g(\omega_i)$ must vanish at $\omega_i = 0$:

$$[\dot{g}(\omega_i)]_{\omega_i=0} = 0.$$

It is useful to introduce the dimensionless parameter c_α which relates the characteristic area α to the product of the "direction-dependent" scales of fluctuation $\theta^{(1)}$ and $\theta^{(2)}$, as follows:

$$c_\alpha = \frac{\alpha}{\theta^{(1)}\theta^{(2)}}. \tag{6.1.27}$$

The parameter c_α plays an important role in the analysis of 2-D random fields, and is evaluated for some cases in the next section.

6.2 Important Special Cases

The Correlation Structure is Quadrant Symmetric

In applications of two-dimensional random field theory it is useful to choose the coordinate axes t_1 and t_2 so as to render the correlation structure of $X(t_1, t_2)$ *quadrant symmetric* (see Sec. 3.2). It then suffices to deal with the quadrant of positive lags ($\tau_1, \tau_2 \geq 0$) and the quadrant of positive frequencies ($\omega_1, \omega_2 \geq 0$). $G(\omega_1, \omega_2)$ is the s.d.f. defined for positive frequencies only, and $g(\omega_1, \omega_2) = \sigma^{-2} G(\omega_1, \omega_2)$ is the associated normalized (unit-area) spectrum. If $X(t_1, t_2)$ is quadrant symmetric (in the weak sense), the correlation measure α may be expressed as follows:

$$\alpha = \pi^2 g(0,0) = 4 \int_0^\infty \int_0^\infty \rho(\tau_1, \tau_2)\, d\tau_1\, d\tau_2. \tag{6.2.1}$$

Likewise, the moments of the 2-D correlation function become

$$\alpha_k^{(i)} = 4 \int_0^\infty \int_0^\infty \tau_i^k \rho(\tau_1, \tau_2)\, d\tau_1\, d\tau_2, \quad i = 1, 2,\ k = 1, 2, \ldots, \tag{6.2.2}$$

and

$$\alpha_{kl} = 4 \int_0^\infty \int_0^\infty \tau_1^k \tau_2^l \rho(\tau_1, \tau_2)\, d\tau_1\, d\tau_2, \quad k,\ l = 1, 2, \ldots. \tag{6.2.3}$$

A number of important special cases, all satisfying the condition of the quadrant symmetry , are examined next.

The Correlation Structure is Separable

In case the correlation structure is separable, the correlation, spectral density, and variance functions can all be expressed as products of the respective one-dimensional functions:

$$\begin{aligned}\rho(\tau_1, \tau_2) &= \rho(\tau_1)\rho(\tau_2),\\ g(\omega_1, \omega_2) &= g(\omega_1)g(\omega_2), \\ \gamma(T_1, T_2) &= \gamma(T_1)\gamma(T_2).\end{aligned} \tag{6.2.4}$$

The correlation measure α also equals the product of the direction-dependent scales of fluctuation:

$$\alpha = \theta^{(1)}\theta^{(2)}. \tag{6.2.5}$$

Hence, from Eq. (6.1.27),

$$c_\alpha = 1. \tag{6.2.6}$$

Consider below two specific examples of separable correlation functions.

Case 1. The random field is obtained by local averaging of an isotropic purely random (uncorrelated) field, the averaging domain being a rectangle moving parallel to the coordinate axes. Denoting by $2a_1$ and $2a_2$ the sides of the rectangle, the correlation function becomes

$$\rho(\tau_1, \tau_2) = \left(1 - \frac{|\tau_1|}{a_1}\right)\left(1 - \frac{|\tau_2|}{a_2}\right), \quad |\tau_1| \leq a_1, \ |\tau_2| \leq a_2. \tag{6.2.7}$$

The direction-dependent scales of fluctuation,

$$\theta^{(i)} = a_i, \qquad i = 1, 2, \tag{6.2.8}$$

characterize the 1-D correlation functions $\rho(\tau_1, 0)$ and $\rho(0, \tau_2)$. Eq. (6.2.7) provides a 2-D extension of the 1-D triangular correlation function [see Eq. (5.1.7)]. Note that the derived random field is anisotropic, even when $a_1 = a_2$.

Case 2. The correlation function is given by

$$\rho(\tau_1, \tau_2) = \exp\left\{-\left(\frac{\tau_1}{b_1}\right)^2 - \left(\frac{\tau_2}{b_2}\right)^2\right\}, \tag{6.2.9}$$

an extension of Eq. (5.1.13). The univariate scales of fluctuation are

$$\theta^{(i)} = \sqrt{\pi}\, b_i, \qquad i = 1, 2. \tag{6.2.10}$$

If $b_1 = b_2$, the correlation structure becomes isotropic as well as separable.

Isotropic Correlation Structure

If the random field is isotropic, its correlation structure can be expressed in terms of the "radial" correlation function

$$\rho^R(\tau) = \rho(\tau, 0) = \rho(0, \tau) = \rho(\tau_1, \tau_2), \tag{6.2.11}$$

where $\tau = \sqrt{\tau_1^2 + \tau_2^2}$. The characteristic area α becomes

$$\alpha = 2\pi \int_0^\infty \tau \rho^R(\tau)\, d\tau = \pi\, \theta_1^R, \tag{6.2.12}$$

in which θ_k^R denotes the kth moment of $\rho^R(\tau)$:

$$\theta_k^R = 2 \int_0^\infty \tau^k \rho^R(\tau)\, d\tau, \quad k = 0, 1, 2, \ldots; \tag{6.2.13}$$

note that $\theta_0^R \equiv \theta^R$. Since α cannot be negative $[\alpha = \pi^2 g(0,0)]$, it may be concluded from Eq. (6.2.12) that for a correlation function to be acceptable as a model for two-dimensional isotropic random variation, its first moment (for $\tau \geq 0$) must be nonnegative, that is, $\theta_1^R \geq 0$. The scales of fluctuation $\theta^{(i)}$, $i = 1, 2$, do not depend on direction, and we may write

$$\theta^{(1)} = \theta^{(2)} = \theta^R. \tag{6.2.14}$$

Inserting Eqs. (6.2.13) and (6.2.14) into the definition of c_α [Eq. (6.1.27)] yields

$$c_\alpha = \frac{\pi\, \theta_1^R}{(\theta^R)^2}. \tag{6.2.15}$$

Note the relationship $c_\alpha = \pi k_1$, where the dimensionless parameter k_1 is defined in Eq. (5.1.39). Table 6.2 lists the values of c_α for a family of correlation models introduced below, while the parameter $k_1 = c_\alpha/\pi$ is evaluated for four types of correlation functions in Sec. 5.1; all the values of c_α are found to lie between 1 and $\pi/2 \simeq 1.57$.

Ellipsoidal Correlation Structure

By appropriate scaling and rotation of the coordinate axes, random fields with ellipsoidal correlation structure can be made into isotropic random fields. In particular, if the coordinate axes t_1 and t_2 are parallel to the principal axes of the iso-correlation ellipses, the correlation structure of $X(t_1, t_2)$ becomes quadrant symmetric, and the dimensionless parameter

Table 6.2 Coefficient $c_\alpha = \pi k_1$ and scale-related parameter for the family of correlation models defined by Eq. (6.2.16).

m	$c_\alpha = \pi k_1$	$\dfrac{\theta^{(i)}}{(\pi/b_i)}$
$\frac{3}{2}$	$\frac{\pi}{2} \simeq 1.57$	1
2	1.2727	1
3	1.1313	1.5
4	1.0861	1.875
5	1.0639	2.1875
7	1.0421	2.7070
10	1.0277	3.3385
15	1.0176	4.1845

c_α is the same as that of the isotropic process. The value of c_α depends only on the shape of the radial correlation function [Eq. (6.2.15)].

Example. (A Family of Autoregressive Models) It is possible to obtain "closed form" analytical results for an entire family of *autoregressive (Markovian) models* whose spectra are given by

$$G(\omega_1, \omega_2) = G_0 \left[1 + \left(\frac{\omega_1}{b_1} \right)^2 + \left(\frac{\omega_2}{b_2} \right)^2 \right]^{-m}, \quad m > 1. \tag{6.2.16}$$

The case $m = 3/2$ corresponds to the exponential (marginal) correlation function. Integrating over all values of ω_1 (from 0 to ∞) yields the marginal s.d.f.

$$G(\omega_2) = \frac{G_0 b_1}{2} \frac{\Gamma(\frac{1}{2})\Gamma(m - \frac{1}{2})}{\Gamma(m)} \left[1 + \left(\frac{\omega_2}{b_2} \right)^2 \right]^{-m+1/2}. \tag{6.2.17}$$

The s.d.f. $G(\omega_1)$ has the same form. The variance σ^2 is obtained by integrating Eq. (6.2.17) over ω_2, yielding

$$\sigma^2 = \frac{G_0 b_1 b_2}{4} \frac{[\Gamma(\frac{1}{2})]^2 \Gamma(m-1)}{\Gamma(m)} = \frac{\pi b_1\, b_2\, G_0}{4(m-1)}, \quad m > 1, \tag{6.2.18}$$

and the scale of fluctuation $\theta^{(i)}$, $i = 1, 2$, equals

$$\theta^{(i)} = \frac{2\sqrt{\pi}}{b_i}\frac{(m-1)\Gamma(m-\frac{1}{2})}{\Gamma(m)}, \quad m > 1. \tag{6.2.19}$$

If m is an integer, Eq. (6.2.19) may also be written as follows:

$$\theta^{(i)} = \frac{\pi}{b_i}\frac{1\cdot 3\cdot 5\cdots(2m-3)}{(m-2)!\,2^{m-2}}, \quad m = 2, 3, \ldots. \tag{6.2.20}$$

The ratio $(\theta^{(i)}b_i/\pi)$ is given for a set of values of m in the last column of Table 6.2. The 2-D correlation measure is

$$\alpha = \frac{\pi^2 G_0}{\sigma^2} = \frac{4\pi(m-1)}{b_1 b_2}, \quad m > 1, \tag{6.2.21}$$

and combining Eqs. (6.2.19) through (6.2.21) yields

$$c_\alpha = \frac{\alpha}{\theta^{(1)}\theta^{(2)}} = \begin{cases} \dfrac{[\Gamma(m)]^2}{(m-1)\,[\Gamma(m-\frac{1}{2})]^2}, & m > 1, \\[2ex] \dfrac{4}{\pi}\dfrac{(m-1)!\,(m-2)!\,2^{2m-4}}{[1\cdot 3\cdot 5\cdots(2m-3)]^2}, & m = 2, 3, \ldots. \end{cases} \tag{6.2.22}$$

Some representative values of c_α are listed in Table 6.2.

6.3 Conditional Variance Functions and Scales of Fluctuation

Conditional Variance Functions

The operation of averaging a two-dimensional random field $X(t_1, t_2)$ over a rectangular area $A = T_1T_2$ may be carried out in two steps, as illustrated in Fig. 6.2. The first step is to integrate $X(t_1, t_2)$ over the distance T_1 along the t_1-axis. This generates a new one-dimensional random function which is the local average of $X(t_1, t_2)$ within a band of width T_1 parallel to the t_2-axis:

$$X_{T_1}(t_2) = X_{T_1}(t_2; t_1) = \frac{1}{T_1}\int_{t_1-T_1/2}^{t_1+T_1/2} X(t_1, t_2)\, dt_1. \tag{6.3.1}$$

The variance of $X_{T_1}(t_2)$ equals the product of the "point variance" σ^2 and the 1-D variance function $\gamma(T_1)$,

$$\mathrm{Var}\,[X_{T_1}] \equiv \sigma^2\gamma(T_1). \tag{6.3.2}$$

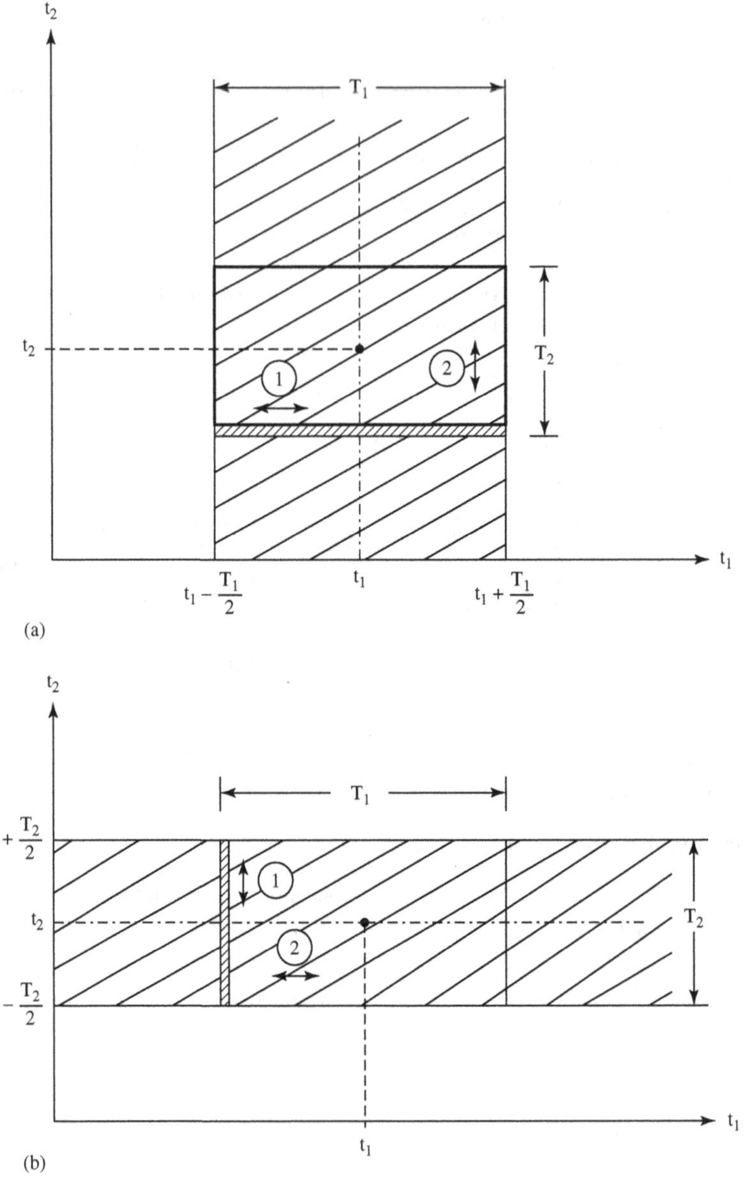

Fig. 6.2 Two ways of averaging $X(t_1, t_2)$ over the rectangular area $A = T_1T_2$: (a) average first with respect to t_1 over a bandwidth T_1, yielding $X_{T_1}(t_2)$, or (b) average first with respect to t_2 over a bandwidth T_2, yielding $X_{T_2}(t_1)$.

The second step is to average $X_{T_1}(t_2)$ over the distance T_2 along the t_2-axis. This leads to further variance reduction, expressed by

$$\sigma_A^2 = \sigma^2 \gamma(T_1)\gamma(T_2|T_1), \tag{6.3.3}$$

where $\gamma(T_2|T_1)$ is the variance function of $X_{T_1}(t_2)$, or the *conditional variance function* of X given prior averaging over the distance T_1 along the t_1-axis. Note that the "bivariate" variance function may be expressed as the product of a "marginal" and a "conditional" variance function:

$$\gamma(T_1, T_2) = \gamma(T_1)\gamma(T_2|T_1). \tag{6.3.4}$$

The two-step procedure may of course be reversed, by averaging first along the t_2-axis and then along the t_1-axis, hence

$$\gamma(T_1, T_2) = \gamma(T_2)\gamma(T_1|T_2) = \gamma(T_1)\gamma(T_2|T_1). \tag{6.3.5}$$

Similar relations exist between the variance functions defined in terms of local integrals:

$$\Delta(T_1, T_2) = \Delta(T_2)\Delta(T_1|T_2) = \Delta(T_1)\Delta(T_2|T_1). \tag{6.3.6}$$

If the correlation between of $X(t_1, t_2)$ is separable, the conditional variance function becomes equal to the "marginal" variance function; for example,

$$\gamma(T_1|T_2) = \gamma(T_1). \tag{6.3.7}$$

Also, the 2-D variance function can then be expressed as

$$\gamma(T_1, T_2) = \gamma(T_1)\gamma(T_2). \tag{6.3.8}$$

Note the striking similarity between Eqs. (6.3.4) through (6.3.8) and the familiar relationships linking the marginal, conditional, and joint probability density functions of two random variables. Furthermore, in this context, the last two equations, applicable when the correlation structure is separable, evoke the concept of statistical independence between two random variables.

Conditional Scales of Fluctuation

The principal correlation parameter of the random process $X_{T_1}(t_2)$ is the conditional scale of fluctuation $\theta_{T_1}^{(2)}$, a characteristic of the (one-dimensional) conditional variance function $\gamma(T_2|T_1)$. If $T_1 = 0$, there is no averaging along the t_1-axis and the conditional scale reduces to $\theta_0^{(2)} \equiv \theta^{(2)}$.

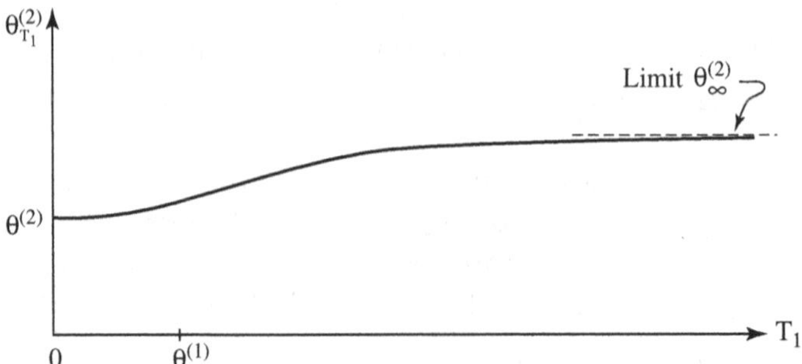

Fig. 6.3 Scale of fluctuation $\theta_{T_1}^{(2)}$ of the derived random process $X_{T_1}(t_2)$ in function of the averaging bandwidth T_1.

As Fig. 6.3 indicates, as T_1 gets larger, $\theta_{T_1}^{(2)}$ tends to increase gradually, converging toward an asymptotic limit $\theta_\infty^{(2)}$. Using Eq. (6.3.4), this limit can be evaluated by letting the intervals T_1 and T_2 become large enough so that each variance function can be replaced its asymptotic form:

$$\frac{\alpha}{T_1 T_2} = \frac{\theta^{(1)}}{T_1}\frac{\theta_{T_1}^{(2)}}{T_2}. \tag{6.3.9}$$

The asymptotic value of $\theta_{T_1}^{(2)}$ becomes, for $T_1 \to \infty$,

$$\lim_{T_1\to\infty} \theta_{T_1}^{(2)} \to \theta_\infty^{(2)} = \frac{\alpha}{\theta^{(1)}} = c_\alpha \theta^{(2)}. \tag{6.3.10}$$

Likewise,

$$\lim_{T_2\to\infty} \theta_{T_2}^{(1)} \to \theta_\infty^{(1)} = \frac{\alpha}{\theta^{(2)}} = c_\alpha \theta^{(1)}, \tag{6.3.11}$$

where α is the volume under the 2-D correlation function, and c_α is the dimensionless factor defined in Eq. (6.1.27), whose typical values range between 1 and $\pi/2$.

In general, the value of $\theta_{T_1}^{(2)}$ can be obtained from Eq. (6.3.5) by taking the limit $T_2 \to \infty$:

$$\gamma(T_1)\frac{\theta_{T_1}^{(2)}}{T_2} = \frac{\theta^{(2)}}{T_2}, \gamma(T_1|T_2 = \infty), \tag{6.3.12}$$

resulting in

$$\theta_{T_1}^{(2)} = \frac{\gamma(T_1|T_2 = \infty)}{\gamma(T_1)} \theta^{(2)}. \tag{6.3.13}$$

The one-dimensional variance function $\gamma(T_1|T_2 = \infty)$ is characterized by the scale $\theta_\infty^{(1)} = \alpha/\theta^{(2)} = c_\alpha \theta^{(1)}$. To see how $\theta_T^{(2)}$ varies with T, consider a specific analytical model for the 1-D variance functions, $\gamma(T_1|T_2 = \infty)$ and $\gamma(T_1)$, appearing in Eq. (6.3.13). Adopting the simplest (idealized) model, given by Eq. (5.1.30), means inserting the expressions

$$\gamma(T_1) = \begin{cases} 1, & T_1 \leq \theta^{(1)}, \\ \dfrac{\theta^{(1)}}{T_1} & T_1 \geq \theta^{(1)}, \end{cases} \tag{6.3.14}$$

and

$$\gamma(T_1|T_2 = \infty) = \begin{cases} 1, & T_1 \leq c_\alpha \theta^{(1)}, \\ \dfrac{c_\alpha \theta^{(1)}}{T_1}, & T_1 \geq c_\alpha \theta^{(1)}, \end{cases} \tag{6.3.15}$$

into Eq. (6.3.13). This results in the following piecewise-linear solution (illustrated in Fig. 6.4) for the conditional scale of fluctuation of $X_{T_1}(t_2)$:

$$\theta_{T_1}^{(2)} = \begin{cases} \theta^{(2)}, & T_1 \leq \theta^{(1)}, \\ \dfrac{T_1 \theta^{(2)}}{\theta^{(1)}}, & \theta^{(1)} \leq T_1 \leq c_\alpha \theta^{(1)}, \\ c_\alpha \theta^{(2)}, & T_1 \geq c_\alpha \theta^{(1)}. \end{cases} \tag{6.3.16}$$

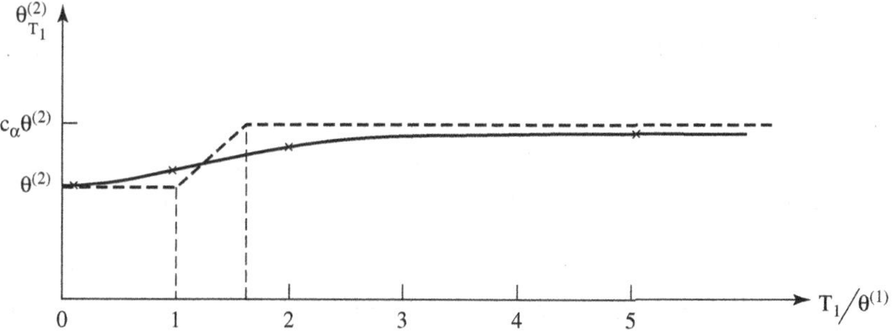

Fig. 6.4 Piecewise-linear approximation of the scale of fluctuation $\theta_{T_1}^{(2)}$. The coefficient c_α equals $\pi/2 \approx 1.57$ for the exponential correlation function (the crosses indicate actual numerical results). The asymptotic upper bound is $\theta_\infty^{(2)} = c_\alpha \theta^{(2)}$.

These values of $\theta_{T_1}^{(2)}$ may now be inserted into that same idealized model for the conditional variance function, namely

$$\gamma(T_2|T_1) = \begin{cases} 1, & T_2 \le \theta_{T_1}^{(2)}, \\ \dfrac{\theta_{T_1}^{(2)}}{T_2}, & T_2 \ge \theta_{T_1}^{(2)}. \end{cases} \tag{6.3.17}$$

Finally, the corresponding approximation for 2-D variance function $\gamma(T_1, T_2)$ is found by multiplying $\gamma(T_1)$ and $\gamma(T_2|T_1)$ (given by Eqs. (6.3.14) and (6.3.17), respectively). Table 6.3 shows the solution, for an isotropic 2-D random field, for different ranges of the averaging intervals T_1 and T_2.

Table 6.3 Piecewise-linear approximation of the 2-D variance function in case the random field $X(t_1, t_2)$ is isotropic.

	$T_1 \le \theta$	$\theta \le T_1 \le c_\alpha \theta$	$T_1 \ge c_\alpha \theta$
$T_2 \le \theta$	1	$\dfrac{\theta}{T_1}$	$\dfrac{\theta}{T_1}$
$\theta \le T_2 \le c_\alpha \theta$	$\dfrac{\theta}{T_2}$	$\dfrac{\theta}{T_2}$ if $T_1 \le T_2$ $\dfrac{\theta}{T_1}$ if $T_1 \ge T_2$	$\dfrac{\theta}{T_1}$
$T_2 \ge c_\alpha \theta$	$\dfrac{\theta}{T_2}$	$\dfrac{\theta}{T_2}$	$\dfrac{\alpha}{T_1 T_2}$

In general, if a single analytical model is imposed on both the conditional and the marginal variance functions, one must expect a slightly different result for $\gamma(T_1, T_2)$ depending on the order of the computation, that is, $\gamma(T_1)\gamma(T_2|T_1)$ versus $\gamma(T_2)\gamma(T_1|T_2)$. To enhance computational accuracy, one might average the results of the two computations, as follows:

$$\gamma(T_1, T_2) = \frac{1}{2}[\gamma(T_1)\gamma(T_2|T_1) + \gamma(T_2)\gamma(T_1|T_2)]. \tag{6.3.18}$$

All the quantities on the right side of Eq. (6.3.18) could be evaluated (approximately) using the same (one-dimensional) analytical model for the variance function, for example [see Eq. (5.1.39)],

$$\gamma(T) = \begin{cases} 1 - k_1 \dfrac{T}{\theta}, & 0 \le T \le \theta, \\ \dfrac{\theta}{T}(1 - k_1 \dfrac{\theta}{T}), & T \ge \theta, \end{cases} \tag{6.3.19}$$

where $k_1 = c_\alpha/\pi$ may be set at 1/3 [the default value corresponding to the triangular correlation function, for which Eq. (6.3.19) is exact]. $\gamma(T_1)$ will depend on $\theta^{(1)}$, $\gamma(T_2|T_1)$ on $\theta^{(2)}_{T_1}$, and so on. The conditional scale $\theta^{(2)}_{T_1}$ itself may be computed by combining Eqs. (6.3.13) and (6.3.19).

More Formal Second-Order Descriptions

For completeness we now derive the equations that relate the conditional variance function $\gamma(T_1|T_2)$ to the other second-order descriptions of the derived process $X_{T_2}(t_1)$, first, the conditional correlation function $\rho(\tau_1|T_2)$ and, second, the conditional spectral density function $g(\omega_1|T_2)$.

Conditional Correlation Function

Starting from Eq. (6.1.6), dividing by $\sigma^2\gamma(T_2)$, and accounting for Eq. (6.3.4) yields the following expression for the conditional variance function of the random process $X_{T_2}(t_1)$:

$$\gamma(T_1|T_2) = \frac{1}{T_1}\int_{-T_1}^{+T_1}\left(1 - \frac{|\tau_1|}{T_1}\right)\rho(\tau_1|T_2)\,d\tau_1, \tag{6.3.20}$$

in which

$$\rho(\tau_1|T_2) = \frac{B(\tau_1|T_2)}{\sigma^2\gamma(T_2)}. \tag{6.3.21}$$

The form of these equations implies that $\rho(\tau_1|T_2)$ and $B(\tau_1|T_2)$ are, respectively, the correlation and covariance functions of $X_{T_2}(t_1)$. The conditional scale of fluctuation is

$$\theta^{(1)}_{T_2} = \int_{-\infty}^{+\infty} \rho(\tau_1|T_2)\,d\tau_1. \tag{6.3.22}$$

The following relations involving $B(\tau_1|T_2)$ are notable:

$$\begin{aligned} B(0|\,T_2) &= \sigma^2\gamma(0, T_2) \equiv \sigma^2\gamma(T_2), \\ B(\tau_1|\,0) &= \sigma^2\rho(\tau_1, 0) \equiv \sigma^2\rho(\tau_1). \end{aligned} \tag{6.3.23}$$

Also, for $T_2 \gg \theta^{(2)}$,

$$\rho(\tau_1|\infty) = \frac{1}{\theta^{(2)}} \int_{-\infty}^{+\infty} \rho(\tau_1, \tau_2)\, d\tau_2. \tag{6.3.24}$$

Finally, combining Eqs. (6.3.22) and (6.3.24) confirms

$$\theta_\infty^{(1)} = \frac{1}{\theta^{(2)}} \int_{-\infty}^{+\infty} \int_{-\infty}^{+\infty} \rho(\tau_1, \tau_2)\, d\tau_1\, d\tau_2 = \frac{\alpha}{\theta^{(2)}}. \tag{6.3.25}$$

Conditional Spectral Density Function

A parallel analysis in the frequency domain starts from Eq (6.1.11); dividing by $\sigma^2\gamma(T_2)$ and accounting for Eq. (6.3.4) yields an expression for the conditional variance function:

$$\gamma(T_1|T_2) = \int_{-\infty}^{+\infty} \left[\frac{\sin(\omega_1 T_1/2)}{\omega_1 T_1/2}\right]^2 s(\omega_1|T_2)\, d\omega_1, \tag{6.3.26}$$

where

$$s(\omega_1|T_2) = \frac{S(\omega_1|T_2)}{\sigma^2\gamma(T_2)}. \tag{6.3.27}$$

The form of these equations implies that $s(\omega_1|T_2)$ is the unit-area spectral density function of $X_{T_2}(t_1)$. The associated scale of fluctuation is

$$\theta_{T_2}^{(1)} = 2\pi\, s(\omega_1|T_2)\big|_{\omega_1=0}, \tag{6.3.28}$$

which for $T_2 \gg \theta^{(2)}$ approaches its upper limit:

$$\theta_\infty^{(1)} = 2\pi\, s(0|\infty) = \frac{4\pi^2 s(0,0)}{\theta^{(2)}} = \frac{\alpha}{\theta^{(2)}}. \tag{6.3.29}$$

We conclude that it is possible to obtain, starting from either $B(\tau_1, \tau_2)$ or $S(\omega_1, \omega_2)$, exact expressions for assorted "conditional" second-order properties, such as for the direction-dependent process $X_{T_2}(t_1)$. The correlation function $\rho(\tau_1|T_2)$ and the unit-area s.d.f. $s(\omega_1|T_2)$ constitute a Fourier transform pair and are related to the conditional variance function $\gamma(T_1|T_2)$ by Eqs. (6.3.20) and (6.3.26), respectively. Note that the operations involving variance functions [Eqs. (6.3.4) through (6.3.6)] have the distinct advantage of requiring only simple algebra. Moreover, they have direct physical meaning in terms of variance reduction under local averaging.

6.4 Covariance of Local Averages

Knowledge of the two-dimensional variance function suffices to evaluate, by means of simple algebra, the covariance of local averages (or integrals) of a homogeneous 2-D random field $X(t_1, t_2)$ over rectangular areas whose sides are parallel to the coordinate axes. Consider the areas $A = T_1 T_2$ and $A' = T_1' T_2'$ shown in Fig. 6.5. By direct extension of Eq. (5.3.5), the covariance of the local integrals I_A and $I_{A'}$ can be expressed as a linear combination of values of the 2-D variance function $\Delta(T_1, T_2) = \sigma^2 \gamma(T_1, T_2)$, as follows:

$$\mathrm{Cov}\,[I_A, I_{A'}] = \frac{\sigma^2}{4} \sum_{k=0}^{3} \sum_{l=0}^{3} (-1)^k (-1)^l \Delta(T_{1k,}, T_{2l}). \tag{6.4.1}$$

The intervals T_{1k} and T_{2l} have the same meaning as in the one-dimensional case. The proof of Eq. (6.4.1) is entirely similar to that of Eq. (5.3.5): it rests on the algebraic identity [Eq. (7.5.3) for $n = 2$] to which the expectation operation is applied. The covariance of the *local averages* $X_A \equiv I_A / A$ and $X_{A'} \equiv I_{A'} / A'$ is

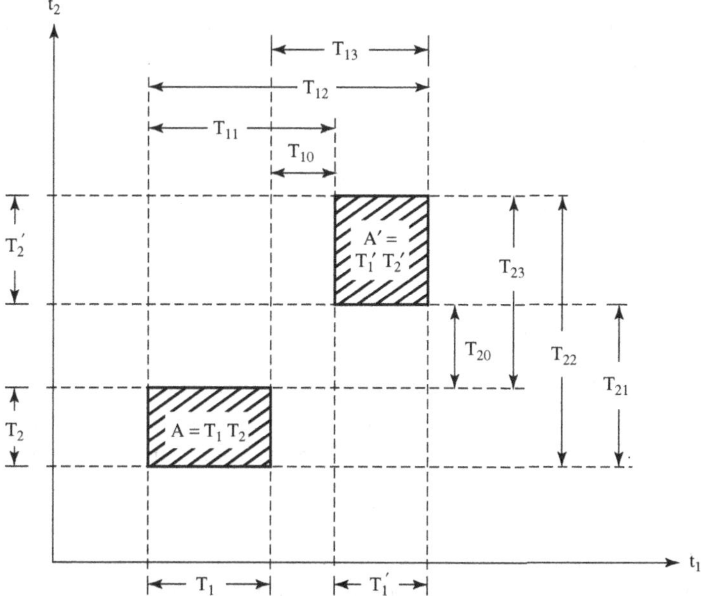

Fig. 6.5 Distances characterizing the relative location of the rectangular areas A and A' in the two-dimensional (orthogonal) parameter space (t_1, t_2).

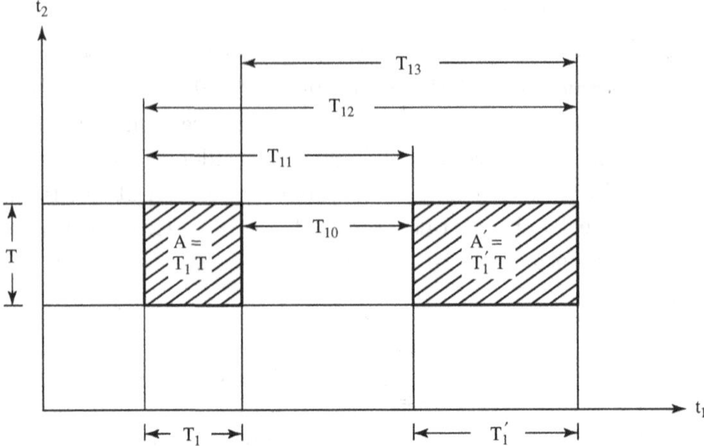

Fig. 6.6 The two averaging areas A and A' share a common dimension.

$$\mathrm{Cov}\,[X_A, X_{A'}] = \frac{1}{(AA')^2}\,\mathrm{Cov}\,[I_A, I_{A'}]. \tag{6.4.2}$$

If the two rectangles are located as shown in Fig. 6.6, namely, if their positions differ only with respect to the coordinate t_2, Eq. (6.4.1) can be simplified to

$$\mathrm{Cov}\,[I_A, I_{A'}] = \frac{\sigma^2 \Delta(T_1)}{2} \sum_{l=0}^{3} (-1)^l \Delta(T_{2l}|T_1). \tag{6.4.3}$$

The same result can be obtained more directly by applying Eq. (5.3.5) to the derived 1-D process $I_{T_1}(t_2)$.

Stochastic Finite Element Analysis

Repeated use of Eq. (6.4.1) permits evaluation of the covariances of "element properties" (i.e., local spatial averages) for each element of a 2-D finite element mesh, as illustrated in Fig. 6.7. Significant computational savings can be realized if all elements are rectangles of equal size and shape. In case the mesh has $m \cdot n$ finite elements, the covariance matrix has $m^2 \cdot n^2$ elements, but only $m \cdot n$ of these differ, and only $m \cdot n$ different values of the 2-D variance function are needed to evaluate them.

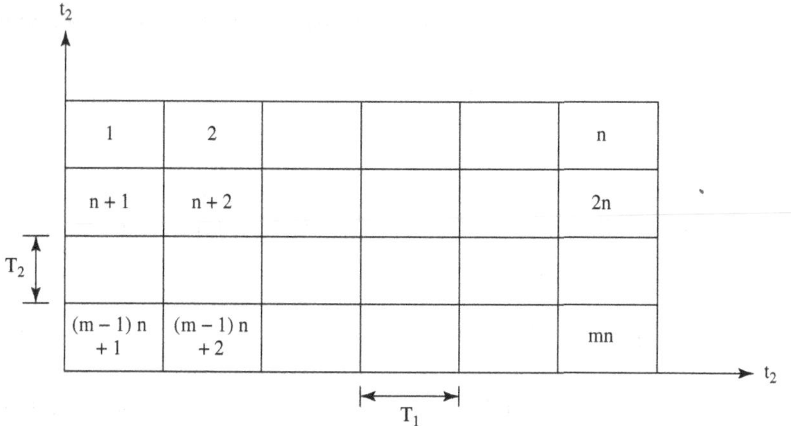

Fig. 6.7 A finite element mesh. In stochastic finite element analysis, key quantities needed are the covariances of the element properties (local averages across each element of the randomly varying property).

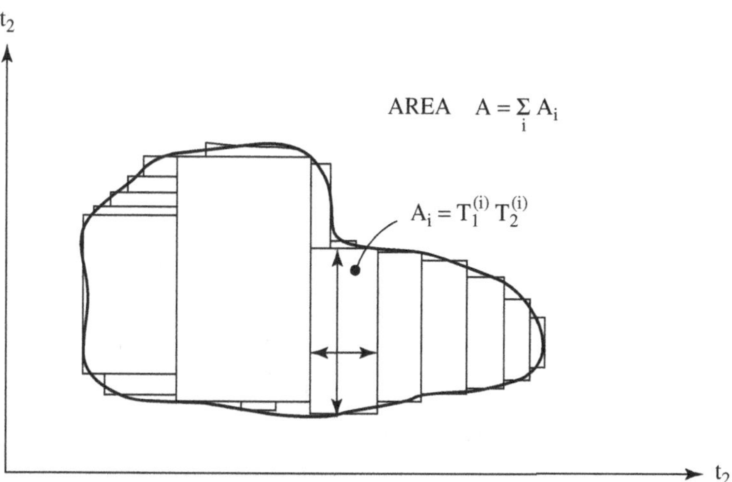

Fig. 6.8 Evaluation of the variance of the integral of a (two-dimensional) random field over an arbitrarily shaped area.

The solution for the covariance of spatial averages over rectangles [Eq. (6.4.1)] serves as a basis for numerical evaluation of: (1) the variances and covariances of spatial averages or local integrals of $X(t_1, t_2)$ over non-rectangular areas; and (2) the variances and covariances of all kinds of "linear transformations" of $X(t_1, t_2)$. Non-rectangular areas may be replaced (to within a prescribed degree of numerical accuracy) by a collection of rectangles, as illustrated in Fig. 6.8. Linear transformations of $X(t_1, t_2)$ may be expressed as follows:

$$Y = \int_A h(t_1, t_2) X(t_1, t_2)\, dt_1\, dt_2 = \sum_{i=1}^{m} h_i X_{A_i}, \tag{6.4.4}$$

where $h(t_1, t_2)$ is a deterministic linear operator. The numerical (finite element) procedure requires that the area A be divided into m nonoverlapping rectangles with areas A_i. Eq. (6.4.4) expresses Y as a linear combination of the vector of spatial averages X_{A_i} whose covariances can be found by means of Eq. (6.4.1). The covariance between two such linear combinations, Y_1 and Y_2, may be obtained in the same way. This mode of *stochastic finite element analysis*, not restricted to linear transformations, is applicable to many types of problems involving (locally homogeneous) 2-D random fields.

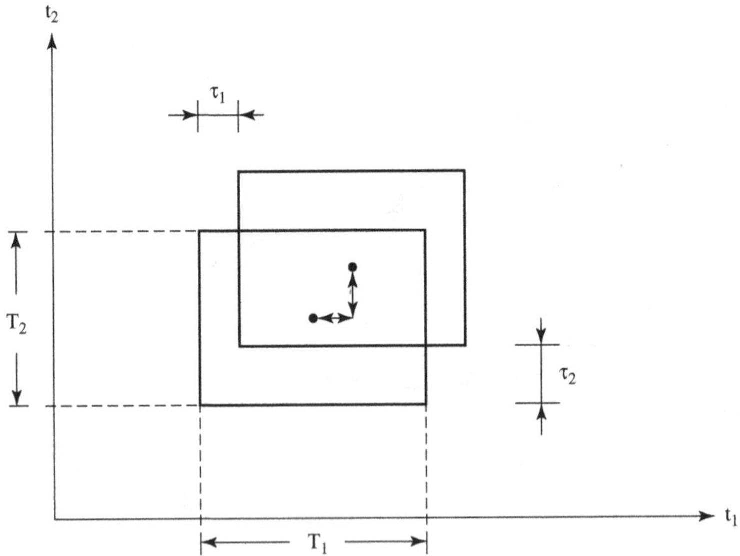

Fig. 6.9 Definition of distances needed to evaluate the covariance function of $X_A(t_1, t_2)$.

Covariance Function of Local Averages

By choosing the intervals T_{1k} and T_{2l} as shown in Fig. 6.9, Eq. (6.4.1) generates an expression for the covariance function of the derived two-dimensional field of local averages $X_A(t_1, t_2)$, where $A = T_1T_2$:

$$\begin{aligned} B_A(\tau_1,\tau_2) &= \frac{\sigma^2}{4T_1^2T_2^2}[\Delta(T_1+\tau_1,T_2+\tau_2)+\Delta(T_1-\tau_1,T_2+\tau_2) \\ &+\Delta(T_1+\tau_1,T_2-\tau_2)+\Delta(T_1-\tau_1,T_2-\tau_2) \\ &-2\Delta(\tau_1,T_2+\tau_2)-2\Delta(\tau_1,T_2-\tau_2)-2\Delta(T_1+\tau_1,\tau_2) \\ &-2\Delta(T_1-\tau_1,\tau_2)+4\Delta(\tau_1,\tau_2)]. \end{aligned} \tag{6.4.5}$$

In the limit for T_1, $T_2 \to 0$, Eq. (6.4.5) yields the inverse relationship between the variance function $\gamma(\tau_1, \tau_2)$ and the covariance function $B(\tau_1, \tau_2)$ of a continuous-parameter homogeneous two-dimensional random field:

$$B(\tau_1,\tau_2) = \frac{\sigma^2}{4}\frac{\partial^4}{\partial\tau_1^2\tau_2^2}\Delta(\tau_1,\tau_2). \tag{6.4.6}$$

The derivation requires re-arranging Eq. (6.4.6) in a form similar to the one-dimensional equivalent, Eq. (5.3.24).

Taking $\tau_1 = 0$ in Eq. (6.4.5) leads to the covariance function of the direction-dependent random process $X_A(t_2)$, as well as its slope variance:

$$\mathrm{Var}\,[\dot{X}_A^{(2)}] = \frac{2\sigma^2\gamma(T_1)}{T_2^2}[1-\rho(T_2|T_1)], \tag{6.4.7}$$

where $\sigma^2\gamma(T_1)$ is the variance of the direction-dependent process $X_{T_1}(t_2)$, while $\rho(\tau_2|T_1)$ is its correlation function evaluated at $\tau_2 = T_2$. Note that Eq. (6.4.7) is a direct extension of Eq. (5.4.11).

Joint Spectral Moment

The most important two-dimensional spectral parameter of the random field $X_A(t_1, t_2)$ is its joint spectral moment

$$\lambda_{11,A} = -\left[\frac{\partial^2 B_A(\tau_1,\tau_2)}{\partial\tau_1\partial\tau_2}\right]_{\tau_1=\tau_2=0}. \tag{6.4.8}$$

We show below that $\lambda_{11,A} = 0$ if the dimensions of the rectangular averaging window differ from zero ($T_1 \neq 0$ and $T_2 \neq 0$), that is, if at least *some* local averaging of $X(t_1, t_2)$ is permitted. $\lambda_{11,A}$ also equals the covariance of the first-order partial derivative $\dot{X}_A^{(1)}$ and $\dot{X}_A^{(2)}$ at a given point (t_1, t_2); and $\lambda_{11,A} = 0$ means that these derivatives are uncorrelated. The proof starts by expressing the series expansion of the two-dimensional covariance function of $X(t_1, t_2)$ near $\tau_1 = \tau_2 = 0$ as follows:

$$B(\tau_1, \tau_2) = \sigma^2 \left[1 + b_1\tau_1 + b_2\tau_2 - d_1\frac{\tau_1^2}{2} - d_2\frac{\tau_2^2}{2} - d_{12}\tau_1\tau_2 + \cdots\right]. \quad (6.4.9)$$

Inserting Eq. (6.4.9) into Eq. (6.1.6) yields the corresponding series for $\gamma(T_1, T_2)$ near $T_1 = T_2 = 0$:

$$\gamma(T_1, T_2) = 1 + \frac{b_1T_1}{3} + \frac{b_2T_2}{3} - \frac{c_1T_1^2}{6} - \frac{c_2T_2^2}{6} - \frac{c_{12}T_1T_2}{9} + \cdots. \quad (6.4.10)$$

A similar analysis in the one-dimensional case [see Eqs. (5.4.2)-(5.4.8)] leads to the conclusion that the term linear in τ vanishes in the corresponding series expansion of $B_T(\tau)$ for $T \neq 0$. Applied to the direction-dependent covariance functions $B_A(\tau_1, 0)$ and $B_A(0, \tau_2)$, this result implies that the terms linear τ_1 and τ_2 in the series expansion of $B_A(\tau_1, \tau_2)$ must also vanish. Furthermore, the constants preceding τ_1^2 and τ_2^2 become the mean square partial derivatives $\mathrm{Var}\,[\dot{X}_A^{(i)}]$, $i = 1, 2$. To see what happens to terms involving $\tau_1\tau_2$, we insert Eq. (6.4.10) into Eq. (6.4.5), accounting $\Delta(T_1, T_2) = T_1^2T_2^2\gamma(T_1, T_2))$, and take the mixed second-order derivative of $B_A(\tau_1, \tau_2)$, as called for on the right side of Eq. (6.4.8). Only the term involving the coefficient d_{12} remains:

$$\frac{\partial^2 B_A(\tau_1, \tau_2)}{\partial\tau_1\partial\tau_2} = \frac{d_{12}\,\sigma^2}{T_1^2\,T_2^2}(\tau_1^2 + 2\tau_1T_1)(\tau_2^2 + 2\tau_2T_2) + \cdots. \quad (6.4.11)$$

If $T_1 \neq 0$ and $T_2 \neq 0$, taking $\tau_1 = \tau_2 = 0$ in Eq. (6.4.11) yields, in combination with Eq. (6.4.8), the desired result:

$$\lambda_{11,A} = 0. \quad (6.4.12)$$

In conclusion, for $T_1 \neq 0$ and $T_2 \neq 0$ the series expansion of the 2-D covariance function of $X_A(t_1, t_2)$ has the following form near $\tau_1 = \tau_2 = 0$:

$$B_A(\tau_1, \tau_2) = \sigma_A^2 - \frac{1}{2}\sum_{i=1}^{n}\tau_i^2\ \mathrm{Var}\,[\dot{X}_A^{(i)}] + \cdots. \quad (6.4.13)$$

The linear terms and the term involving $\tau_1\tau_2$ vanish. For $T_i \to 0$ and $\tau_i \to 0$ ($i = 1, 2$), Eq. (6.4.11) produces the indeterminate result (0/0). In

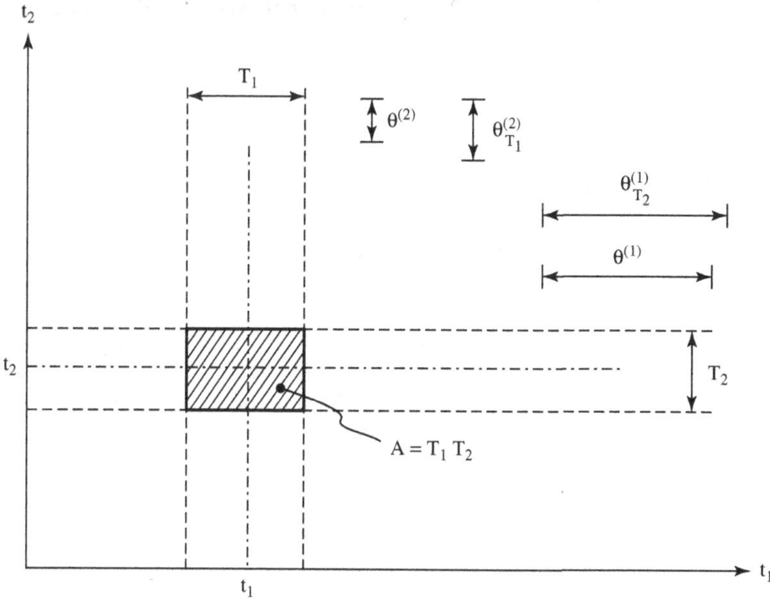

Fig. 6.10 Scales of fluctuation associated with the direction-dependent process $X_{T_1}(t_2)$ and $X_{T_2}(t_1)$.

this case, there is no local averaging, and the joint second-order spectral moment, provided it exists, is given $\lambda_{11} = \lambda_{11,A=0}$.

6.5 Statistics of Level Excursions and Extremes

The theory developed in Chap. 4 for two-dimensional mean square differentiable random processes can now be confidently applied to the random fields $I_A(t_1, t_2)$ and $X_A(t_1, t_2)$ obtained by local aggregation or averaging of the homogeneous random field $X(t_1, t_2)$. These derived random fields will satisfy the conditions for existence of the mean square partial derivatives and (under mild conditions imposed by the Central Limit Theorem) tend to become Gaussian as the size of averaging domain increases.

In this section, interest focuses on excursions of the field $X_A(t_1, t_2)$ above a threshold level b. We first derive some relevant statistics for the direction-dependent one-dimensional processes $X_A(t_1)$ and $X_A(t_2)$, generated by fixing one of the coordinates (see Fig. 6.10).

Level Crossings of Direction-Dependent Local Average Processes

In the same way as the random process $X(t)$ generates the local average process $X_T(t)$, the process $X_{T_2}(t_1)$ gives rise to

$$X_A(t_1) = \frac{1}{T_1}\int_{t_1-T_1/2}^{t_1+T_1/2} X_{T_2}(t_1)\,dt_1, \tag{6.5.1}$$

a single observation of which is a local average of $X(t_1,t_2)$ over the rectangular area $A = T_1T_2$. The variance of $X_A(t_1)$ is

$$\text{Var}\,[X_A^{(1)}] = \text{Var}\,[X_A] = \sigma^2\gamma(T_2)\gamma(T_1|T_2) = \sigma^2\gamma(T_1,T_2), \tag{6.5.2}$$

and, by direct extension of Eq. (5.4.11), its mean square derivative is

$$\begin{aligned}\text{Var}\,[\dot{X}_A^{(1)}] &= \frac{2}{T_1^2}\sigma^2\gamma(T_2)[1-\rho(T_1|T_2)]\\ &= 2\ \text{Var}\,[X_A]\left\{\frac{1-\rho(T_1|T_2)}{\Delta(T_1|T_2)}\right\}.\end{aligned} \tag{6.5.3}$$

If the distances T_1 and T_2 are much larger than their respective scales of fluctuation, $\theta^{(1)}$ and $\theta^{(2)}$, that is, if $T_i \gg \theta^{(i)}$ for $i = 1, 2$, the variances of X_A and the partial derivatives $\dot{X}_A^{(i)}$ $(i = 1, 2)$ become:

$$\text{Var}\,[X_A] = \frac{\sigma^2\alpha}{A}, \qquad \text{and} \tag{6.5.4}$$

$$\text{Var}\,[\dot{X}_A^{(i)}] = \frac{2\ \text{Var}\,[X_A]}{c_\alpha T_i\,\theta^{(i)}}, \quad T_i \gg \theta^{(i)},\ i = 1,2. \tag{6.5.5}$$

Based on the theory presented in Sec. 4.4, we can now evaluate the mean rate of b-upcrossings and the mean lengths of excursion above a given level b by the processes $X_A(t_1)$ and $X_A(t_2)$. In particular, the mean rate of b-upcrossings by the process $X_A(t_i)$, $i = 1, 2$, is

$$\nu_{b,A}^{(i)} = f_A(b)\left(\frac{2}{\pi}\ \text{Var}\,[\dot{X}_A^{(i)}]\right)^{1/2}, \quad i = 1,2, \tag{6.5.6}$$

and the mean length (measured along the t_i axis) of excursions above b by the process $X_A(t_i)$, $i = 1, 2$, is

$$E[\mathcal{T}_{b,A}^{(i)}] = (2\pi\ \text{Var}\,[\dot{X}_A^{(i)}])^{1/2}\frac{F_A^c(b)}{f_A(b)}, \quad i = 1,2, \tag{6.5.7}$$

where $f_A(\cdot)$ and $F_A^c(\cdot)$ denote, respectively, the p.d.f. and the complementary c.d.f. of the local average X_A.

Mean Size of Regions of Excursion above High Levels

For high threshold levels, one pictures the two-dimensional pattern of excursions (regions where $X_A \geq b$) as a collection of isolated patches. As the level b rises, the patches become more isolated and their sizes shrink. The one-dimensional excursion lengths may be crudely interpreted as the dimensions, parallel to the respective coordinate axes, of the area $\mathscr{A}_b \equiv \mathscr{A}_{b,A}$ of each isolated region of excursion. Since the first-order partial derivatives $\dot{X}_A^{(1)}$ and $\dot{X}_A^{(2)}$ at a given point are uncorrelated ($\lambda_{11,A} = 0$), the mean size of isolated excursion regions above a high level b can be expressed approximately as follows:

$$E[\mathscr{A}_{b,A}] \simeq E[\mathscr{T}_{b,A}^{(1)}]E[\mathscr{T}_{b,A}^{(2)}] = \left(\frac{F_A^c(b)}{f_A(b)}\right)^2 \frac{2\pi}{\sigma_{\dot{X}_A}^2}, \tag{6.5.8}$$

where $\sigma_{\dot{X}_A}^2$ is the geometric mean of the mean-square partial derivatives:

$$\sigma_{\dot{X}_A}^2 = (\mathrm{Var}\,[\dot{X}_A^{(1)}]\mathrm{Var}\,[\dot{X}_A^{(2)}])^{1/2}, \tag{6.5.9}$$

which can also be expressed as in Eq. (6.5.3).

Similar results, paralleling Eq. (4.5.8), can be obtained for the mean area of excursion above b for the "envelope" field $R_A(t_1, t_2)$ (associated with $X_A(t_1, t_2)$):

$$E[\mathscr{A}_{b,R_A}] \simeq \left(\frac{F_{R_A}^c(b)}{f_{R_A}(b)}\right)^2 \frac{2\pi}{\sigma_{\dot{X}_A}|\mathbf{\Delta}_A|^{1/2}}, \tag{6.5.10}$$

in which the determinant of $\mathbf{\Delta}_A$ depends on the bandwidth factors $\delta_A^{(1)}$ and $\delta_A^{(2)}$ of the direction-dependent processes $X_A(t_i)$, $i = 1, 2$.

Mean Frequency of Local Maxima above High Levels

In the same way as the pattern of excursions along any line parallel to a coordinate axis approaches a (one-dimensional) Poisson process, the pattern of high-level excursions in the plane must tend toward a 2-D Poisson process characterized by $\mu_{b,A}$, the mean number of excursions per unit area. From

Eq. (4.6.2), replacing X by X_A and taking $|\mathbf{V}_A| = 1$, we obtain

$$\mu_{b,A} \simeq \frac{1}{2\pi}[f_A(b)]^2[F_A^c(b)]^{-1}\sigma^2_{\dot{X}_A}. \tag{6.5.11}$$

The mean rate of excursions above a high level b by the corresponding envelope process $R_A \equiv R_A(t_1, t_2)$ follows from Eq. (4.6.4):

$$\mu_{b,R_A} = \frac{1}{2\pi}[f_{R_A}(b)]^2[F_{R_A}^c(b)]^{-1}\sigma_{\dot{X}_A}|\mathbf{\Delta}_A|^{1/2}. \tag{6.5.12}$$

These estimates for the 2-D excursion statistics can now be used to approximate the probability distribution of the maximum value of $X_A(t_1, t_2)$ within an area of given size, in the manner outlined in Sec. 4.6. The general solution for the multi-dimensional case is presented in Sec. 7.7.

6.6 Invariants for 2-D Homogeneous Random Fields

Invariance under Linear Transformation

Assume that a homogeneous random field $X(t_1, t_2)$ is linearly transformed into a new homogeneous random field $Y(t_1, t_2)$. The linear system is characterized by the impulse response function $h(t_1, t_2)$ or the frequency response function $H(\omega_1, \omega_2)$. These two system functions form a Fourier transform pair. In particular, we have

$$H(0,0) = \int_{-\infty}^{+\infty}\int_{-\infty}^{+\infty} h(t_1, t_2)\, dt_1\, dt_2. \tag{6.6.1}$$

Also,

$$S_Y(\omega_1, \omega_2) = |H(\omega_1, \omega_2)|^2 S_X(\omega_1, \omega_2). \tag{6.6.2}$$

The expressions for the mean and the variance of $Y(t_1, t_2)$ are, respectively,

$$m_Y = H(0,0)\, m_X, \tag{6.6.3}$$

and

$$\sigma_Y^2 = \int_{-\infty}^{+\infty}\int_{-\infty}^{+\infty} |H(\omega_1, \omega_2)|^2 S_X(\omega_1, \omega_2)\, d\omega_1\, d\omega_2. \tag{6.6.4}$$

By taking $\omega_1 = \omega_2 = 0$ in Eq. (6.6.2), the respective 2-D correlation parameters α_Y and α_X can be related as follows:

$$\alpha_Y \sigma_Y^2 = |H(0,0)|^2 \alpha_X \sigma_X^2 = \frac{m_Y^2}{m_X^2}\alpha_X \sigma_X^2, \tag{6.6.5}$$

assuming $m_X \neq 0$. The sought-after invariance relation follows:

$$\alpha_Y \left(\frac{\sigma_Y}{m_Y}\right)^2 = \alpha_X \left(\frac{\sigma_X}{m_X}\right)^2. \tag{6.6.6}$$

In case the transformation keeps the mean unchanged (so that $H(0,0) = 1$), or when $m_X = 0$, the invariance relation becomes:

$$\alpha_Y \sigma_Y^2 = \alpha_X \sigma_X^2. \tag{6.6.7}$$

Eqs. (6.6.6) and (6.6.7) express a principle of "conservation of (density of) uncertainty", analogous in form to the 1-D case presented in Sec. 5.6.

Invariance Under Local Spatial Averaging

A special case is $Y \equiv X_A$, when the system function $h(t_1, t_2)$ is a rectangular box covering an area of size $A = T_1T_2$ and having unit volume. It follows from Eq. (6.6.7) that the correlation parameter of the local average process $X_A(t_1, t_2)$ equals

$$\alpha_A = \frac{\alpha\sigma^2}{\sigma_A^2} = \frac{\alpha}{\gamma(T_1, T_2)}. \tag{6.6.8}$$

If the dimensions T_1 and T_2 vanish, $\alpha_A \to \alpha$, whereas for large averaging areas, when $T_i \gg \theta^{(i)}$, $i = 1, 2$, the correlation parameter becomes $\alpha_A \to T_1T_2 = A$. The implication is that repeated local averaging causes the direction-dependent scales of fluctuation $\theta_A^{(i)}$, $i = 1, 2$, to increase gradually; and it also leads to increasing values for the spectral bandwidth measures $\delta_A^{(i)}$ and $\varepsilon_A^{(i)}$, $i = 1, 2$, which approach their respective unit upper bounds. One concludes that local averaging produces these quantifiable effects: (1) reduction of the "point variance"; (2) proportional increase in the correlation measures; and (3) increase in the degree of disorder, as measured by the spectral bandwidth factors.

2-D White Noise

The case of two-dimensional white noise is of special interest. Its spectral density function is the same at all points in the 2-D frequency domain:

$$S(\omega_1, \omega_2) = S_0. \tag{6.6.9}$$

To analyze its behavior, we view it as the limit, for $\omega_1^* \to +\infty$ and $\omega_2^* \to \infty$, of a random field with a "band-limited" 2-D spectral density function,

$$S(\omega_1, \omega_2) = S_0, \quad -\omega_1^* \le \omega_1 \le \omega_1^*, \; -\omega_2^* \le \omega_2 \le \omega_2^*, \tag{6.6.10}$$

whose variance is found by integrating $S(\omega_1, \omega_2)$ over all frequencies:

$$\sigma^2 = 4\, S_0\, \omega_1^* \omega_2^*. \tag{6.6.11}$$

The normalized or unit-volume s.d.f. is

$$s(\omega_1, \omega_2) = \frac{S(\omega_1, \omega_2)}{\sigma^2} = \frac{1}{4\, \omega_1^* \omega_2^*}. \tag{6.6.12}$$

From Eq. (6.1.10), the "correlation area" is

$$\alpha = 4\pi^2 s(0,0) = \frac{\pi^2}{\omega_1^* \omega_2^*}. \tag{6.6.13}$$

The ideal-white-noise characteristics emerge when ω_1^* and ω_2^* are both allowed to increase without limit,

$$\sigma^2 \to \infty, \quad \alpha \to 0, \tag{6.6.14}$$

while keeping the product $\alpha\sigma^2$ constant:

$$\alpha\, \sigma^2 = 4\pi^2 S_0 = \pi^2 G_0, \tag{6.6.15}$$

where $G_0 = 4S_0$ is the spectral density defined for positive frequencies only. Combining Eqs. (6.6.7) and (6.6.15), note that if a field $Y(t_1, t_2)$ represents the homogeneous response of a linear system to zero-mean white noise excitation with spectral intensity S_0, then

$$\alpha_Y\, \sigma_Y^2 = 4\, \pi^2 S_0. \tag{6.6.16}$$

The 2-D correlation parameter of the response field is easily found from the relation $\alpha_Y = 4\pi^2 S_0/\sigma_Y^2$ once the response variance σ_Y^2 is known.

6.7 Space-Time Processes: Frequency-Dependent Scale of Fluctuation

If the two-dimensional homogeneous random field under study is a *space-time process* $X(u, t)$, one may choose to describe the time variation in the frequency domain without similarly transforming the spatial coordinate u. The nature of the temporal variation is often very different from that of the variation in space, especially in applications (e.g., earthquakes, wind, sea waves) involving waves in random media. The basic "mixed transform" relations are presented in Sec. 3.8. Throughout this section and the next, the correlation structure of the space-time process is assumed to be *quadrant symmetric* (q.s.). This facilitates physical interpretation and practical application of the results, and is easily relaxed when the need arises.

Background

In Sec. 3.8 the space-time covariance function $B(\nu, \tau)$ is introduced as the covariance between two observations at points (u, t) and (u', t') separated by the distance $\nu = u - u'$ and the time lag $\tau = t - t'$. By one-step Fourier transformation [Eq. (3.8.5)] the time lag τ is converted into the frequency ω, thus yielding the space-time cross-spectral density function $C(\nu, \omega)$. This function reduces to the familiar point s.d.f. if the locations u and u' coincide, or $\nu = 0$, namely $C(0, \omega) \equiv S(\omega)$. By "normalizing" $C(\nu, \omega)$ with respect to its value at $\nu = 0$, we obtain the *frequency-dependent spatial correlation function*:

$$\rho_\omega(\nu) = \frac{C(\nu, \omega)}{S(\omega)}, \qquad \text{(from 3.8.8)}$$

where ω is made a subscript to emphasize the dependence on the separation distance ν. The function $\rho_\omega(\nu)$ quantifies the degree of spatial correlation associated with individual sinusoidal components of the space-time process $X(u, t)$. The "composite" spatial correlation function $\rho(\nu)$ may be expressed as follows:

$$\rho(\nu) = \int_{-\infty}^{+\infty} \rho_\omega(\nu) s(\omega)\, d\omega \overset{\text{q.s.}}{\underset{\downarrow}{=}} \int_0^{\infty} \rho_\omega(\nu) g(\omega)\, d\omega, \qquad \text{(from 3.8.10)}$$

where $g(\omega)$ is the one-sided unit-area s.d.f. (representing random variation with time at a fixed location). The interpretation is that $\rho(\nu)$ is a weighted combination of the frequency-dependent correlation functions $\rho_\omega(\nu)$, the weights, $g(\omega)\, d\omega$, being the fractional contributions to the variance σ^2.

If the $\nu \to k$ transformation is made in addition to the $\tau \to \omega$ transformation, $C(\nu, \omega)$ generates the two-dimensional s.d.f. $S(k, \omega)$, and $\rho_\omega(\nu)$ the frequency-dependent wave number spectrum $s_\omega(k)$. We may write:

$$s_\omega(k) = \frac{S(k, \omega)}{S(\omega)} = \frac{s(k, \omega)}{s(\omega)}, \qquad \text{(from 3.8.15)}$$

where $s(k,\omega)$ and $s(\omega)$ are normalized with respect to the variance. Also,

$$s(k) = \int_{-\infty}^{+\infty} s_\omega(k)s(\omega)\,d\omega \overset{\text{q.s.}}{\underset{\downarrow}{=}} \int_0^\infty s_\omega(k)g(\omega)\,d\omega, \qquad \text{(from 3.8.17)}$$

where $s(k)$ is the unit-area wave number spectrum (characterizing spatial variation at a point in time).

Frequency-Dependent Spatial Scale of Fluctuation

A two-dimensional space-time process $X(u,t)$, observed at a given instant t_0, becomes a random spatial pattern $X(u,t_0)$ characterized (like any stationary random function) by its mean m, variance σ^2, and scale of fluctuation θ^u. Provided it exists, θ^u may be expressed in terms of the correlation function $\rho(\nu)$, where $\nu = u - u'$, as follows:

$$\theta^u = \int_{-\infty}^{+\infty} \rho(\nu)\,d\nu = 2\int_0^\infty \rho(\nu)\,d\nu. \qquad (6.7.1)$$

Alternately, in terms of the unit-area wave number spectrum $s(k)$, where k is the wave number (or spatial circular frequency), we have

$$\theta^u = 2\pi s(k)|_{k=0}. \qquad (6.7.2)$$

The *frequency-dependent spatial scale of fluctuation* can now be defined, first in terms of $\rho_\omega(\nu)$:

$$\theta^u_\omega = 2\int_0^\infty \rho_\omega(\nu)\,d\nu. \qquad (6.7.3)$$

The basic relationship between the scales θ^u and θ^u_ω is obtained by combining Eqs. (6.7.1), (6.7.3), and (3.8.10):

$$\theta^u = 2\int_0^\infty d\nu \int_0^\infty \rho_\omega(\nu)g(\omega)\,d\omega = \int_0^\infty \theta^u_\omega\, g(\omega)\,d\omega. \qquad (6.7.4)$$

In words, the spatial scale of fluctuation θ^u can be expressed as a weighted combination of the frequency-dependent scales θ^u_ω, the weighting function being the unit-area s.d.f. $g(\omega)$. Also, from Eq. (3.8.19)

$$s_\omega(k) = \frac{1}{\pi}\int_0^\infty \rho_\omega(\nu)\cos\nu k\,d\nu. \qquad (6.7.5)$$

Taking $k = 0$ and introducing the definition of θ^u_ω [Eq. (6.7.3)], we obtain

$$\theta^u_\omega = 2\pi s_\omega(k)|_{k=0}, \qquad (6.7.6)$$

and it is easy to confirm:

$$\theta^u = 2\pi \int_0^\infty s_\omega(0) g(\omega)\, d\omega = \int_0^\infty \theta_\omega^u\, g(\omega)\, d\omega. \tag{6.7.7}$$

Physical Interpretation

The random process $X(u,t)$ may be expressed as a sum of J independent sinusoidal components $X_j(u,t)$ (assuming zero mean, for convenience):

$$X(u,t) = \sum_{j=1}^{J} X_j(u,t) = \sum_{j=1}^{J} C_j(u) \cos[\omega_j t - \Phi_j(u)]. \tag{6.7.8}$$

The nth component in this expansion is a sinusoid with frequency ω_j, random amplitude $C_j(u)$, and random phase angle $\Phi_j(u)$. At a given instant $t = t_0$, Eq. (6.7.8) expresses the pattern of spatial variation, $X(u,t_0)$, as a sum of independent random contributions $X_j(u,t_0)$. The variance of the nth sinusoidal component is

$$E[X_j^2] = \frac{1}{2} E[C_j^2(u)] = G(\omega_j)\Delta\omega = \sigma^2 g(\omega_j)\Delta\omega. \tag{6.7.9}$$

The scale of fluctuation of the component $X_j(u,t_0)$ is $\theta_{\omega_j}^u$, while the composite scale of fluctuation is θ^u. In general, the scale of fluctuation of a sum of independent random functions may be expressed as a linear combination of the component scales of fluctuation. The coefficients in the linear relationship are the fractional contributions each component makes to the total variance. In the case at hand, the total variance is σ^2 and the fractional contribution of the jth component is $g(\omega_j)\Delta\omega$, and the composite scale θ^u is related to the component scales $\theta_{\omega_j}^u$ as follows:

$$\theta^u = \sum_{j=1}^{J} \theta_{\omega_j}^u g(\omega_j)\Delta\omega. \tag{6.7.10}$$

This relationship is equivalent to Eq. (6.7.4) and converges to it when the limits $J \to \infty$ and $\Delta\omega \to 0$ are taken.

Behavior of the Frequency-Dependent Spatial Scale

How exactly does the scale θ_ω^u depend on the frequency ω? Intuitively, since θ_ω^u characterizes components of spatial variation associated with frequency ω, one expects $\theta_\omega^u \propto \omega^{-1}$. However, for $\omega \to 0$, this inverse proportionality law would imply $\theta^u_\omega \to \infty$, which is inconsistent with the formula [Eq. (6.7.4)] linking θ^u_ω and θ^u.

Combining Eqs. (6.7.2), (6.7.3) and (3.8.20) yields the following expression for the ratio of spatial scales:

$$\frac{\theta^u_\omega}{\theta^u} = \left.\frac{s(k,\omega)}{s(\omega)s(k)}\right|_{k=0}, \tag{6.7.11}$$

where $s(k,\omega)$, $s(\omega)$, and $s(k)$ are all normalized spectra. This result permits evaluation of θ^u_ω directly in terms of $s(k,\omega)$. (Recall that the spectra $s(\omega)$ and $s(k)$ can be obtained by partial integration of $s(k,\omega)$.) Examples of the use of Eq. (6.7.11) are presented at the end of this section.

Behavior near the Frequency Origin

To see what happens at $\omega = 0$, recall that the two-dimensional spectrum $s(k,\omega)$ is related to the 2-D correlation measure α as follows:

$$\alpha = \int_{-\infty}^{+\infty}\int_{-\infty}^{+\infty} \rho(\nu,\tau)\,d\nu\,d\tau = (2\pi)^2 s(k,\omega)|_{k=\omega=0}. \tag{6.7.12}$$

Also, the scale of temporal fluctuation of the random process $X(u_0,t)$ is

$$\theta^t = 2\pi s(\omega)|_{\omega=0}. \tag{6.7.13}$$

Hence for $k = \omega = 0$, Eq. (6.7.11) yields

$$\frac{\theta^u_0}{\theta^u} = \left.\frac{s(k,\omega)}{s(k)s(\omega)}\right|_{k=\omega=0} = \frac{\alpha}{\theta^u\,\theta^t} = c_\alpha, \tag{6.7.14}$$

where c_α is the dimensionless 2-D correlation parameter of the space-time process $X(u,t)$. This establishes the following intriguing connection:

$$\lim_{\omega\to 0}\theta^u_\omega = c_\alpha\theta^u \equiv \lim_{T\to\infty}\theta^u_T, \tag{6.7.15}$$

where θ^u_T is the scale of *spatial* fluctuation of the *temporal* average, over the time interval T, of the space-time process $X(u,t)$.

Approximate Functional Form

Further relevant information about the behavior of θ^u_ω at low frequencies is obtainable from the series expansions of $s(k,\omega)$ and $s(\omega)$. From

Eq. (6.1.20), *mutatis mutandis*, we obtain

$$s(k,\omega)|_{k=0} = \frac{\alpha}{(2\pi)^2} - \frac{1}{2(2\pi)^2}\alpha_2^t\omega^2 + \cdots, \tag{6.7.16}$$

where, from Eq. (6.1.18),

$$\alpha_2^t = \int_{-\infty}^{+\infty}\int_{-\infty}^{+\infty} \tau^2 \rho(\nu,\tau)\, d\nu\, d\tau. \tag{6.7.17}$$

Likewise, based on Eq. (5.2.12), the series expansion of $s(\omega)$ is

$$s(\omega) = \frac{\theta^t}{2\pi} - \frac{1}{4\pi}\theta_2^t\omega^2 + \cdots, \tag{6.7.18}$$

where

$$\theta_2^t = \int_{-\infty}^{+\infty} \tau^2 \rho(\tau)\, d\tau. \tag{6.7.19}$$

Combining Eqs. (6.7.11), (6.7.16) and (6.7.18) enables one to express the first few terms of the series expansion for the frequency-dependent spatial scale as follows:

$$\theta_\omega^u = \theta_{\omega=0}^u \left\{1 - \frac{1}{2}\left(\frac{\omega}{\Omega_1^u}\right)^2 + \cdots\right\}, \tag{6.7.20}$$

in which $\theta_{\omega=0}^u = c_\alpha\theta^u$ [see Eq. (6.7.14)] and

$$\Omega_1^u = \left[\frac{\alpha_2^t}{\alpha} - \frac{\theta_2^t}{\theta^t}\right]^{-1/2}, \tag{6.7.21}$$

is a parameter having the dimensions of frequency.

The functional form expressed below, and depicted in Fig. 6.12, for the frequency-dependent spatial scale θ_ω^u is consistent with the information just given, and in particular with the (first few terms in the) above series expansion:

$$\theta_\omega^u = c_\alpha\theta^u \left[1 + \left(\frac{\omega}{\Omega_1^u}\right)^2\right]^{-1/2}, \tag{6.7.22}$$

in which Ω_1^u, given by Eq. (6.7.21), is the "corner frequency" that marks the transition between two modes of functional dependence ($\theta_\omega^u \simeq c_\alpha\theta^u$ and $\theta_\omega^u \propto \omega^{-1}$) of the frequency-dependent scale θ_ω^u. Figure 6.11 indicates that

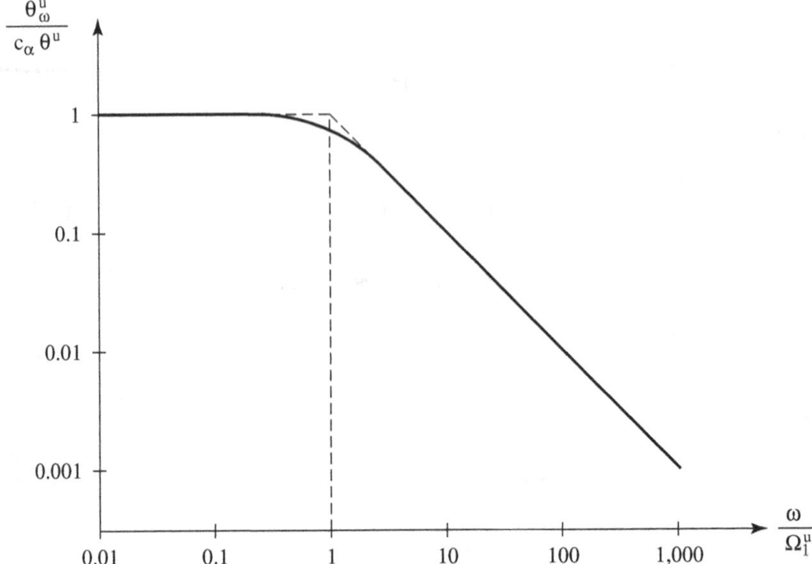

Fig. 6.11 Frequency-dependent scale of fluctuation θ^u_ω of the space-time process $X(u, t)$.

the function is relatively flat until ω reaches Ω^u_1. At higher frequencies, the scale θ^u_ω tends to vary in inverse proportion to ω.

It should be noted that this type of analysis is meaningful only when the 2-D correlation structure is *not* separable. As shown in Case 3 below, the corner frequency Ω^u_1, given by Eq. (6.7.21), becomes infinite when the correlation structure is separable, and in this case $\theta^u_\omega = \theta^u$ for all $\omega \geq 0$. If the correlation structure is not separable, Ω^u_1 is expected to be a low multiple of $1/\theta^t$; a default value, $\Omega^u_1 = 2/\theta^t$, happens to be exact for Case 1 presented below.

Examples of 2-D Space-Time Correlation Structures

Case 1. Exponential Correlation Structure

Consider the exponential correlation function

$$\rho(\nu, \tau) = \exp\{-[(a\nu)^2 + (b\tau)^2]^{1/2}\}, \tag{6.7.23}$$

which corresponds to the 2-D wave number-frequency (k–ω) spectrum

$$s(k,\omega) = \frac{1}{2\,\pi\,a\,b}\left[1+\left(\frac{k}{a}\right)^2+\left(\frac{\omega}{b}\right)^2\right]^{-3/2}. \tag{6.7.24}$$

Integrating over ω yields

$$s(k) = \frac{1}{\pi a}\left[1+\left(\frac{k}{a}\right)^2\right]^{-1}. \tag{6.7.25}$$

Likewise,

$$s(\omega) = \frac{1}{\pi b}\left[1+\left(\frac{\omega}{b}\right)^2\right]^{-1} \equiv \frac{1}{2}g(\omega). \tag{6.7.26}$$

The scales of spatial and temporal fluctuation are, respectively,

$$\theta^u = \frac{2}{a}, \quad \theta^t = \frac{2}{b}, \tag{6.7.27}$$

and the 2-D correlation parameter is

$$\alpha = \frac{2\pi}{a\,b}. \tag{6.7.28}$$

Hence,

$$c_\alpha = \frac{\alpha}{\theta^u\theta^t} = \frac{\pi}{2} \simeq 1.57. \tag{6.7.29}$$

The expression for the frequency-dependent scale of fluctuation is obtained by inserting Eqs. (6.7.24) and (6.7.26) into Eq. (6.7.11). The result is

$$\begin{aligned}\theta^u_\omega &= \left.\frac{2\pi s(k,\omega)}{s(\omega)}\right|_{k=0} = \frac{\pi}{a}\left[1+\left(\frac{\omega}{b}\right)^2\right]^{-1/2}\\ &= c_\alpha\theta^u\left[1+\left(\frac{\omega}{b}\right)^2\right]^{-1/2},\end{aligned} \tag{6.7.30}$$

agreeing exactly with Eq. (6.7.22), with $\Omega^u_1 = b$.

We will now confirm that the result $\Omega^u_1 = b$ can also be obtained directly from Eq. (6.7.21). Since the correlation structure of $X(u,t)$ is ellipsoidal, the parameters needed to evaluate Ω^u_1 can be expressed in terms of the moments of the radial correlation $\rho^R(\tau) = \rho(0,\tau)$ with scale of fluctuation θ^t. In terms of the moments

$$\theta^R_k = 2\int_0^\infty \tau^k\rho^R(\tau)\,d\tau, \tag{6.7.31}$$

the expression Eq. (6.7.21), for the spectral parameter Ω_1^u can be restated as follows:

$$\Omega_1^u = \left[\frac{\theta_3^R}{2\theta_1^R} - \frac{\theta_2^R}{\theta_0^R}\right]^{-1/2}. \tag{6.7.32}$$

In the case at hand, $\rho^R(\tau) = \rho(0,\tau) = \exp\{-b\tau\}$, so that

$$\theta_k^R = 2\int_0^\infty \tau^k \exp\{-b\tau\}\,d\tau = \frac{2k!}{b^{k+1}}, \quad k = 0, 1, 2, \ldots. \tag{6.7.33}$$

Combining Eqs. (6.7.32) and (6.7.33) yields the sought-after confirmation:

$$\Omega_1^u = \left[\frac{6/b^4}{2/b^2} - \frac{2/b^3}{1/b}\right]^{-1/2} = b. \tag{6.7.34}$$

Case 2. Autoregressive Correlation Structure

Consider again the family of autoregressive (Markovian) correlation models examined in Sec. 6.2. To make it fit the notation used to describe 2-D space-time processes, it suffices to replace (ω_1/b_1) and (ω_2/b_2) by (k/a) and (ω/b), respectively, in Eq. (6.2.16):

$$S(k,\omega) = S_0\left[1 + \left(\frac{k}{a}\right)^2 + \left(\frac{\omega}{b}\right)^2\right]^{-m}, \quad m > 1, \tag{6.7.35}$$

where the case $m = 3/2$ corresponds to the 2-D exponential correlation structure just considered. It is easy to show that the frequency-dependent spatial scale θ_ω^u has exactly the same functional form as in Eq. (6.7.30), so that $\Omega_1^u = b$ for this entire family of models.

Case 3. Separable Correlation Structure

When $\rho(\nu,\tau) = \rho(\nu)\rho(\tau)$, it follows from the Wiener-Khinchine relations that the cross-spectral density function $C(\nu,\omega)$ is also separable:

$$C(\nu,\omega) = S(\omega)\rho(\nu). \tag{6.7.36}$$

Hence, based on Eq. (3.8.8),

$$\rho_\omega(\nu) = \rho(\nu), \tag{6.7.37}$$

and the scale θ_ω^u does not depend on frequency, $\theta_\omega^u = \theta^u$. Also $c_\alpha = \alpha/(\theta^u\theta^t) = 1$, and the expression, Eq. (6.7.21), for the corner frequency

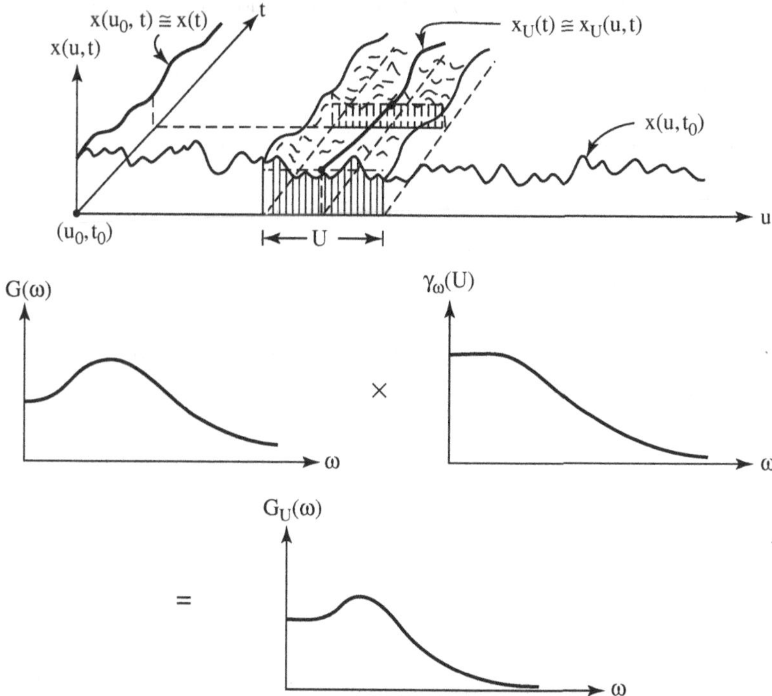

Fig. 6.12 The frequency-dependent variance function $\gamma_\omega(U)$ serves to connect the spectral density function of $X(t)$ and $X_U(t)$. $G_U(\omega)$ equals the product of $G(\omega)$ and $\gamma_\omega(U)$.

yields:

$$\Omega_1^u = \left[\frac{\theta_2^t\,\theta^u}{\theta^t\,\theta^u} - \frac{\theta_2^t}{\theta^t}\right]^{-1/2} = \infty. \tag{6.7.38}$$

Separability implies the lack of any linkage between the patterns of the random variation in time and space.

6.8 Space-Time Processes: Frequency-Dependent Variance Function

In problems involving time-dependent random fields, the analyst may seek information about the spectral content of spatial averages (or integrals) of space-time processes. This section is devoted to the development of a theoretical framework for formulating and solving such problems. The approach is a direct extension of the methodology developed in Chap. 5. The spatial variance function, now permitted to depend on frequency, describes the effect of spatial averaging on components of the "point" variance associated with different narrow bands of frequency. The basic concepts and relations, introduced here for a 2-D space-time process $X(u,t)$, are subsequently extended (in Chap. 7) to time-varying random fields $X(\mathbf{u},t)$, where $\mathbf{u} = (u_1, u_2)$ or $\mathbf{u} = (u_1, u_2, u_3)$. Consider the local spatial average of $X(u,t)$ over a fixed distance U centered at u (see Fig. 6.12):

$$X_U(u,t) = \frac{1}{U}\int_{u-U/2}^{u+U/2} X(u',t)\,du'. \tag{6.8.1}$$

The derived space-time random field $X_U(u,t)$ may be expressed as a sum of sinusoids, in the same way as $X(u,t)$ in Eq. (6.7.8):

$$X_U(u,t) = \sum_{j=1}^{J} X_{U,j}(u,t), \tag{6.8.2}$$

where

$$X_{U,j}(u,t) = \frac{1}{U}\int_{u-U/2}^{u+U/2} X_j(u',t)\,du'. \tag{6.8.3}$$

Clearly, the component processes $X_{U,j}(u,t)$ and $X_j(u,t)$ have frequencies in the same narrow spectral band $\Delta\omega$ centered at ω_j. Also, $X_j(u,t)$ is characterized by the variance $G(\omega_j)\Delta\omega$ and the scale of fluctuation $\theta^u_{\omega_j}$. The variance of the frequency-specific spatial average $X_{U,j}(u,t)$ may be expressed as follows:

$$G_U(\omega_j)\Delta\omega = [G(\omega_j)\Delta\omega]\,\gamma_{\omega_j}(U), \tag{6.8.4}$$

where $\gamma_{\omega_j}(U)$ is by definition the variance function of $X_n(u,t)$. Taking $\omega_j = \omega$, one obtains the frequency-dependent variance function $\gamma_\omega(U)$ that indicates how the "point" spectral density function $G(\omega)$ changes when $X(u,t)$ undergoes local spatial averaging. From Eq. (6.8.4), the point s.d.f. of $X_U(u,t)$ is

$$G_U(\omega) = G(\omega)\,\gamma_\omega(U). \tag{6.8.5}$$

This relationship is illustrated in Fig. 6.12. The variance of $X_U(u,t)$ is found by integrating $G_U(\omega)$ over all frequencies,

$$\sigma_U^2 = \sum_{j=1}^{J} G_U(\omega_n)\Delta\omega \to \int_0^\infty G_U(\omega)\,d\omega = \int_0^\infty G(\omega)\gamma_\omega(U)\,d\omega, \tag{6.8.6}$$

and can also be expressed directly in terms of the variance function $\gamma(U)$ of the composite process,

$$\sigma_U^2 = \sigma^2\gamma(U). \tag{6.8.7}$$

Combining Eqs. (6.8.6) and (6.8.7) yields the important relationship:

$$\gamma(U) = \int_0^\infty \gamma_\omega(U)g(\omega)\,d\omega. \tag{6.8.8}$$

For $U \to \infty$, this equation reduces to the expression linking the overall and frequency-dependent spatial scales, θ^u and θ^u_ω [see Eq. (6.7.4)].

Finally, the variance function $\gamma_\omega(U)$ can be evaluated in the usual way (as for one-dimensional random processes studied in Chap. 5) by adopting a simple (approximate) functional form that depends only on the scale of fluctuation, in this case θ^u_ω.

There are many opportunities for practical application of the concepts and methodology presented in Secs. 6.7 and 6.8, for instance, in the field of stochastic dynamics of structures subjected to wind forces, earthquake ground motions or ocean waves. A sampling of references to applications is found in Sec. 7.6 on stochastic finite element analysis.

Chapter 7

Multi-Dimensional Local Average Processes

Fate, Time, Chance and Change?
— To these All Things are subject.

Shelley, Prometheus Unbound

This chapter generalizes the results obtained in Chaps. 5 and 6 for an n-dimensional homogeneous random field $X(\mathbf{t})$, where $\mathbf{t} = (t_1, \ldots, t_n)$ represents a point in the parameter space. Special attention is devoted to the analysis of "random media," spatial random fields in three dimensions. In most instances it is no more difficult to state the general rather than the specific ($n = 3$) results, but the concepts can more easily be interpreted in the tangible, physical context of three-dimensional space.

An n-dimensional space-time random process is characterized by a vector of $(n - 1)$ spatial coordinates $\mathbf{u}$ and time t. Although it can be analyzed by means of the general multi-dimensional theory, there are significant dividends to an approach that accounts for the special role of the time parameter, for instance, by analyzing the frequency-dependence of the spatial correlation structure, as is done in Secs. 7.3 and 7.4.

7.1 Variance Function and Correlation Measures

Consider a weakly homogeneous n-dimensional random field $X(\mathbf{t})$, where $\mathbf{t} = (t_1, t_2, \ldots, t_n)$, with mean $m = 0$ and variance σ^2, and denote by

$$D = T_1 T_2 \cdots T_n = \prod_{i=1}^{n} T_i \tag{7.1.1}$$

the size of a "domain" in the parameter space over which the field is locally averaged. The domain is a (generalized) rectangular box whose sides are parallel to the coordinate axes, as illustrated in Fig. 7.1 for $n = 3$. The derived random field of spatial averages is

$$X_D(\mathbf{t}) = \frac{1}{D} \int_{\mathbf{0}}^{\mathbf{T}} X(\mathbf{t}) d\mathbf{t}, \tag{7.1.2}$$

where $\mathbf{T} = (T_1, T_2, \ldots, T_n)$ is the vector of averaging dimensions T_i, $i = 1$, $2, \ldots, n$. The mean of $X_D(\mathbf{t})$ is zero and its variance is

$$\sigma_D^2 \equiv \text{Var}[X_D(\mathbf{t})] = \sigma^2 \gamma(\mathbf{T}) \equiv \sigma^2 \gamma(T_1, T_2, \ldots, T_n), \tag{7.1.3}$$

where $\gamma(\mathbf{T}) = \gamma(T_1, T_2, \ldots, T_n)$ is the (multi-dimensional) variance function of $X(\mathbf{t})$. Another form of the variance function,

$$\Delta(\mathbf{T}) = D^2 \gamma(\mathbf{T}), \tag{7.1.4}$$

is associated with the random field of "local integrals," $I_D(\mathbf{t}) = DX_D(\mathbf{t})$, over the same generalized rectangular domain.

Generalized Wiener-Khinchine Relations

The correlation function of $X(\mathbf{t})$ is denoted by $\rho(\boldsymbol{\tau})$ and its normalized spectral density function by $s(\boldsymbol{\omega})$. The generalized Wiener-Khinchine relations link these two functions [Eqs. (3.4.6) and (3.4.7)], as follows:

$$s(\boldsymbol{\omega}) = \frac{1}{(2\pi)^n} \int_{-\infty}^{+\infty} \rho(\boldsymbol{\tau}) \cos \boldsymbol{\omega} \cdot \boldsymbol{\tau} \, d\boldsymbol{\tau}, \tag{7.1.5}$$

$$\rho(\boldsymbol{\tau}) = \int_{-\infty}^{+\infty} s(\boldsymbol{\omega}) \cos \boldsymbol{\omega} \cdot \boldsymbol{\tau} \, d\boldsymbol{\omega}, \tag{7.1.6}$$

where the integrals are over all components of the lag space and the frequency domain, respectively. Taking $\boldsymbol{\tau} = \mathbf{0}$ in Eq. (7.1.6) confirms that $s(\boldsymbol{\omega})$ is a normalized (unit-integral) spectrum:

$$\rho(\mathbf{0}) = \int_{-\infty}^{+\infty} s(\boldsymbol{\omega}) \, d\boldsymbol{\omega} = 1. \tag{7.1.7}$$

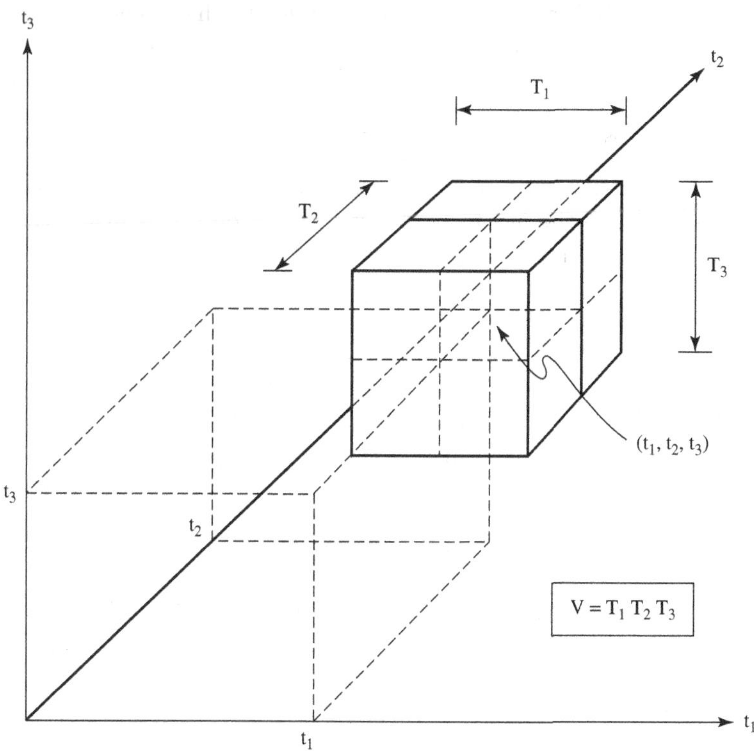

Fig. 7.1 Averaging domain with volume $V = T_1 T_2 T_3$ in the 3-dimensional parameter space (t_1, t_2, t_3).

Integral Correlation Parameters

The correlation parameter η of the random field $X(\mathbf{t})$, provided it exists, may be expressed in three alternate ways, as follows:

$$\eta = \int_{-\infty}^{+\infty} \rho(\boldsymbol{\tau})\, d\boldsymbol{\tau} = (2\pi)^n s(\mathbf{0}) = \lim_{\mathbf{T}\to\infty} D\,\gamma(\mathbf{T}). \tag{7.1.8}$$

The quantity η refers to a general (n-dimensional) field, but will also be used to represent the particular case $n = 4$. If the field has exactly one, two, or three coordinates, the quantity η is instead denoted by:

$$\eta \equiv \begin{cases} \theta, & \text{if } n = 1, \\ \alpha, & \text{if } n = 2, \\ \beta, & \text{if } n = 3. \end{cases} \tag{7.1.9}$$

These same symbols will also be used to refer to the moments of the n-dimensional correlation function (with appropriate addscripts).

Conditions of Existence of the Correlation Measures

The validity of the triad of interpretations in Eq. (7.1.8) is subject to some conditions on the moments of $\rho(\boldsymbol{\tau})$. These conditions insure that the variance function approaches its simple asymptotic form

$$\gamma(\mathbf{T}) = \eta/D \tag{7.1.10}$$

when the dimensions $T_1, T_2, \ldots, T_n$ of the averaging domain become very large. In the frequency domain the basic condition is that the partial first derivatives of $s(\boldsymbol{\omega})$ evaluated at $\boldsymbol{\omega} = \mathbf{0}$ be equal to zero:

$$\left[\frac{\partial s(\boldsymbol{\omega})}{\partial \omega_i}\right]_{\boldsymbol{\omega}=\mathbf{0}} = 0, \quad i = 1, 2, \ldots, n. \tag{7.1.11}$$

The equivalent condition in the space-time domain is that all second-order moments of $\rho(\boldsymbol{\tau})$ be finite, for i, $j = 1, \ldots, n$,

$$|\eta_2^{(i)}| < \infty \quad \text{and} \quad \eta_{11}^{(ij)} < \infty, \tag{7.1.12}$$

where

$$\begin{aligned} \eta_2^{(i)} &= \int_{-\infty}^{+\infty} \tau_i^2 \rho(\boldsymbol{\tau})\, d\boldsymbol{\tau}, \\ \eta_{11}^{(ij)} &= \int_{-\infty}^{+\infty} \tau_i \tau_j \rho(\boldsymbol{\tau})\, d\boldsymbol{\tau}. \end{aligned} \tag{7.1.13}$$

$\eta_{11}^{(ij)} = 0$ if the correlation structure of $X(\mathbf{t})$ is quadrant symmetric with respect to the pair of coordinates, i and j. In general, the series expansion of $s(\boldsymbol{\omega})$ near the frequency origin $\boldsymbol{\omega} = \mathbf{0}$ can be expressed as follows:

$$s(\boldsymbol{\omega}) = \frac{\eta}{(2\pi)^n} - \frac{1}{2(2\pi)^n}\left[\sum_{i=1}^{n} \omega_i^2 \eta_2^{(i)} - 2\sum_{i=1}^{n} \sum_{\substack{j\neq i \\ j=1}}^{n} \omega_i \omega_j \eta_{11}^{(ij)}\right] + \cdots, \tag{7.1.14}$$

which reduces to Eq. (6.1.20) for $n = 2$.

The Isotropic Case

If the field is isotropic, $\rho(\boldsymbol{\tau})$ will depend on the radial lag $\tau = [\sum_i \tau_i^2]^{1/2}$ only, and η can then be expressed in terms of the moments of the radial correlation function $\rho^R(\tau) = \rho(\tau, 0, \ldots, 0)$, as follows:

$$\eta = \begin{cases} 2(2\pi)^{n-1}\theta_{n-1}^R, & n = 2, 3, \ldots, \\ \theta_0^R, & n = 1, \end{cases} \tag{7.1.15}$$

where

$$\theta_k^R = 2\int_0^\infty \tau^k \rho^R(\tau)\, d\tau, \quad k = 0, 1, 2, \ldots. \tag{7.1.16}$$

Eq. (7.1.12), the condition for the validity of Eqs. (7.1.8) and (7.1.14), becomes

$$|\theta_{n+1}^R| < \infty. \tag{7.1.17}$$

The condition $\eta \geq 0$ is a constraint on the validity of correlation models for an n-dimensional isotropic random field. In particular, for $n = 3$, Eq. (7.1.15) implies that θ_2^R must be positive. This further implies that, for $n = 3$, the one-dimensional spectrum $G(\omega)$ must have a local maximum at $\omega = 0$. [see Eq. (5.2.12)], and is consistent with Eq. (3.4.44) which states that $G(\omega)$ must decrease monotonically with frequency to be acceptable as a model for 3-D isotropic random fields.

Lower-Order Correlation Measures

Random fields of lower dimensionality are obtained by keeping one or more of the coordinates of the n-dimensional random field $X(\mathbf{t})$ fixed. Each field so obtained is characterized by its own integral correlation measure. The following cases are of particular interest:

1. The correlation measure that characterizes the random variation of $X(\mathbf{t})$ if only the coordinate t_i is fixed,

$$\eta^{(i)} = \int_{-\infty}^{+\infty} \rho(\boldsymbol{\tau}_i)\, d\boldsymbol{\tau}_i, \tag{7.1.18}$$

where $\boldsymbol{\tau}_i$ is a vector (of size $n-1$) comprising all elements of $\boldsymbol{\tau}$ except τ_i.

2. The parameter that characterizes the correlation structure of $X(\mathbf{t})$ when exactly two coordinates, t_i and t_j, are fixed,

$$\eta^{(ij)} = \int_{-\infty}^{+\infty} \rho(\boldsymbol{\tau}_{ij})\, d\boldsymbol{\tau}_{ij}, \tag{7.1.19}$$

where $\boldsymbol{\tau}_{ij}$ is a vector (of size $n-2$) comprising all elements of $\boldsymbol{\tau}$ except τ_i and τ_j.
3. The scale of fluctuation characterizing one-dimensional variation along lines parallel to the t_i-axis,

$$\theta^{(i)} = \int_{-\infty}^{+\infty} \rho(\tau_i)\, d\tau_i. \tag{7.1.20}$$

4. The correlation measure characterizing two-dimensional random variation on a plane parallel to the axes t_i and t_j,

$$\alpha^{(ij)} = \int_{-\infty}^{+\infty}\int_{-\infty}^{+\infty} \rho(\tau_i, \tau_j)\, d\tau_i\, d\tau_j. \tag{7.1.21}$$

Furthermore, assorted *dimensionless* coefficients may be defined, the most useful of which are

$$c_\eta^{(i)} = \frac{\eta}{\theta^{(i)}\eta^{(i)}}, \quad \text{and} \quad c_\eta = \eta^{n-1}\left[\prod_{i=1}^{n} \eta^{(i)}\right]^{-1}. \tag{7.1.22}$$

For $n = 2$, in terms of the correlation parameters encountered in Chap. 6, we have $\eta \equiv \alpha^{(12)}$, $\eta^{(1)} = \theta^{(2)}$ and $\eta^{(2)} = \theta^{(1)}$. The preceding results will now be specialized for the case $n = 3$.

The Three-Dimensional Case

For $n = 3$, the size D of the n-dimensional domain becomes the volume $V = T_1T_2T_3$ of a rectangular box. (The same role is played by the area A for $n = 2$ and by the interval T for $n = 1$.) The correlation measure η, referred to by β for $n = 3$ [see Eq. (7.1.9)], is given by

$$\eta \equiv \beta = \begin{cases} \displaystyle\int_{-\infty}^{+\infty}\int_{-\infty}^{+\infty}\int_{-\infty}^{+\infty} \rho(\tau_1, \tau_2, \tau_3)\, d\tau_1\, d\tau_2\, d\tau_3, \\ (2\pi)^3 s(0,0,0), \\ \lim V\gamma(T_1, T_2, T_3), \quad \text{when } T_1, T_2, T_3 \to \infty. \end{cases} \tag{7.1.23}$$

Lower-Order Correlation Measures

Some useful lower-order correlation measures are defined next. (For parameters not explicitly mentioned, expressions similar to those given below are obtained by straightforward substitution of indexes.) The scale of fluctuation of one-dimensional random variation parallel to the t_1-axis is

$$\theta^{(1)} \equiv \beta^{(2\,3)} = \int_{-\infty}^{+\infty} \rho(\tau_1, 0, 0)\, d\tau_1 = \int_{-\infty}^{+\infty} \rho(\tau_1)\, d\tau_1. \tag{7.1.24}$$

Two-dimensional random variation in a plane perpendicular to the t_1-axis is characterized by

$$\begin{aligned} \beta^{(1)} \equiv \alpha^{(2\,3)} &= \int_{-\infty}^{+\infty}\int_{-\infty}^{+\infty} \rho(0, \tau_2, \tau_3)\, d\tau_2\, d\tau_3 \\ &= \int_{-\infty}^{+\infty}\int_{-\infty}^{+\infty} \rho(\tau_2, \tau_3)\, d\tau_2\, d\tau_3. \end{aligned} \tag{7.1.25}$$

Also, for $n = 3$ the dimensionless factor defined by Eq. (7.1.22) is

$$c_\beta = \frac{(\beta)^2}{\alpha^{(2\,3)}\alpha^{(1\,2)}\alpha^{(1\,3)}}. \tag{7.1.26}$$

The Isotropic Case

For *isotropic* 3-D fields, $\rho(\tau_1, \tau_2, \tau_3) = \rho^R(\tau)$, where $\tau = (\tau_1^2 + \tau_2^2 + \tau_3^2)^{1/2}$, and the "correlation volume" [see Eq. (7.1.23)] becomes

$$\beta = (2\pi)^2 \int_0^\infty \tau^2 \rho^R(\tau)\, d\tau = 2\pi^2 \theta_2^R, \tag{7.1.27}$$

subject to the necessary and sufficient condition [Eq. (7.1.17)]:

$$|\theta_4^R| < \infty \tag{7.1.28}$$

or [see Eq. (7.1.11)]:

$$\left[\frac{\partial s(\omega_1, \omega_2, \omega_3)}{\partial \omega_1}\right]_{\omega_1=\omega_2=\omega_3=0} = 0. \tag{7.1.29}$$

Only the following two lower-order correlation measures are needed for 3-D isotropic random fields:

$$\begin{aligned} \theta &= \theta^{(i)} \equiv \beta^{(ij)}, \\ \alpha &= \alpha^{(ij)} \equiv \beta^{(i)}, \quad \text{for all } i \text{ and } j. \end{aligned} \tag{7.1.30}$$

Also, the dimensionless parameters defined by Eq. (7.1.22) become $c_\beta^{(i)} = \beta/(\alpha\theta)$, $i = 1, 2, 3$ and $c_\beta = \beta^2/\alpha^3$, respectively.

Separable Correlation Structure

The correlation structure of a 3-D random field is *fully separable* if

$$\rho(\tau_1, \tau_2, \tau_3) = \rho(\tau_1)\rho(\tau_2)\rho(\tau_3). \tag{7.1.31}$$

In this case, we have

$$\eta \equiv \beta = \theta^{(1)}\theta^{(2)}\theta^{(3)}, \tag{7.1.32}$$

and the dimensionless coefficients, like c_β, are all equal to one. Evidently, only three correlation parameters are needed: $\theta^{(i)}$, $i = 1, 2, 3$.

Partial Isotropy or Separability

In some important practical applications, the correlation function may have the following form:

$$\rho(\tau_1, \tau_2, \tau_3) = \rho\left(\sqrt{\tau_1^2 + \tau_2^2}\right)\rho(\tau_3), \tag{7.1.33}$$

indicating the correlation structure of the 3-D random field $X(t_1, t_2, t_3)$ is isotropic with respect to t_1 and t_2, and is also *partially separable.* As mentioned in Sec. 3.1, such a model may be appropriate in case t_1 and t_2 represent geographical coordinates while t_3 denotes time, such as when the rate of rainfall is modeled as a (locally homogeneous) space-time random field. A further useful extension may be to make the (t_1, t_2) correlation structure ellipsoidal by scaling the t_1-axis and t_2-axis differently.

Invariance under Linear Transformation

We now generalize some important and useful results concerning to quantities that remain invariant under linear transformation. They were introduced for one-dimensional random functions in Sec. 5.6 and extended to 2-D random fields in Sec. 6.6. If an n-dimensional homogeneous random field $X(\mathbf{t})$ is transformed linearly into a new such field $Y(\mathbf{t})$, in a way that preserves the mean value, $m_X = m_Y$, then the product $\eta\sigma^2$ will remain invariant:

$$\eta_X \sigma_X^2 = \eta_Y \sigma_Y^2. \tag{7.1.34}$$

More generally, by extension of Eqs. (5.6.6) and (6.6.7), if $m_X \neq m_Y$ the quantity $\eta\sigma^2/m^2$ remains invariant. Local averaging over a (generalized) rectangular domain D is an example of a linear a transformation that preserves the mean, in which case we may write:

$$\eta\sigma^2 = \eta_D \sigma_D^2 = \eta_D \sigma^2 \gamma(\mathbf{T}). \tag{7.1.35}$$

It follows that the n-dimensional correlation measure of the local average process $X_D(\mathbf{t})$ may be expressed as:

$$\eta_D = \eta/\gamma(\mathbf{T}). \tag{7.1.36}$$

For $T_i \gg \theta^{(i)}$, $i = 1, \ldots, n$, we have $\gamma(\mathbf{T}) \to \eta/D$, so that the integral correlation measure η_D of the local average field becomes equal to $D = T_1 T_2 \ldots T_n$, the size of the averaging domain:

$$\eta_D \to D, \quad T_i \gg \theta^{(i)}. \tag{7.1.37}$$

From an n-dimensional homogeneous random field $X(\mathbf{t})$, one can generate numerous fields of lower dimension, by keeping a subset of coordinates fixed, or by averaging or integrating over a limited range of a subset of coordinates. The invariance relations apply to these lower-dimensional random processes as well. For example, if all the coordinates except t_i are held constant, the n-dimensional field $X(\mathbf{t})$ reduces to a one-dimensional process $X(t_i)$ with mean m, variance σ^2, and scale of fluctuation $\theta^{(i)}$. If the process $X(t_i)$ is transformed linearly into $Y(t_i)$ in a way that preserves its mean, the invariance relation becomes:

$$\theta_X^{(i)} \sigma_X^2 = \theta_Y^{(i)} \sigma_Y^2. \tag{7.1.38}$$

If the transformation involves just local averaging over a distance T_i along the ith coordinate axis, Eq. (7.1.36) will reduce to

$$\theta_{T_i}^{(i)} = \theta^{(i)}/\gamma(T_i). \tag{7.1.39}$$

For $T_i \gg \theta^{(i)}$, substituting $\gamma(T_i) = \theta^{(i)}/T_i$ into Eq. (7.1.39) yields

$$\theta_{T_i}^{(i)} \to T_i, \qquad T_i \gg \theta^{(i)}. \tag{7.1.40}$$

As is also expressed by Eq. (7.1.37), information about the details of the correlation structure of the original random field gets lost in the process of long-range local averaging, (i.e., when $T_i \gg \theta^{(i)}$ for $i = 1, 2, \ldots, n$), only the scales of the fluctuation matter.

Composite and Multi-Scale Random Fields

Assume that a homogeneous field can be expressed as the sum of M independent random fields:

$$X(\mathbf{t}) = X_1(\mathbf{t}) + \cdots + X_M(\mathbf{t}). \tag{7.1.41}$$

The mean and variance of $X(\mathbf{t})$ are, respectively,

$$\begin{aligned} m &= m_1 + \cdots + m_M, \\ \sigma^2 &= \sigma_1^2 + \cdots + \sigma_M^2. \end{aligned} \tag{7.1.42}$$

The spectral density, correlation and variance functions of $X(\mathbf{t})$ can all be expressed in terms of the ratios $q_k = \sigma_k^2/\sigma^2$, namely the fractional contributions to the total variance (which are nonnegative and sum to one):

$$\begin{aligned} g(\boldsymbol{\omega}) &= q_1 g_1(\boldsymbol{\omega}) + \cdots + q_M g_M(\boldsymbol{\omega}), \\ \rho(\boldsymbol{\tau}) &= q_1 \rho_1(\boldsymbol{\tau}) + \cdots + q_M \rho_M(\boldsymbol{\tau}), \\ \gamma(\mathbf{T}) &= q_1 \gamma_1(\mathbf{T}) + \cdots + q_M \gamma_M(\mathbf{T}), \end{aligned} \tag{7.1.43}$$

and the various correlation measures (such as $\eta, \theta^{(i)}$) and spectral moments (such as $\lambda_2^{(i)}, \lambda_{kl}^{(ij)}$) can all be similarly expressed:

$$\begin{aligned} \theta^{(i)} &= q_1 \theta_1^{(i)} + \cdots + q_M \theta_M^{(i)}, \quad i = 1, \ldots, n, \\ \eta &= q_1 \eta_1 + \cdots + q_M \eta_M, \\ \lambda_k^{(i)} &= q_1 \lambda_{k,1}^{(i)} + \cdots + q_M \lambda_{k,M}^{(i)}, \quad i = 1, \ldots, n, \quad k = 0, 1, 2, \ldots. \end{aligned} \tag{7.1.44}$$

The local average field $X_D(\mathbf{t})$ also becomes a sum of M independent component fields:

$$X_D(\mathbf{t}) = X_{D_1}(\mathbf{t}) + \cdots + X_{D_M}(\mathbf{t}). \tag{7.1.45}$$

Its variance and mean-square partial derivatives can all be obtained straightforwardly in terms of the composite quantities expressed above.

Scale Spectrum

Composite random field modeling is particularly useful when the component random fields have different scales of fluctuation (and related integral correlation measures). As explained in Sec. 5.9, one can then express by means of the *scale spectrum* each component's contribution to the total variance of the multi-scale random field. Properties of the scale spectrum (a plot of the fractional contributions of each component as a function of its scale fluctuation) are presented in Sec. 5.9.

In multi-scale stochastic modeling of the properties of *random media*, it is often useful to express the overall spatial variation as a sum of statistically independent components, each with its own variance and scale of fluctuation. In the case of a heterogeneous particulate medium such as a granular material, comprised of particles and voids, an efficient stochastic description of the complex microstructure yields an estimate of the smallest scale of fluctuation, expressed in terms of statistics of the void-solid microstructure (Vanmarcke [142]). A modest amount of local spatial averaging – over distances that are low multiples of this scale of fluctuation, typical of test specimens of the material – will greatly reduce the relative size of the microstructure's contribution to the residual variability (of the locally averaged medium). This quantitatively links local-average-based stochastic continuum models to the underlying (discontinuous, two-phased) particulate random medium.

7.2 Conditional Variance Functions and Correlation Measures

The three-dimensional variance function $\gamma(\mathbf{T}) = \gamma(T_1, T_2, T_3)$ may be expressed as the product of three one-dimensional variance functions:

$$\gamma(T_1, T_2, T_3) = \gamma(T_1)\,\gamma(T_2|T_1)\,\gamma(T_3|T_1, T_2). \tag{7.2.1}$$

This equation states in two different ways how much reduction of the point variance σ^2 occurs when the random field $X(\mathbf{t})$ is averaged inside a rectangular box with dimensions T_1, T_2 and T_3 (and volume $V = T_1T_2T_3$). The right side of the equation reflects the sequence of steps in the averaging operation: it implies that the field is first averaged over the distance T_1 along the t_1-axis; the resulting 2-D process, conditioned by T_1, is then averaged over the distance T_2 along the t_2-axis; finally, the derived random function of t_3, conditioned by T_1 and T_2, is averaged over the distance T_3. A similar, yet different, sequence of steps is reflected in

$$\gamma(T_1, T_2, T_3) = \gamma(T_1, T_2)\,\gamma(T_3|T_1, T_2) = \gamma(T_3)\,\gamma(T_1, T_2|T_3). \tag{7.2.2}$$

Eqs. (7.2.1) and (7.2.2) may of course also be stated in terms of variance functions denoted by Δ, which relate to local integrals (or aggregations) instead of local averages.

Each of the variance functions on the right side of Eq. (7.2.1) describes the correlation structure of some 1-D random process, and is therefore characterized by a scale of fluctuation. The various links are depicted below:

Variance function		Random process		Scale
$\gamma(T_3\|T_1, T_2)$	$\leftrightarrow$	$X_{T_1T_2}(t_3)$	$\leftrightarrow$	$\theta^{(3)}_{T_1T_2}$
$\gamma(T_2\|T_1)$	$\leftrightarrow$	$X_{T_1}(t_2)$	$\leftrightarrow$	$\theta^{(2)}_{T_1}$
$\gamma(T_1)$	$\leftrightarrow$	$X(t_1)$	$\leftrightarrow$	$\theta^{(1)}$.

For example, as Fig. 7.2 shows, the random variable $X_{T_1T_2}(t_3^*)$ is obtained by averaging $X(t_1, t_2, t_3)$ over a rectangular area T_1T_2 in the plane (for which $t_3 = t_3^*$) perpendicular to the t_3-axis. The random process $X_{T_1T_2}(t_3)$ is characterized by the variance $\sigma^2\gamma(T_1, T_2)$ and the scale of fluctuation $\theta^{(3)}_{T_1T_2}$ (the superscript refers to the direction of the one-dimensional variation). The corresponding local integral process $I_{T_1T_2}(t_3)$ differs from local average process $X_{T_1T_2}(t_3)$ by the factor $(T_1T_2)^2 = (V/T_3)^2$.

Related Scales of Fluctuation

The question arises as to how the scale of fluctuation $\theta^{(3)}_{T_1T_2}$ depends on the dimensions T_1 and T_2. To arrive at a general solution (in which T_1 and T_2 can take arbitrary positive values), it is instructive to consider a sequence of increasingly complex special cases.

Case 1. $T_1 = T_2 = 0$. There is no local averaging, hence

$$\theta^{(3)}_{T_1T_2} = \theta^{(3)}. \tag{7.2.3}$$

Case 2. $T_1 \gg \theta^{(1)}$, $T_2 \gg \theta^{(2)}$. When $T_i \gg \theta^{(i)}$, $i = 1, 2, 3$, each of the variance functions in the following relationship:

$$\gamma(T_1, T_2, T_3) = \gamma(T_1, T_2)\,\gamma(T_3|T_1, T_2), \tag{7.2.4}$$

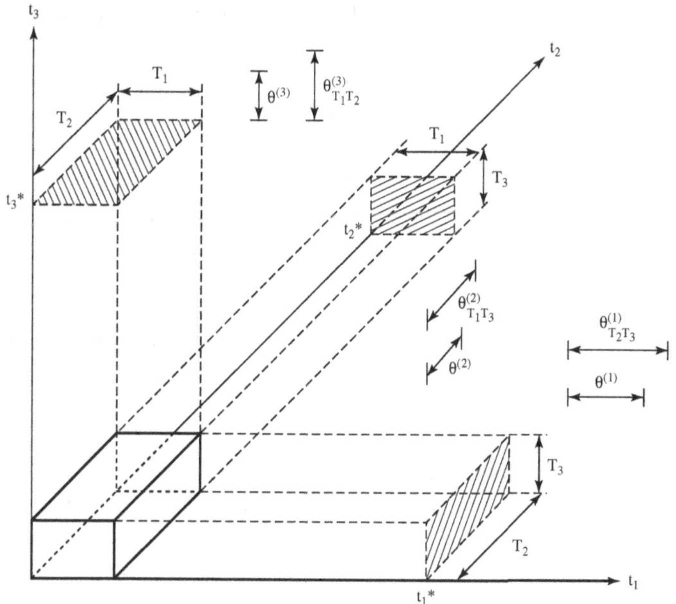

Fig. 7.2 Scales of fluctuation of the direction-dependent random processes; for example, $\theta^{(3)}_{T_1T_2}$ is the scale of the process $X^{(3)}_{T_1T_2}(t_3)$.

may be replaced by its asymtotic form, yielding:

$$\frac{\beta}{T_1\,T_2\,T_3} = \frac{\beta^{(3)}}{T_1T_2}\,\frac{\theta^{(3)}_{T_1T_2}}{T_3}. \tag{7.2.5}$$

Hence, the scale of fluctuation of $X_{T_1T_2}(t_3)$ is given by

$$\theta^{(3)}_{T_1T_2} = \frac{\beta}{\beta^{(3)}} = c^{(3)}_{\beta}\theta^{(3)}. \tag{7.2.6}$$

Case 3. T_1 arbitrary, $T_2 = 0$. First, note that $\theta^{(3)}_{T_1T_2}$ is now identical to $\theta^{(3)}_{T_1}$, the scale of fluctuation representing 1-D variation (parallel to the t_3-axis) in a plane perpendicular to the t_2-axis. From the theory of two-dimensional fields [Eq. (6.3.13)], we have

$$\theta^{(3)}_{T_1T_2} \equiv \theta^{(3)}_{T_1} = \theta^{(3)}\,\frac{\gamma(T_1|T_3=\infty)}{\gamma(T_1)}, \tag{7.2.7}$$

where the variance functions $\gamma(T_1|T_3=\infty)$ and $\gamma(T_1)$ depend on the scales $\alpha^{(1\,3)}/\theta^{(3)}$ and $\theta^{(1)}$, respectively.

Case 4. T_1 arbitrary, $T_2 \gg \theta^{(2)}$. The starting point is again a relationship between variance functions, namely:

$$\gamma(T_2, T_3)\,\gamma(T_1|T_2, T_3) = \gamma(T_2)\,\gamma(T_1|T_2)\,\gamma(T_3|T_1, T_2). \tag{7.2.8}$$

Taking $T_2 \to \infty$ and $T_3 \to \infty$ leads to

$$\theta^{(3)}_{T_1 T_2} = \frac{\beta^{(1)}}{\theta^{(2)}} \frac{\gamma(T_1|T_2 = \infty, T_3 = \infty)}{\gamma(T_1|T_2 \to \infty)}. \tag{7.2.9}$$

The variance function $\gamma(T_1|T_2 = \infty, T_3 = \infty)$ is controlled by the scale $\beta/\beta^{(1)} = c^{(1)}_{\beta}\theta^{(1)}$, and the variance function $\gamma(T_1|T_2 = \infty)$ by the scale $\alpha^{(1\,2)}/\theta^{(2)}$.

Case 5. T_1 and T_2 arbitrary, the general case. Starting again from two alternative expansions of the 3-D variance function, and letting $T_3 \to \infty$, one obtains:

$$\gamma^{(1)}(T_1)\gamma^{(2)}(T_2|T_1)\frac{\theta^{(3)}_{T_1 T_2}}{T_3} = \frac{\theta^{(3)}}{T_3}\,\gamma(T_1|T_3 = \infty)\,\gamma(T_2|T_1, T_3 = \infty). \tag{7.2.10}$$

Hence,

$$\theta^{(3)}_{T_1 T_2} = \theta^{(3)} \frac{\gamma(T_1|T_2 = \infty)}{\gamma(T_1)} \frac{\gamma(T_2|T_1, T_3 = \infty)}{\gamma(T_2|T_1)}. \tag{7.2.11}$$

Accounting for Eq. (7.2.7), we can also write:

$$\theta^{(3)}_{T_1 T_2} = \theta^{(3)}_{T_1} \frac{\gamma(T_2|T_1, T_3 = \infty)}{\gamma(T_2|T_1)}. \tag{7.2.12}$$

Each 1-D variance function in Eqs. (7.2.7) - (7.2.12) depends on the quotient of its argument (either T_1 or T_2) and some conditional scale of fluctuation. In order to evaluate $\theta^{(3)}_{T_1 T_2}$, a specific analytical form must be chosen for the 1-D variance function. Using the simplest model throughout, namely Eq. (6.3.16), yields the piecewise-linear approximation presented in Table 7.1 for the isotropic case. Adopting instead the default model, namely Eq. (5.1.39), for the 1-D variance function throughout, would result in a much better approximation for $\theta^{(3)}_{T_1 T_2}$. Note also that the scale $\theta^{(3)}_{T_1 T_2}$ can of course also be calculated directly from $\rho(\tau_1, \tau_2, \tau_3)$, provided the latter is available; the relevant methodology parallels that presented in Sec. 6.3.

Table 7.1 Piecewise-linear approximation for the scale of fluctuation $\theta^{(3)}_{T_1 T_2}$ in case the random field $X(t_1, t_2, t_3)$ is isotropic.

	Value of T_2			
Value of T_2	$T_1 \leq \theta$	$\theta \leq T_1 \leq \dfrac{\alpha}{\theta}$	$\dfrac{\alpha}{\theta} \leq T_1 \leq \dfrac{\beta}{\alpha}$	$T_1 \geq \dfrac{\beta}{\alpha}$
$T_2 \leq \theta$	θ	T_1	$\dfrac{\alpha}{\theta}$	$\dfrac{\alpha}{\theta}$
$\theta \leq T_2 \leq \dfrac{\alpha}{\theta}$	T_2	T_1 if $T_2 \leq T_1$ T_2 if $T_1 \leq T_2$	$\dfrac{\alpha}{\theta}$	$\dfrac{\alpha}{\theta}$
$\dfrac{\alpha}{\theta} \leq T_2 \leq \dfrac{\beta}{\alpha}$	$\dfrac{\alpha}{\theta}$	$\dfrac{\alpha}{\theta}$	T_2 if $T_2 \leq T_1$ T_1 if $T_1 \geq T_2$	T_2
$T_2 \geq \dfrac{\beta}{\alpha}$	$\dfrac{\alpha}{\theta}$	$\dfrac{\alpha}{\theta}$	T_1	$\dfrac{\beta}{\alpha}$

Generalization for n-Dimensional Fields

In analyzing an n-dimensional homogeneous random field $X(\mathbf{t})$, it is useful to introduce two types of derived local average processes. First, averaging $X(\mathbf{t})$ along the ith coordinate axis over the distance T_i produces a random process having $n - 1$ parameters:

$$X_{T_i}(\mathbf{t}_i) = \frac{1}{T_i} \int_0^{T_1} X(\mathbf{t})\, dt_i, \tag{7.2.13}$$

where $\mathbf{t}_i$ denotes a vector (of size $n-1$) comprising all elements of $\mathbf{t}$ except t_i. The variance of this process is

$$\operatorname{Var}[X_{T_i}] = \sigma^2 \gamma(T_i), \tag{7.2.14}$$

and its variance function is

$$\frac{\operatorname{Var}[X_D]}{\operatorname{Var}[X_{T_i}]} \equiv \gamma(\mathbf{T}_i | T_i), \tag{7.2.15}$$

where $\mathbf{T}_i$ denotes a vector (of size $n-1$) comprising all elements of $\mathbf{T}$ except T_i. The integral correlation measure of the field $X_{T_i}(\mathbf{t}_i)$ is denoted by $\eta^{(i)}_{\mathbf{T}_i}$.

A second derived process results from averaging $X(\mathbf{t})$ along all the coordinate axes except the t_i-axis:

$$X_{\mathbf{T}_i}(t_i) = \frac{T_i}{D} \int_{\mathbf{0}}^{\mathbf{T}_i} X(\mathbf{t})\, d\mathbf{t}_i. \tag{7.2.16}$$

This is a one-dimensional random process whose variance is

$$\mathrm{Var}\,[X_{\mathbf{T}_i}] = \sigma^2\, \gamma(\mathbf{T}_i), \tag{7.2.17}$$

and whose variance function depends on T_i (and is conditioned on $\mathbf{T}_i$):

$$\frac{\mathrm{Var}\,[X_D]}{\mathrm{Var}\,[X_{\mathbf{T}_i}]} \equiv \gamma(T_i|\mathbf{T}_i). \tag{7.2.18}$$

The latter is characterized by the scale of fluctuation $\theta^{(i)}_{\mathbf{T}_i}$.

The second-order properties of the direction-dependent process $X_{\mathbf{T}_i}(t_i)$ can be described not only by the variance function $\gamma(T_i|\mathbf{T}_i)$ but also by the correlation function $\rho(\tau_i|\mathbf{T}_i)$ and the unit-area spectral density function $s(\omega_i|\mathbf{T}_i)$.

The two derived processes [defined by Eqs. (7.2.13) and (7.2.16), respectively] are in a sense complementary. Each process can, through further averaging, generate the n-dimensional derived random field $X_D(\mathbf{t})$, defined in Eq. (7.1.3). The relationship between the variance functions,

$$\gamma(\mathbf{T}) = \gamma(T_i)\, \gamma(\mathbf{T}_i|T_i) = \gamma(\mathbf{T}_i)\, \gamma(T_i|\mathbf{T}_i), \tag{7.2.19}$$

quite clearly indicates the sequence of fractional reductions of the original "point variance" σ^2, leading to $\sigma^2_{\mathbf{D}} = \sigma^2\gamma(\mathbf{T})$. Examples of multidimensional correlation models, and derivations of their parameters, are presented in Sec. 7.4.

Limiting Forms of the Correlation Structure

If the dimensions of a rectangular domain D all exceed their respective scales of fluctuation, *the correlation structure of $X_D(\mathbf{t})$ tends to become separable*, and the integral correlation measure η_D can be expressed as a product of lower-dimension correlation parameters,

$$\eta_D = \prod_{i=1}^{n} \theta^{(i)}_D. \tag{7.2.20}$$

This follows directly from the fact that the derived local average process has a (new) correlation measure $\eta_D \to D$ that equals the size of the rectangular averaging domain $D = T_1T_2 \ldots T_n$. Moreover, if $T_i \gg \theta^{(i)}$, $i = 1, \ldots, n$, the correlation function of the directon-dependent processes $X_D(t_i)$ tends to become triangular, with scale $\theta_D^{(i)} = T_i$ [see Eqs. (5.1.7), (5.1.8) and (5.1.23)], so the multi-dimensional correlation function approaches

$$\rho_D(\boldsymbol{\tau}) = \prod_{i=1}^{n} \rho_D(\tau_i) = \prod_{i=1}^{n} \left(1 - \frac{|\tau_i|}{T_i}\right). \tag{7.2.21}$$

The associated (separable, hence quadrant symmetric) spectral density function is

$$g_D(\boldsymbol{\omega}) = \frac{D}{\pi^n} \prod_{i=1}^{n} \left[\frac{\sin(\omega_i T_i/2)}{\omega_i T_i/2}\right]^2 \equiv 2^n s_D(\boldsymbol{\omega}). \tag{7.2.22}$$

The corresponding multi-dimensional variance function takes the form:

$$\gamma_D(\mathbf{T}') = \prod_{i=1}^{n} \gamma_D(T_i'), \tag{7.2.23}$$

in which each one-dimensional variance function tends toward the triangular correlation function with scale of fluctuation $\theta_D^{(i)} = T_i$, for $i = 1, 2, \ldots, n$, namely:

$$\gamma_D(T_i') = \begin{cases} 1 - \dfrac{T_i'}{3\,T_i}, & T_i' \leq T_i, \\[2ex] \left(\dfrac{T_i}{T_i'}\right)\left(1 - \dfrac{T_i}{3\,T_i'}\right), & T_i' \geq T_i. \end{cases} \tag{7.2.24}$$

Besides characterizing the asymptotically exact form of the correlation structure of $X_D(\mathbf{t})$ (when $\theta^{(i)} \gg T_i$), these relations provide simple approximate models for the correlation, spectral density, and variance functions of separable n-dimensional homogeneous random fields.

It is important to note that the shape of the averaging domain – a generalized rectangle – matters. If the averaging region were a (generalized) sphere with radius R (hence, a circle in 2-D space), the correlation structure of the local average field would become isotropic when $R \gg \theta^{(i)}$, $i = 1, 2, \ldots, n$. In conclusion, for an averaging region whose dimensions are large compared to the field's correlation distances, the region's geometry itself determines the correlation structure of the derived local-average field.

7.3 Frequency-Dependent Spatial Random Variation

A homogeneous n-dimensional space-time process $X(\mathbf{u}, t)$ observed at a given instant t_0 becomes a spatial random pattern $X(\mathbf{u}, t_0)$ in $(n - 1)$ dimensions. If the process is isotropic in space, the spatial correlation structure (when $t = t_0$) is characterized by the one-, two-, and three-dimensional correlation measures θ^u, α^u, and β^u, respectively, while the temporal scale of fluctuation at a given point in space is denoted by θ^t. The space-time processes $X(u, t)$, $X(u_1, u_2, t)$ and $X(u_1, u_2, u_3, t)$ are further characterized by the integral correlation measures α, β, and η, respectively.

The nature of the temporal variation is often very different from that of the spatial variation, and it makes sense in many applications to analyze the time variation in the frequency domain without similarly transforming the spatial coordinates. As in Secs. 6.7 and 6.8, which deal with 2-D space-time processes, it is assumed here that the correlation function $\rho(\boldsymbol{\nu}, \tau)$ (where $\boldsymbol{\nu} = \mathbf{u} - \mathbf{u}'$ and $\tau = t - t'$) is quadrant symmetric. The underlying second-order stochastic theory is presented in Sec. 3.8.

Frequency-Dependent Spatial Correlation Measures

The space-time process $X(\mathbf{u}, t)$ may be expressed as a sum of J independent random components $X_j(\mathbf{u}, t)$ associated with narrow, nonoverlapping intervals on the frequency axis:

$$X(\mathbf{u}, t) = \sum_{j=1}^{J} X_j(\mathbf{u}, t) = \sum_{j=1}^{J} C_j(\mathbf{u}) \cos(\omega_j t - \phi_j(\mathbf{u})). \tag{7.3.1}$$

The component $X_j(\mathbf{u}, t)$, representing contributions from a narrow frequency band $\Delta\omega$ centered at ω_j, has random amplitude $C_j(\mathbf{u})$ and random phase angle $\Phi_j(\mathbf{u})$. At a given instant t_0, Eq. (7.3.1) expresses the spatial pattern $X(\mathbf{u}, t_0)$ as a sum of independent random contributions with component variances

$$E[X_j^2] = \frac{1}{2} E[C_j^2(\mathbf{u})] = G(\omega_j)\Delta\omega = \sigma^2 g(\omega_j)\Delta\omega. \tag{7.3.2}$$

The spatial correlation structure of $X_j(\mathbf{u}, t_0)$, now frequency-dependent, is characterized by the frequency-dependent correlation function $\rho_\omega(\boldsymbol{\nu})$ or the wave number spectrum $s_{\omega_j}(\mathbf{k})$, given by Eqs. (3.8.8) and (3.8.20), respectively. The frequency-dependent spatial correlation measures of $X_j(\mathbf{u}, t_0)$

are defined in much the same way as their composite counterparts, but are now subscripted. Taking $\omega_j = \omega$, we express:

1. the frequency-dependent scale of fluctuation ($n = 2$),

$$\theta^u_\omega = \begin{cases} 2\int_0^\infty \rho_\omega(\nu)\,d\nu, \\ 2\pi s_\omega(k)|_{k=0}, \end{cases} \tag{7.3.3}$$

2. the frequency-dependent areal correlation measure ($n = 3$),

$$\alpha^u_\omega = \begin{cases} \int_{-\infty}^{+\infty}\int_{-\infty}^{+\infty} \rho_\omega(\nu_1,\nu_2)\,d\nu_1\,d\nu_2, \\ (2\pi)^2 s_\omega(k_1,k_2)|_{k_1=k_2=0}, \end{cases} \tag{7.3.4}$$

3. the frequency-dependent volumetric correlation measure ($n = 4$),

$$\beta^u_\omega = \begin{cases} \iint_{-\infty}^{+\infty}\int \rho_\omega(\boldsymbol{\nu})\,d\boldsymbol{\nu}, \\ (2\pi)^3 s_\omega(\mathbf{k})|_{\mathbf{k}=0}, \end{cases} \tag{7.3.5}$$

where $\boldsymbol{\nu} = (\nu_1, \nu_2, \nu_3)$ and $\mathbf{k} = (k_1, k_2, k_3)$.

Each frequency-dependent correlation measure is related to the corresponding composite integral measure, as in Eq. (6.7.4), namely:

$$\theta^u = \sum_{j=1}^{J} \theta^u_{\omega_j} g(\omega_j)\Delta\omega \rightarrow \int_0^\infty \theta^u_\omega g(\omega)\,d\omega. \tag{7.3.6}$$

Likewise,

$$\alpha^u = \sum_{j=1}^{J} \alpha^u_{\omega_j} g(\omega_j)\Delta\omega \rightarrow \int_0^\infty \alpha^u_\omega g(\omega)\,d\omega, \tag{7.3.7}$$

and

$$\beta^u = \sum_{j=1}^{J} \beta^u_{\omega_j} g(\omega_j)\Delta\omega \rightarrow \int_0^\infty \beta^u_\omega g(\omega)\,d\omega. \tag{7.3.8}$$

The question dealt with next is, "how do the spatial correlation measures α^u_ω and β^u_ω depend on ω, the frequency?"

Functional Form of Frequency-Dependent Correlation Measures

Two Spatial Dimensions

The frequency-dependent spatial correlation measures α_ω^u and β_ω^u may be analyzed in much the same way as the frequency-dependent scale θ_ω^u in Sec. 6.7. Consider first the 2-D measure α_ω^u. Combining Eqs. (7.3.4) and (3.8.15), we obtain

$$\alpha_\omega^u = \left.\frac{(2\pi)^2 s(k_1, k_2, \omega)}{s(\omega)}\right|_{k_1=k_2=0}. \tag{7.3.9}$$

Taking $\omega = 0$ yields

$$\lim_{\omega\to 0} \alpha_\omega^u = \frac{\beta}{\theta^t} = c_2^u \alpha^u \equiv \lim_{T\to\infty} \alpha_T^u, \tag{7.3.10}$$

in which $c_2^u = \beta/(\theta^t \alpha^u)$ is a constant defined, *mutatis mutandis*, as in Eq. (7.1.22) [replacing η by β and c_α by c_2^u, with the subscript 2 indicating the number of spatial coordinates], and α_T^u is the spatial correlation measure of the temporal average, over the time interval T, of the space-time process $X(u_1, u_2, t)$. Eq. (7.3.10) is analogous to Eq. (6.8.8).

The analysis in Sec. 6.7 reveals that θ_ω^u will vary in inverse proportion to frequency ω when the latter exceeds the corner frequency Ω_1^u. This information, along with the relation $\alpha^u \propto [\theta^u]^2$, indicates the following asymptotic behavior:

$$\alpha_\omega^u \propto \left(\frac{\Omega_2^u}{\omega}\right)^2, \quad \omega \gg \Omega_2^u, \tag{7.3.11}$$

where Ω_2^u refers to another (yet to be determined) corner frequency. The relationship presented below [paralleling Eq. (6.7.22) with $c_\alpha \equiv c_1^u$] is suggested for the ratio of spatial correlation measures:

$$\frac{\alpha_\omega^u}{\alpha^u} = c_2^u \left[1 + \left(\frac{\omega}{\Omega_2^u}\right)^2\right]^{-1}. \tag{7.3.12}$$

It meets expectations as to how α_ω^u behaves at high frequencies [Eq. (7.3.11)] as well as at $\omega = 0$ [Eq. (7.3.10)]. The value of the corner frequency Ω_2^u can be determined by replacing the spectra in Eq. (7.3.9), the relationship defining α_ω^u, by their series expansions, as follows:

$$\alpha_\omega^u = \frac{\beta - \frac{1}{2}\beta_2^t\omega^2 + \cdots}{\theta^t - \frac{1}{2}\theta_2^t\omega^2 + \cdots} = c_2^u \left\{1 - \left(\frac{\omega}{\Omega_2^u}\right)^2 + \cdots\right\}, \tag{7.3.13}$$

in which

$$\Omega_2^u = \left[\frac{\beta_2^t}{2\beta} - \frac{\theta_2^t}{2\theta^t}\right]^{-1/2}, \tag{7.3.14}$$

presuming that the quantity inside the square root will be positive. Examples are provided in Sec. 7.4.

Three Spatial Dimensions

The expression for the 3-D frequency-dependent spatial correlation measure β_ω^u is obtained from Eqs. (7.3.5) and (3.8.20):

$$\beta_\omega^u = \left.\frac{(2\pi)^3 s(k_1, k_2, k_3, \omega)}{s(\omega)}\right|_{\mathbf{k}=\mathbf{0}}. \tag{7.3.15}$$

Taking $\omega = 0$ yields

$$\lim_{\omega\to 0} \beta_\omega^u = \frac{\eta}{\theta^t} = c_3^u \beta^u \equiv \lim_{T\to\infty} \beta_T^u, \tag{7.3.16}$$

where c_3^u is a constant [like that defined by Eq. (7.1.22)] and β_T^u is the spatial correlation measure of the temporal average, over the interval T, of the space-time process $X(u_1, u_2, u_3, t)$.

At frequencies exceeding the corner frequency Ω_3^u, the function β_ω^u will vary in proportion to ω^{-3},

$$\beta_\omega^u \propto \left(\frac{\Omega_3^u}{\omega}\right)^3, \quad \omega \gg \Omega_3^u. \tag{7.3.17}$$

The following expression is consistent with Eqs. (7.3.16) and (7.3.17), and parallels Eq. (7.3.12):

$$\frac{\beta_\omega^u}{\beta^u} = c_3^u \left[1 + \left(\frac{\omega}{\Omega_3^u}\right)^2\right]^{-3/2}, \tag{7.3.18}$$

in which

$$c_3^u = \frac{\eta}{\beta^u \theta^t}, \tag{7.3.19}$$

and

$$\Omega_3^u = \left[\frac{\eta_2^t}{3\eta} - \frac{\theta_2^t}{3\theta^t}\right]^{-1/2}. \tag{7.3.20}$$

This expression for Ω_3^u is found by equating the coefficients of ω^2 in the series expansions of Eq. (7.3.18) and (the spectra defining β_ω^u in) Eq. (5.3.13).

Ellipsoidal Correlation Structure

If the correlation structure of an n-dimensional space-time process $X(\mathbf{u}, t)$ is ellipsoidal, the corner frequency Ω_{n-1}^u may be expressed in terms of the moments of the radial correlation function with scale θ^t [see Eqs. (6.7.28)–(6.7.30) for $n = 2$]. We may write:

$$\Omega_{n-1}^u = \left\{ \frac{1}{n-1} \left[\frac{\theta_{n+1}^R}{n\,\theta_{n-1}^R} - \frac{\theta_2^R}{\theta_0^R} \right] \right\}^{-1/2}, \quad n = 2, 3, 4, \tag{7.3.21}$$

where $n - 1$ is the number of spatial coordinates. Eq. (7.3.21) properly reduces to Eq. (6.7.32) when $n = 2$.

Separable Spatial and Temporal Correlation Structure

If the patterns of spatial and temporal correlation are *separable*, so that the overall correlation function can be expressed as a product of space- and time-related factors, then

$$c_{n-1}^u = 1, \quad n = 2, 3, 4, \quad \text{and} \tag{7.3.22}$$

$$\Omega_{n-1}^u = \infty, \quad n = 2, 3, 4. \tag{7.3.23}$$

All the "corner frequencies" become infinite and none of the spatial correlation measures depend on frequency.

Frequency-Dependent Variance Function

Denote by $X_A(u_1, u_2, t)$ the spatial average of $X(u_1, u_2, t)$ over a rectangular area $A = U_1 U_2$. At a given instant, $t = t_0$, this averaging operation becomes identical to that considered in Chap. 6. Indeed, the Chap. 6 approach may be applied to the individual sinusoidal components of $X(u_1, u_2, t)$ and $X_A(u_1, u_2, t)$, representing the same narrow frequency band $\Delta\omega$ centered at ω. The variances of these components are related as follows:

$$G_{X_A}(\omega)\Delta\omega \equiv G_A(\omega)\Delta\omega = [G(\omega)\Delta\omega]\,\gamma_\omega(U_1, U_2), \tag{7.3.24}$$

where $G_{X_A}(\omega) \equiv G_A(\omega)$ is the "point" spectral density function of $X_A(u_1, u_2, t)$, and $\gamma_\omega(U_1, U_2)$ is the frequency-dependent 2-D variance function.

Integration over all frequencies yields the variance of X_A:

$$\sigma_A^2 = \int_0^\infty G_A(\omega)\, d\omega = \int_0^\infty \gamma_\omega(U_1, U_2) G(\omega)\, d\omega. \tag{7.3.25}$$

The variance of X_A can also be expressed in terms of the composite variance function $\gamma(U_1, U_2)$:

$$\sigma_A^2 = \sigma^2 \gamma(U_1, U_2). \tag{7.3.26}$$

It follows that the frequency-dependent and composite variance functions are related as follows:

$$\gamma(U_1, U_2) = \int_0^\infty \gamma_\omega(U_1, U_2) g(\omega)\, d\omega. \tag{7.3.27}$$

The unit-area s.d.f. of X_A is

$$g_A(\omega) = g(\omega) \frac{\gamma_\omega(U_1, U_2)}{\gamma(U_1, U_2)}. \tag{7.3.28}$$

The variance function $\gamma_\omega(U_1, U_2)$ depends on the correlation measures $\alpha_\omega, \theta_\omega^{(1)}$, and $\theta_\omega^{(2)}$ introduced earlier. The theory in Chap. 6 involving the 2-D variance function $\gamma(T_1, T_2)$, the conditional variance function $\gamma(T_1|T_2)$, the correlation measures α, $\theta^{(1)}$, $\theta^{(2)}$, $\theta_T^{(1)}$, and so on, can be now extended with minimum effort. It suffices to replace T by U and add the reference to frequency ω. Also, the variances σ^2, σ_A^2, and $\sigma_{I_A}^2$ may be replaced by $G(\omega)\Delta\omega$, $G_A(\omega)\Delta\omega$ and $G_{I_A}(\omega)\Delta\omega$, respectively.

The analysis of the 4-D process $X(\mathbf{u}, t)$, where $\mathbf{u} = (u_1, u_2, u_3)$, proceeds in much the same way. Spatial averaging now occurs inside a rectangular box with volume $V = U_1 U_2 U_3$, and the theory permits evaluation of the spectral density function of the derived space-time process $X_V(\mathbf{u}, t)$. The latter could represent, for instance, time-dependent pollutant concentrations in a random medium (air, water, or soil), with V defining the volume inside which the concentrations are measured.

Generalization

Spatial averaging of a space-time process $X(\mathbf{u}, t)$ over a region D in space (a distance U, an area $A = U_1 U_2$ or a volume $V = U_1 U_2 U_3$) generates a new space-time process $X_D(\mathbf{u}, t)$. This process may be viewed as a sum of sinusoids, in the same way as $X(\mathbf{u}, t)$ in Eq. (7.3.1):

$$X_D(\mathbf{u}, t) = \sum_{j=1}^{J} X_{D,j}(\mathbf{u}, t), \tag{7.3.29}$$

where

$$X_{D,j}(\mathbf{u}, t) = \frac{1}{D} \int_D X_j(\mathbf{u}, t)\, d\mathbf{u}. \tag{7.3.30}$$

The variance of this component process is

$$G_D(\omega_j)\Delta\omega = \gamma_{\omega_j}(\mathbf{U}) G(\omega_j)\Delta\omega, \tag{7.3.31}$$

where $\gamma_{\omega_j}(\mathbf{U})$ is the frequency-dependent variance function of $X_j(\mathbf{u}, t)$, and $\Delta\omega$ denotes the width of a narrow band of frequencies centered on ω_j. Summing the variances of all these components yields the composite variance of $X_D(\mathbf{u}, t)$:

$$\sigma_D^2 = \sum_{j=1}^{J} G_D(\omega_j)\Delta\omega = \sigma^2 \sum_{j=1}^{J} \gamma_{\omega_j}(\mathbf{U}) g(\omega_j)\Delta\omega = \sigma^2\gamma(\mathbf{U}). \tag{7.3.32}$$

For $\Delta\omega \to 0$ and $J \to \infty$, replacing ω_j by ω, Eq. (7.3.31) becomes

$$G_D(\omega) \equiv G_{X_D}(\omega) = G(\omega)\gamma_\omega(\mathbf{U}). \tag{7.3.33}$$

Also,

$$\gamma(\mathbf{U}) = \int_0^\infty \gamma_\omega(\mathbf{U}) g(\omega)\, d\omega. \tag{7.3.34}$$

It is clear that the frequency-dependent correlation parameters previously expressed in terms of $\rho_\omega(\boldsymbol{\nu})$ and $s_\omega(\mathbf{k})$ can also be stated in terms of $\gamma_\omega(\mathbf{U})$. If the subscript ω is added, the relationships have essentially the same form as those introduced in Sec. 7.1. For example, we may write

$$\gamma_\omega(U_1, U_2) = \frac{\alpha_\omega^u}{U_1 U_2}, \quad U_i \gg \theta_\omega^{(i)},\ i = 1, 2, \tag{7.3.35}$$

and

$$\gamma_\omega(U_1, U_2, U_3) = \frac{\beta_\omega^u}{U_1 U_2 U_3}, \quad U_i \gg \theta_\omega^{(i)},\ i = 1, 2, 3. \tag{7.3.36}$$

Also, the multi-dimensional variance functions – either composite or dependent on frequency – may be expressed as products of one-dimensional variance functions, for example:

$$\gamma_\omega(U_1, U_2) = \gamma_\omega(U_1)\gamma_\omega(U_2|U_1). \tag{7.3.37}$$

The results presented in Sec. 7.2 remains valid; it suffices to replace elements of **T** by elements of **U** and to add the subscript ω throughout.

7.4 Some Tractable Space-Time Correlation Models

A Family of 3-D Space-Time Processes

The three-dimensional wave number-frequency spectrum of a family of autoregressive space-time processes $X(u_1, u_2, t)$, isotropic in space, is

$$S(k_1, k_2, \omega) = S_0 \left[1 + \left(\frac{k_1}{a}\right)^2 + \left(\frac{k_2}{a}\right)^2 + \left(\frac{\omega}{b}\right)^2\right]^{-m}, \quad m > \frac{3}{2}. \tag{7.4.1}$$

The correlation structure is ellipsoidal since the associated covariance function depends on $\tau^* = [(a\nu_1)^2 + (a\nu_2)^2 + (b\tau)^2]^{1/2}$. (The case $m = 2$ corresponds to the exponential correlation model.) The wave number spectrum $S(k_1, k_2)$ is obtained by integrating the above 3-D spectrum over all frequencies:

$$S(k_1, k_2) = S_0\, b\sqrt{\pi}\,\frac{\Gamma(m-\frac{1}{2})}{\Gamma(m)}\left[1 + \left(\frac{k_1}{a}\right)^2 + \left(\frac{k_2}{a}\right)^2\right]^{-m+1/2}. \tag{7.4.2}$$

Integration of $S(k_1, k_2, \omega)$ over the wave number space (k_1 and k_2) yields the (temporal) spectral density function

$$S(\omega) = S_0\, a^2\, \pi\, \frac{\Gamma(m-1)}{\Gamma(m)}\left[1 + \left(\frac{\omega}{b}\right)^2\right]^{m+1}. \tag{7.4.3}$$

Further integration of Eq. (7.4.2) (over all wave numbers) or Eq. (7.4.3) (over all frequencies) generates the "point variance" of $X(u_1, u_2, t)$:

$$\sigma^2 = S_0\, a^2 b\, \pi^{3/2} \frac{\Gamma(m-\frac{3}{2})}{\Gamma(m)}, \quad m > \frac{3}{2}. \tag{7.4.4}$$

The 3-D correlation measure is

$$\beta = (2\pi)^3 s(0,0,0) = \frac{8\pi^{3/2}}{a^2 b}\frac{\Gamma(m)}{\Gamma(m-\frac{3}{2})}. \tag{7.4.5}$$

Other pertinent correlation parameters are

$$\alpha^u = (2\pi)^2 s(k_1, k_2)|_{\mathbf{k}=0} = \frac{4\pi\,\Gamma(m-1/2)}{\alpha^2\,\Gamma(m-3/2)} \tag{7.4.6}$$

and

$$\theta^t = 2\pi s(0) = \frac{2\pi^{1/2}}{b}\frac{\Gamma(m-1)}{\Gamma(m-3/2)}. \tag{7.4.7}$$

The spatial scale θ^u has the same form as Eq. (7.4.7); it suffices to replace b by a. The following dimensionless factors are of interest:

$$c_\alpha^u = \frac{\alpha^u}{(\theta^u)^2} = \frac{\Gamma(m-1/2)\,\Gamma(m-3/2)}{[\Gamma(m-1)]^2} \tag{7.4.8}$$

and

$$c_\beta^u \equiv c_2^u = \frac{\beta}{\alpha^u\theta^t} = \frac{\Gamma(m)\,\Gamma(m-3/2)}{\Gamma(m-1/2)\,\Gamma(m-1)}. \tag{7.4.9}$$

The frequency-dependent spatial correlation measures are

$$\theta_\omega^u = 2\pi\left.\frac{S(k,\omega)}{S(\omega)}\right|_{k=0} = \frac{2\pi^{1/2}}{a}\frac{\Gamma(m-1/2)}{\Gamma(m-1)}\left[1+\left(\frac{\omega}{b}\right)^2\right]^{-1/2} \tag{7.4.10}$$

and

$$\alpha_\omega^u = (2\pi)^2\left.\frac{S(k_1,k_2,\omega)}{S(\omega)}\right|_{\mathbf{k}=0} = \frac{4\pi}{a^2}\frac{\Gamma(m)}{\Gamma(m-1)}\left[1+\left(\frac{\omega}{b}\right)^2\right]^{-1}. \tag{7.4.11}$$

It is notable that, for this family of space-time correlation models, the corner frequencies Ω_1^u and Ω_2^u are equal:

$$\Omega_1^u = \Omega_2^u = b. \tag{7.4.12}$$

Also, the dimensionless ratio of the frequency-dependent spatial correlation measures,

$$\frac{\alpha_\omega^u}{(\theta_\omega^u)^2} = \frac{\Gamma(m)\,\Gamma(m-1)}{[\Gamma(m-1/2)]^2}, \tag{7.4.13}$$

does not depend on frequency.

A Family of 4-D Space-Time Processes

The analysis of a homogeneous 4-D space-time process $X(u_1, u_2, u_3, t)$ with a spectrum of the autoregressive (Markovian) type,

$$S(\mathbf{k}, \omega) = S_0 \left[1 + \sum_{i=1}^{3} \left(\frac{k_i}{a_i}\right)^2 + \left(\frac{\omega}{b}\right)^2\right]^{-m}, \quad m > 2, \tag{7.4.14}$$

proceeds along parallel lines. Note that $a_1 = a_2 = a_3$ would imply that the random field is isotropic in space. The wave number spectrum is obtained by integration over ω:

$$S(\mathbf{k}) = S_0 \, b \sqrt{\pi} \, \frac{\Gamma(m - \frac{1}{2})}{\Gamma(m)} \left[1 + \sum_{i=1}^{3} \left(\frac{k_i}{a_i}\right)^2\right]^{-m+1/2}. \tag{7.4.15}$$

Threefold integration over the wave numbers yields

$$S(\omega) = S_0 a_1 a_2 a_3 \, \pi^{3/2} \, \frac{\Gamma(m - \frac{3}{2})}{\Gamma(m)} \left[1 + \left(\frac{\omega}{b}\right)^2\right]^{-m+3/2}. \tag{7.4.16}$$

The variance of $X(\mathbf{u}, t)$ is

$$\sigma^2 = S_0 \, a_1 \, a_2 \, a_3 \, b \pi^2 \, \frac{\Gamma(m - 2)}{\Gamma(m)} = \frac{S_0 \, \pi^2 \, a_1 \, a_2 \, a_3 \, b}{(m-1)(m-2)}, \quad m > 2. \tag{7.4.17}$$

The following mixed spectra are also needed to evaluate the frequency-dependent correlation parameters:

$$S(k_1, k_2, \omega) = S_0 \, a_3 \sqrt{\pi} \, \frac{\Gamma(m - 1/2)}{\Gamma(m)} \left[1 + \sum_{i=1}^{2} \left(\frac{k_i}{a_i}\right)^2 + \left(\frac{\omega}{b}\right)^2\right]^{-m+1/2} \tag{7.4.18}$$

and

$$S(k_1, \omega) = S_0 \, a_2 a_3 \, \pi \, \frac{\Gamma(m - 1)}{\Gamma(m)} \left[1 + \left(\frac{k_1}{a_1}\right)^2 + \left(\frac{\omega}{b}\right)^2\right]^{-m+1}. \tag{7.4.19}$$

Assuming isotropy in space, the frequency-dependent spatial correlation measures are

$$\beta_\omega^u = (2\pi)^3 \left.\frac{S(\mathbf{k}, \omega)}{S(\omega)}\right|_{\mathbf{k}=0} = \frac{8\pi^{3/2}}{a^3} \, \frac{\Gamma(m)}{\Gamma(m - 3/2)} \left[1 + \left(\frac{\omega}{b}\right)^2\right]^{-3/2}, \tag{7.4.20}$$

$$\alpha_\omega^u = (2\pi)^2 \left.\frac{S(k_1, k_2, \omega)}{S(\omega)}\right|_{k_1=k_2=0} = \frac{4\pi}{a^2} \frac{\Gamma(m - 1/2)}{\Gamma(m - 3/2)} \left[1 + \left(\frac{\omega}{b}\right)^2\right]^{-1} \tag{7.4.21}$$

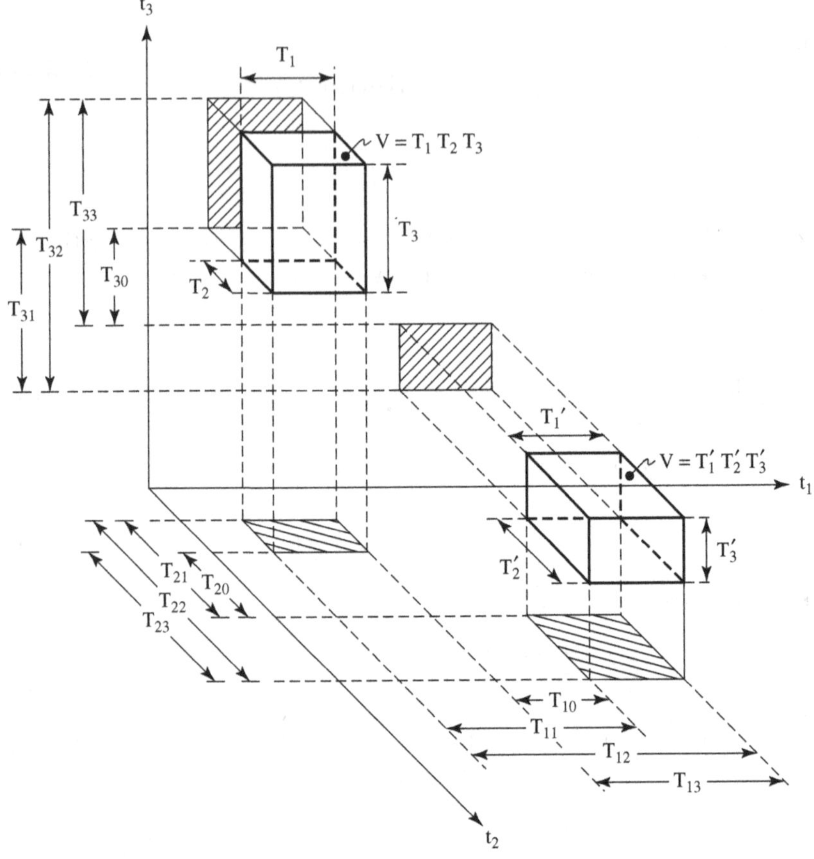

Fig. 7.3 Distances characterizing the relative location of the volumes V and V' in the three-dimensional parameter space.

and

$$\theta_\omega^u = 2\pi \frac{S(k_1, \omega)}{S(\omega)}\bigg|_{k_1=0} = \frac{2\pi^{1/2}}{a} \frac{\Gamma(m-1)}{\Gamma(m-3/2)} \left[1 + \left(\frac{\omega}{b}\right)^2\right]^{-1/2} . \qquad (7.4.22)$$

For this family of random fields, the corner frequencies appearing in the expressions for the frequency-dependent correlation measures are all the same: $\Omega_{n-1}^u = b$, for $n = 2, 3, 4$. Various dimensionless measures can easily be found by inserting Eqs. (7.4.14) - (7.4.22) into their definitions.

7.5 Covariance of Local Averages

General Formulation

An algebraic relation similar to (and extending) Eq. (5.3.3) exists in reference to a function $X(t_1, \ldots, t_n)$ of two or more orthogonal coordinates. Consider two generalized "rectangular" domains, $D = T_1 T_2 \cdots T_n$ and $D' = T'_1 T'_2 \cdots T'_n$, with sides parallel to the coordinate axes, as illustrated for $n = 2$ in Fig. 6.5 and for $n = 3$ in Fig. 7.3. Along the ith coordinate axis the dimensions T_i and T'_i and the distances T_{i0}, T_{i1}, T_{i2}, and T_{i3} are defined exactly as in the one-dimensional case. The relative location of the two domains D and D' (in n-dimensional parameter space) is fully characterized by the matrix

$$\mathbf{T} = \begin{bmatrix} T_{10} & T_{11} & T_{12} & T_{13} \\ T_{20} & T_{21} & T_{22} & T_{23} \\ \vdots & \vdots & \vdots & \vdots \\ T_{n0} & T_{n1} & T_{n2} & T_{n3} \end{bmatrix}. \tag{7.5.1}$$

Each row of this matrix consists of four distances along one of the coordinate axes. From this matrix a total of 4^n different vectors of size n can be constructed by choosing exactly one element from each row. Specifically, if k_i denotes the second subscript of the element selected in row i, then the index vector $\mathbf{k} = (k_1, \ldots, k_i, \ldots, k_n)$ fully defines each selection $\mathbf{T_k} = (T_{1k_1}, \ldots, T_{ik_i}, \ldots, T_{nk_n})$, whose components are the dimensions of a "rectangular" domain $D_\mathbf{k}$. The generalized algebraic identity can now be stated in the following form:

$$2^n I_D I_{D'} = \sum_{\mathbf{k}} \left[\prod_{i=1}^{n} (-1)^{k_i} \right] I^2_{D_\mathbf{k}}, \tag{7.5.2}$$

where I_D, $I_{D'}$, and $I_{D_\mathbf{k}}$ represent integrals of $X(t_1, \ldots, t_n)$ over the rectangular domains D, D', and $D_\mathbf{k}$, respectively. The summation in Eq. (7.5.2) is over all possible choices (4^n) for the vector $\mathbf{k}$, and every coefficient in the sum is either $+1$ or -1. Eq. (7.5.2) reduces to Eq. (5.3.3) when $n = 1$.

The covariance between X_D and $X_{D'}$ can now be expressed as a linear combination of 4^n terms [based on Eq. (7.5.2), to which the expection operation is applied]:

$$\operatorname{Cov}[X_D, X_{D'}] = \frac{\sigma^2}{2^n D D'} \sum_{\mathbf{k}} \prod_{i=1}^{n} (-1)^{k_i} \; \Delta(\mathbf{T_k}), \tag{7.5.3}$$

where $\Delta(\mathbf{T_k}) = D^2\,\gamma(\mathbf{T}) = \Delta(T_{1k_1},\ldots,T_{ik_i},\ \ldots,\ T_{nk_n})$. Again, the summation is over the set of all possible choices (4^n) of the vector $\mathbf{k}$.

Eq. (7.5.3) permits computationally efficient second-order stochastic analysis of random fields – in particular, stochastic finite element analysis, in which the variances and covariances of local averages across finite elements are needed – based directly on the knowledge of, or a model for, the n-dimensional variance function $\Delta(\mathbf{T}) = D^2\gamma(\mathbf{T})$.

The Three-Dimensional Case

Fig. 7.3 shows the locations of two rectangular boxes with volumes V and V' in the parameter space of a three-dimensional random field $X(\mathbf{t})$. The covariance of the local integrals of $X(\mathbf{t})$ inside the volumes V and V' may be expressed as a linear combination of 3-D variance functions. Consistent with Eq. (7.5.3), the expression (for $n = 3$) contains a total of $4^3 = 64$ terms:

$$\operatorname{Cov}\left[I_V, I_{V'}\right] = (VV')^2\ \operatorname{Cov}\left[X_V, X_{V'}\right]$$

$$= \frac{\sigma^2}{8}\sum_{j=0}^{3}\sum_{k=0}^{3}\sum_{l=0}^{3}(-1)^j(-1)^k(-1)^l\Delta(T_{1j}, T_{2k}, T_{3l}). \qquad (7.5.4)$$

The definition of the distances along each coordinate axis is the same as in the one-dimensional case, and $X_V = I_V/V$ refers to the local average.

Covariance Function of Local Average Fields

To obtain the covariance function of the derived random field $X_D(\mathbf{t})$,

$$B_D(\boldsymbol{\tau}) = \operatorname{Cov}\left[X_D(\mathbf{t}), X_D(\mathbf{t}')\right], \qquad (7.5.5)$$

where $D = T_1 \cdot T_2 \cdots T_n$ and $\boldsymbol{\tau} = \mathbf{t} - \mathbf{t}'$ (that is, $\tau_1 = t_1 - t_1'$, $\tau_2 = t_2 - t_2'$, and so on), it suffices to choose the following form for the matrix $\mathbf{T}$:

$$\mathbf{T} = \begin{bmatrix} T_1 - \tau_1 & \tau_1 & T_1 + \tau_1 & \tau_1 \\ \vdots & \vdots & \vdots & \vdots \\ T_i - \tau_i & \tau_i & T_i + \tau_i & \tau_i \\ \vdots & \vdots & \vdots & \vdots \\ T_n - \tau_n & \tau_n & T_n + \tau_n & \tau_n \end{bmatrix}. \qquad (7.5.6)$$

For $\mathbf{T} \to \mathbf{0}$ the field $X_D(\mathbf{t})$ reduces to $X(\mathbf{t})$, and Eq. (7.5.3) will generate the relationship between $B(\boldsymbol{\tau})$ and $\Delta(\mathbf{T})$, namely:

$$B(\boldsymbol{\tau}) = \frac{\sigma^2}{2^n}\left[\frac{\partial^{2n}\Delta(\boldsymbol{\tau})}{\partial\tau_1^2\partial\tau_2^2\cdots\partial\tau_n^2}\right], \tag{7.5.7}$$

where $\Delta(\boldsymbol{\tau}) = (\tau_1\tau_2\cdots\tau_n)^2\gamma(\boldsymbol{\tau})$. Eq. (7.5.7) generalizes Eqs. (5.3.24) and (6.4.6).

The approach just outlined may also be used to evaluate the cross spectral density function of two local spatial averages $X_D(t)$ and $X_{D'}(t)$ of a multi-dimensional space-time process $X(\mathbf{u}, t)$. The averaging domains D and D' may be distances, areas, volumes, or just points in space (if either D or $D' \to 0$). Finally, when analyzing the dependence on frequency of various representations of the spatial correlation structure, one needs to replace $\Delta(\mathbf{T})$ by $\Delta_\omega(\mathbf{U})$ and n by $n' = n - 1$ in Eq. (7.5.3).

7.6 Stochastic Finite Element Analysis

In Secs. 6.4 and 7.5 methodology based on the variance function is presented to calculate the covariance matrix of element averages required in stochastic finite element analysis. The theory accounts for the reduction of the point variance and for correlation between element averages as a function of the dimensions of the finite element mesh and the correlation parameters of the underlying random field.

In particular, if the correlation structure of a homogeneous n-dimensional random field $X(\mathbf{t})$ is represented by the variance function $\Delta(\mathbf{T})$, it is possible to calculate, by means of simple algebra [Eq. (7.5.3)], the matrix of covariances of local averages associated with a set of finite elements into which a region in the parameter space is partitioned. (If one attempts to solve the same problem in terms of the correlation function $\rho(\boldsymbol{\tau})$, each covariance calculation would require $2n$-fold numerical integration over the lag space.)

For example, as discussed in Sec. 6.4, repeated use of Eq. (6.4.1) permits evaluation of the covariances of element properties (local spatial averages for each element) for the 2-D finite element mesh shown in Fig. 6.7. Nonrectangular areas may be replaced (to within an arbitrary degree of numerical accuracy) by a collection of rectangles, as shown in Fig. 6.8.

In specific applications the random field could be a spatially varying property (compressibility, permeability, strength) of a distributed disordered system, or it could be a space-time process (for example, wind pressure or earthquake ground acceleration) that excites a system. The technique just described has been applied to finite element analysis of the deformation of beams with randomly-varying rigidity (Vanmarcke and Grigoriu [145]); seepage through a randomly permeable dam foundation (Choot [33]); pore pressures in an embankment dam characterized by random transmissivity (Hachich and Vanmarcke [68]); and pollution migration in porous media (Chaudhuri and Sekhar [30; 31]).

A similar procedure can be used to solve important practical problems in stochastic dynamics where the excitation takes the form of a space-time random process. Based on the concept of the frequency-dependent variance function, one can calculate the matrix of cross-spectral density functions of local spatial averages of a space-time process, providing the input for a second-order stochastic dynamic analysis of an appropriately discretized model of a structure subjected to random excitation.

To solve nonlinear finite element problems numerically, one can make use of the method of enumeration mentioned in Sec. 2.2. Assuming the local averages (associated with the different finite elements) are jointly Gaussian, it is possible to evaluate the probability mass associated with a prespecified range of values of the element properties. A deterministic response analysis can then be carried out for each combination of element properties having non-negligible probability mass. In this way, probability masses can be transferred directly from the space of the random input quantities to that of the random response quantities.

Another all-purpose numerical analysis procedure is Monte Carlo simulation. Given the first- and second-order statistics of the set of element averages, it is possible to simulate sets of correlated observations which serve as input into (repeated) deterministic response analysis. Based on a sufficient number of input realizations and response calculations, sample statistics of pertinent response parameters can then be obtained. Publications describing both methods and applications of numerical simulation of random fields with prescribed correlation structures include Bruining and Van Batenburg [24], Roberto *et al.* [114], Frimpong and Achireko [62], Spanos and Zeldin [123], and Vanmarcke *et al.* [146].

We conclude this discussion about stochastic finite element analysis by stressing the following points: (1) the variance function strongly depends on one or more integral correlation measures and can be closely approximated in terms of only these parameters (mainly the direction-dependent scales of fluctuation); (2) covariances can be expressed as simple linear combinations of the variance function; (3) the approach can be easily extended to situations (common in practice) where correlation decay occurs in stages, when the variance function is expressed as a sum of two or more components with very different scales of fluctuation. Further uses and interpretations are suggested below.

Optimal Linear Estimation

It is useful to re-examine classical linear estimation theory (reviewed in Sec. 2.6) in light of the results of the theory of local averages of random fields. In the estimation problem discussed in Sec. 2.6, the random variable Y_1 is to be predicted based on the observation of the random variable Y_2. The posterior mean and variance of Y_1 given Y_2 are, respectively,

$$E[Y_1|Y_2] = m_1 + B_{12}(Y_2 - m_2)/\sigma_2^2 \tag{7.6.1}$$

and

$$\mathrm{Var}[Y_1|Y_2] = \sigma_1^2(1 - \rho_{1\,2}^2), \tag{7.6.2}$$

where m_i and σ_i^2 are the prior means and variances ($i = 1, 2$), $B_{1\,2}$ is the prior covariance of Y_1 and Y_2, and $\rho_{1\,2} = B_{1\,2}/\sigma_1\sigma_2$ is the coefficient of correlation between Y_1 and Y_2. Eq. (7.6.1) may be used to predict Y_1 based on an actual observation $Y_2 = y_2$, or it may be interpreted as a linear relationship between two random variables, $E[Y_1|Y_2]$ and Y_2. Eq. (7.6.2) quantifies the amount of variance reduction that can be achieved by observing Y_2. Note that the outcome of the observation does not affect the posterior variance. In the random field context, several new applications and interpretations of these results are now suggested:

1. Y_1 and Y_2 represent local averages of a homogeneous random field $X(\mathbf{t})$ over two different finite elements in the parameter space. Recall that the covariance $B_{1\,2}$ may be expressed as a linear combination of variance functions [Eq. (7.5.3)].
2. Y_2 is the local average of a random field across a finite element in the parameter space, and Y_1 is the value of the random field at a point randomly located inside the element. In this case $B_{12} = \sigma_2^2 = \sigma^2\gamma(\mathbf{T})$, $\rho_{12} = [\gamma(\mathbf{T})]^{1/2}$ and $m_1 = m_2 = m$. From Eqs. (7.6.1) and (7.6.2) we obtain:

$$E[Y_1|Y_2] = Y_2, \tag{7.6.3}$$

$$\mathrm{Var}[Y_1|Y_2] = \sigma^2[1 - \gamma(\mathbf{T})]. \tag{7.6.4}$$

3. In a variant of the previous case, Y_1 is the local average across a finite element, and Y_2 is an observation point randomly located inside the finite element. In this case $m_1 = m_2 = m$, $\sigma_1^2 = \sigma^2\gamma(\mathbf{T})$, $\sigma_2^2 = \sigma^2$, and $\rho_{12}^2 = \gamma(\mathbf{T})$. The posterior mean of Y_1 given Y_2 is now a weighted combination of the prior mean m and the observation Y_2:

$$E[Y_1|Y_2] = m[1-\gamma(\mathbf{T})] + Y_2\gamma(\mathbf{T}). \tag{7.6.5}$$

The posterior variance, according to Eq. (7.6.2), is

$$\mathrm{Var}[Y_1|Y_2] = \sigma^2\gamma(\mathbf{T})[1-\gamma(\mathbf{T})]. \tag{7.6.6}$$

4. The stochastic analysis and updating procedures may also be applied to individual sinusoidal components of a space-time process. At various locations in space, measuring devices record time histories that provide the information used to update the stochastic characterization of the space-time process. This methodology provides the basis for *conditional simulation* of (space-time) random fields, whose values (or time histories) at specific points are known or prescribed. For instance, random fields of earthquake ground motion can be simulated that are consistent with recorded ground-motion time histories at accelerograph locations (Vanmarcke *et al.* [143; 146]).

The two-variable case may be extended by considering a random vector $\mathbf{Y}_1$ to be predicted based on the observation of the random vector $\mathbf{Y}_2$. The mean value vectors are $\mathbf{m}_1$ and $\mathbf{m}_2$ and the covariance matrix of all the random variables contained in $\mathbf{Y}_1$ and $\mathbf{Y}_2$ may be partitioned as follows:

$$\mathbf{B_Y} = \left[\begin{array}{c|c} \mathbf{B}_{11} & \mathbf{B}_{12} \\ \hline \mathbf{B}_{12}^t & \mathbf{B}_{22} \end{array}\right], \tag{7.6.7}$$

where $\mathbf{B}_{11} = \mathbf{B}_{\mathbf{Y}_1}$, $\mathbf{B}_{22} = \mathbf{B}_{\mathbf{Y}_2}$, and $\mathbf{B}_{12} = \mathbf{B}_{\mathbf{Y}_1,\mathbf{Y}_2}$. The optimal linear prediction equation gives the vector of posterior means [Eq. (2.6.24)]:

$$E[\mathbf{Y}_1|\mathbf{Y}_2] = \mathbf{m}_1 + \mathbf{B}_{12}\mathbf{B}_{22}^{-1}[\mathbf{Y}_2 - \mathbf{m}_2], \tag{7.6.8}$$

where $\mathbf{B}_{12}\mathbf{B}_{22}^{-1} = \mathbf{A}_0$ is the matrix of optimal coefficients. The posterior covariance matrix of $\mathbf{Y}_1$ given $\mathbf{Y}_2$ [from Eq. (2.6.25)] is

$$\mathbf{B}_{11|2} = \mathbf{B}_{11} - \mathbf{A}_0\mathbf{B}_{22}\mathbf{A}_0^t = \mathbf{B}_{11} - \mathbf{B}_{12}\mathbf{B}_{22}^{-1}\mathbf{B}_{12}^t. \tag{7.6.9}$$

Again, it does not depend on the observation vector $\mathbf{Y}_2$.

The quantities to be observed or estimated may be local averages of the random field under study. The required prior covariance can then be expressed algebraically in terms of the variance function, which is in turn controlled by the principal correlation parameters of the random field. The analysis retains its relative simplicity in case the random field is a composite of several uncorrelated fields with different correlation properties. It thus becomes possible to incorporate into the analysis measurement errors (whether random or systematic). Recall that uncorrelated measurement errors may be modeled as components with near-zero scale, and systematic errors as components with a suitably large scale of fluctuation.

It is useful to categorize data acquisition programs according to the type or level of information they are designed to generate. The objective of a measurement program may be to provide baseline information about the sources of uncertainty (inherent variability, random and systematic measurement errors) and about the amount of uncertainty contributed by each source. Such baseline information is also needed to obtain first estimates of the parameters (m, σ, θ, and δ) of random field models. The baseline program generates background information that can be used to design higher-level data acquisition programs in a more informed and rational manner. The purpose of a higher-level data acquisition may be to improve or update (in a Bayesian sense) parameter estimates obtained from the baseline program, to predict (forecast) values of the random field at specific points, or to update estimates of system response and performance. The writer sought to develop this type of baseline information for geotechnical site characterization (Vanmarcke and Fenton [144]), using extensive data from the (US) National Geotechnical Experimentation Sites (NGES).

7.7 Partial Derivatives of Local-Average Fields

The Three-Dimensional Case

A three-dimensional homogeneous random field $X(t_1, t_2, t_3)$ generates random processes of lower dimensionality when one or two of its coordinates are fixed. Consider, in particular, the one-dimensional processes representing random variation in a direction parallel to a principal axis:

$$X^{(i)}(\mathbf{t}) = \begin{cases} X(t_1, t_2^*, t_3^*), \\ X(t_1^*, t_2, t_3^*), \\ X(t_1^*, t_2^*, t_3), \end{cases} \tag{7.7.1}$$

where t_i^* denotes a specific value of t_i. Each 1-D process possesses the same first-order characteristics as $X(\mathbf{t})$, namely mean m and variance σ^2, while their respective scales of fluctuation are $\theta^{(1)}$, $\theta^{(2)}$, and $\theta^{(3)}$.

We have shown in Chap. 4 that the statistics of level excursions and extremes of the random field $X(\mathbf{t})$ depend on the determinant of $\mathbf{\Lambda}_{11}$, the covariance matrix of the partial second-order derivatives of $X(\mathbf{t})$, and are therefore sensitive to details of the high frequency content of $X(\mathbf{t})$, about which information is usually lacking. This problem can be circumvented by permitting some local averaging, which acts to suppress the microscale fluctuations of $X(\mathbf{t})$, so that $\mathbf{\Lambda}_{11}$ (of the derived random field) becomes a diagonal matrix, with as diagonal elements the mean square partial derivatives. Specifically, a new random field $X_V(\mathbf{t})$ is derived from $X(\mathbf{t})$ by local spatial averaging inside a moving rectangular box whose sides T_1, T_2, and T_3 remain parallel to the respective coordinate axes. The size of the volume over which $X(\mathbf{t})$ is locally averaged equals $V = T_1T_2T_3$. Like the original random process $X(\mathbf{t})$, the derived field $X_V(\mathbf{t}) = X_V(t_1, t_2, t_3)$ gives rise to three uni-directional processes $X_V^{(i)}(t_i)$, $i = 1, 2, 3$, with the same mean and variance, respectively

$$E[X_V] = m, \quad \text{and} \tag{7.7.2}$$

$$\text{Var}\,[X_V] = \sigma^2\gamma(T_1, T_2, T_3) = \sigma^2\gamma(\mathbf{T}). \tag{7.7.3}$$

The derivatives of the direction-dependent 1-D processes are in effect the partial derivatives of $X_V(\mathbf{t})$:

$$\dot{X}_V^{(i)}(\mathbf{t}) = \frac{\partial}{\partial t_i} X_V(\mathbf{t}), \quad i = 1, 2, 3. \tag{7.7.4}$$

The covariance matrix of the partial derivatives $X_V(\mathbf{t})$ at location $\mathbf{t}$ is diagonal for $T_i > 0$, $i = 1, 2, 3$, since

$$E[\dot{X}_V^{(i)}(\mathbf{t})\dot{X}_V^{(j)}(\mathbf{t})] = \lambda_{11,V}^{(ij)} = 0, \quad i \neq j, \tag{7.7.5}$$

while the diagonal elements of the matrix $\mathbf{\Lambda}_{11,V}$,

$$\text{Var}\,[\dot{X}_V^{(i)}(\mathbf{t})] = \lambda_{11,V}^{(ii)} \equiv \lambda_{2,V}^{(i)}, \tag{7.7.6}$$

are the mean-square partial derivatives (for $i = 1, 2, 3$) of $X_V(\mathbf{t})$.

The n-Dimensional Case

Extension to n-dimensional homogeneous random fields is now straightforward. The fluctuations of the field $X_D(\mathbf{t})$ in the ith direction are represented by the one-dimensional process $X_D(t_i)$. The variance of its derivative $\dot{X}_D(t_i)$, the partial derivative of the field $X_D(\mathbf{t})$ with respect to t_i, has the following form, a direct extension of Eqs. (5.4.11) and (5.4.12):

$$\text{Var}\,[\dot{X}_{_D}^{(i)}] = 2\ \text{Var}\,[X_D]\left\{\frac{1-\rho(T_i|\mathbf{T}_i)}{\Delta(T_i|\mathbf{T}_i)}\right\}, \quad i = 1, \ldots, n, \tag{7.7.7}$$

where $\Delta(T_i|\mathbf{T}_i) = T_i^2\gamma(T_i|\mathbf{T}_i)$ and $\rho(\tau_i|\mathbf{T}_i)$ are, respectively, the variance function and the correlation function of the one-dimensional process $X_{\mathbf{T}_i}(t_i)$ introduced in Sec. 7.2. The covariance matrix of the partial derivatives of $X_D^{(i)}(\mathbf{t})$ at a point $\mathbf{t}$ is diagonal, presuming $T_i \neq 0, i = 1, \ldots, n$, as follows:

$$\mathbf{\Lambda}_{11,D} = \begin{bmatrix} \text{Var}\,[\dot{X}_{_D}^{(1)}] & & 0 \\ & \ddots & \\ 0 & & \text{Var}\,[\dot{X}_{_D}^{(n)}] \end{bmatrix}. \tag{7.7.8}$$

If the dimensions of the domain D all exceed their respective marginal scales of fluctuation ($T_i \gg \theta^{(i)}$), the variance of $X_D(\mathbf{t})$ converges toward

$$\sigma_{_D}^2 \equiv \text{Var}\,[X_D] = \eta\sigma^2/D, \tag{7.7.9}$$

and the mean-square partial derivatives take the asymptotic form

$$[\sigma_{\dot{X}_D}^{(i)}]^2 \equiv \text{Var}[\dot{X}_{_D}^{(i)}] = \frac{2\eta^{(i)}\sigma^2}{DT_i} = \frac{2\sigma_{_D}^2}{T_i c_\eta^{(i)}\theta^{(i)}}, \; i = 1, \ldots, n. \tag{7.7.10}$$

The determinant of $\mathbf{\Lambda}_{11,D}$, the product of its diagonal elements, then takes the value

$$|\mathbf{\Lambda}_{11,D}| = [\sigma_{\dot{X}_D}^{(1)}\sigma_{\dot{X}_D}^{(2)}\cdots\sigma_{\dot{X}_D}^{(n)}]^2 = \frac{2^n\eta^{n-1}\sigma^{2n}}{c_\eta D^{n+1}} = \frac{2^n\sigma_{_D}^{2n}}{\eta c_\eta D}, \tag{7.7.11}$$

in which the coefficient c_η is defined in Eq. (7.1.22), and the correlation measure η in Eq. (7.1.8); D is the size of the domain over which $X(\mathbf{t})$ is locally averaged; and σ is the standard derivation of $X(\mathbf{t})$.

7.8 Statistics of High-Level Excursions and Extremes

Classical extreme value distributions, reviewed in Sec. 2.2, represent maxima or minima of a large number of independent, identically distributed random variables. Herein, we derive extreme value distributions that are consistent with the theory of homogeneous random fields, accounting explicitly for statistical dependence and for information about the size and occurrence frequency of isolated regions of excursion above high (or below low) threshold levels.

Five-Step Derivation

Consider a scalar homogeneous random field $X(\mathbf{t})$, where $\mathbf{t} = (t_1, ..., t_n)$ is a point in the n-dimensional parameter space. A single probability density function $f(x)$ characterizes the state of the field at any location $\mathbf{t}$; and the cumulative and complementary cumulative distribution functions are $F(x)$ and $F^c(x) = 1 - F(x)$, respectively. The notation $X(t_i)$, $i = 1, \ldots, n$, is used to refer to a stationary random process on a line parallel to the t_i-axis.

Step 1. For each uni-directional random process $X(t_i)$, the mean rate of 'upcrossings' of a threshold level b is proportional to the p.d.f. of X evaluated at b,

$$\nu_{b,i}^{+} \propto f(b). \tag{7.8.1}$$

This follows from the classical result by Rice [113], stated in Eq. (4.5.15), and is valid if $X(t_i)$ and its derivative are statistically independent; owing to homogeneity (and hence stationarity), they are known to be uncorrelated.

Step 2. The mean length of stay above the the level b by the process $X(t_i)$ can be obtained by solving a system of two equations and two unknowns [see Eqs. (4.5.1) and (4.5.2)], yielding

$$E[\mathscr{T}_{b,i}^{+}] = F^c(b)\,[\nu_{b,i}^{+}]^{-1} \propto F^c(b)\;[f(b)]^{-1}, \tag{7.8.2}$$

where $F^c(b)$ is the complementary c.d.f. of X evaluated at b.

Step 3. The mean size $E[\mathscr{D}_b^{+}]$ of an isolated region where $X(\mathbf{t})$ exceeds a high threshold b, is approximately equal to the product of its 'average dimensions' $E[\mathscr{T}_{b,i}^{+}]$, each of which varies with b in accordance with Eq. (7.8.2). Hence we may write:

$$E[\mathscr{D}_b^{+}] \simeq k_0\,[F^c(b)]^n\,[f(b)]^{-n}, \tag{7.8.3}$$

in which k_0 is a proportionality factor that can in some cases be determined analytically [see e.g. Eq. (7.8.7) below].

Step 4. The mean rate μ_b^+ of occurrence of isolated excursions above a high level b is obtained by expressing the aggregate domain of excursion in a region of total size d in two ways, as follows:

$$E[\mathscr{D}_b^+]\,\mu_b^+ d = F^c(b)\,d. \tag{7.8.4}$$

Combining Eqs. (7.8.3) and (7.8.4), yields

$$\mu_b^+ = \frac{F^c(b)}{E[\mathscr{D}_b^+]} \simeq k_0^{-1}[F^c(b)]^{1-n}[f(b)]^n. \tag{7.8.5}$$

Step 5. The probability mass function of $N_b(d)$, the total number of isolated excursions above a high level b in a domain of size d, tends toward the Poisson law, with mean $\mu_b^+ d$. Since the two events $\{N_b(d) = 0\}$ and $\{\max_d X(t) \leq b\}$ are equivalent, the distribution of the maximum of $X(t)$ over the domain d takes the following asymptotic form, as $b \to \infty$:

$$\begin{aligned} P\left[\max_d X(t) \leq b\right] &= P\left[N_b(d) = 0\right] = \exp\{-\mu_b^+ d\} \\ &\simeq \exp\{-k_0^{-1} d\, [F^c(b)]^{1-n}[f(b)]^n\}. \end{aligned} \tag{7.8.6}$$

Eq. (7.8.6) does not account for several effects that gain importance as the level b decreases. First, there is a non-zero probability that the random field is above b at any location $\mathbf{t}$. Second, the excursion regions are not 'points' in the parameter space, as the Poisson law implies. Third, local maxima tend to be clustered at lower values of the level b. All three effects can be accounted for, as described in Sec. 4.4, yielding an improved estimate (for the extreme value distribution) that will converge toward Eq. (7.8.6) as $b \to \infty$.

The results just presented all depend on the p.d.f. and the complementary c.d.f. of $X(t)$ evaluated at the threshold level b. The effect of the correlation structure and higher-order statistics of $X(t)$ is represented by the factor k_0. Several specific cases are considered below.

Gaussian Random Fields

For zero-mean Gaussian random fields (and their envelopes, considered below), the factor k_0 depends on the determinant of the matrix $\mathbf{\Lambda}_{11}$ of joint spectral moments of second order, which is the covariance matrix of the partial derivatives of the random field at a point in the parameter space. For example, for an n-dimensional *isotropic* Gaussian random field, accounting Eqs. (7.7.2) and (4.7.2), we have:

$$k_0 = (2\pi)^{-n/2}\,[\nu_{0,i}^{+}\,\sigma]^{-n}, \tag{7.8.7}$$

in which $\nu_{0,i}^{+}$ is the zero-upcrossing rate of the random variation along any straight line in the parameter space. A key requirement for $\nu_{0,i}^{+}$, and therefore the factor k_0, to be properly defined is that the random field $X(t)$ must be mean square differentiable.

Local Averages of Gaussian Random Fields

From any zero-mean homogeneous random field $X(t)$, one can derive a family of 'local average' fields $X_D(t)$, where D also denotes the size of the domain over which $X(t)$ is locally averaged. In case the dimensions of the averaging domain are all much greater than their respective correlation scales, the derived random field $X_D(t)$ will tend to become Gaussian with mean $m_D = m = 0$ and variance $\sigma_D^2 = \sigma^2(\eta/D)$, where η is the integral correlation measure and σ^2 the variance of $X(t)$. The quantity σ must now be replaced by σ_D in Eq. (7.8.7), and $\nu_{0,i}^{+}$ takes the following form:

$$\nu_{0,i}^{+} = \sqrt{2}\pi(cD\eta)^{1/(2n)}, \tag{7.8.8}$$

where $c = c_\eta$ is the constant characterizing the n-dimensional isotropic correlation structure. The mean size of individual excursion regions may be expressed as a function of $u = |b - m|/\sigma_D$, for $u \to \infty$:

$$E[\mathscr{D}_b^{+}] = \pi^{n/2}(cD\eta)^{1/2}u^{-n}, \tag{7.8.9}$$

in which $c = c_\eta$ and u are dimensionless; both η and D measure the sizes of regions in the parameter space, and n is the random field's dimension.

Rayleigh Envelope Fields

Similar results are obtainable for the envelope processes $R(\mathbf{t})$ and $R_D(\mathbf{t})$ that correspond, respectively, to $X(\mathbf{t})$ and $X_D(\mathbf{t})$. The probability density function $f(x)$ and its c.d.f. $F(x)$ now become those of the Rayleigh distribution. The expressions for k_0 are readily obtainable from results such as those presented in Secs. 4.4 and 6.6. As mentioned, however, for k_0 to be defined, the underlying random field must be mean square differentiable, which a small amount of local averaging guarantees.

Relation to the Gumbel Distribution of Maximum Values

Assume that $X(t)$ is exponentially distributed, with p.d.f.

$$f(b) = \lambda\, e^{-\lambda b}, \qquad b \geq 0, \tag{7.8.10}$$

and corresponding c.d.f.

$$F^c(b) = e^{-\lambda b}, \qquad b \geq 0. \tag{7.8.11}$$

Inserting these expressions into Eq. (7.8.6) yields the distribution of maximum values attributed to Gumbel [67]:

$$P[\max X(t) \leq b] = \exp\{-k_0^{-1} d\,\lambda^n\, e^{-\lambda b}\} = \exp\{-(d/d_0)\, e^{-\lambda b}\}, \tag{7.8.12}$$

where $d_0 = k_0 \lambda^{-n}$ denotes the size of a characteristic element. The mean size of isolated regions of excursions above the level b is found to be independent of b:

$$E[\mathscr{D}_b^+] = k_0 \left[\frac{F(b)}{f(b)}\right]^n = k_0\, \lambda^{-n} \equiv d_0, \tag{7.8.13}$$

which equals the 'characteristic element' size. This summons the image of randomly ordered cells of the same size, with independent and exponentially distributed cell properties (constant within the region occupied by each cell). Cells are 'randomly selected', with decreasing probability, $e^{-\lambda b}$, as the level b increases. A similar model can be constructed for minimum values, i.e., by treating them as maximum values of the field $X'(t) = -X(t)$.

In conclusion, high-level excursion statistics of homogeneous random fields are shown to depend in a surprisingly simple way on the first-order probability density function. The correlation structure of the field is captured by a single parameter k_0, that can in some cases be determined analytically. In other cases, it may have to be estimated from empirical data. The deviation above [in Eqs. (7.8.10) - (7.8.13)] shows that some classical extreme value models can be derived from, and are consistent with, homogeneous random field theory [138].

Extremes of Multi-Scale Random Fields

The focus in Chaps. 5 through 7 has been on local averages of single-scale random fields. For continuous-parameters processes, even a small amount of local averaging renders mean square derivatives finite and enables one to derive stable, robust estimates of level-crossing and extreme-value statistics. As explained in Sec. 5.9 and at the end of Sec. 7.1, this methodology is readily extendable to multi-component or multi-scale random fields, for which the composite statistics can be expressed as linear combination of the corresponding component statistics. For homogeneous random fields that can be modeled as sums of many independent component fields, representing random variation over a wide range of scales of fluctuation, the scale spectrum and the range of scales (instead of the correlation measures of single components) may be expected to dominate behavior; and this may result in very different level-excursion and extreme-value statistics. This important topic, beyond the scope of this book, is addressed by the writer [138] in the context of a major application of random field theory to modeling the energy density fluctuations in the very early universe, presumably arising during (the inflation phase of) the Big Bang [138]; in this case, the scale spectrum is constant over a very wide range of scales that obey a geometric-growth law, and the corresponding spectral density function is that of band-limited fractional ("$1/f$") noise.

Chapter 8

Overview of Findings

I cannot make my peace with the randomness doctrine: I cannot abide the notion of purposelessness and blind chance in nature. And yet I do not know what to put in its place for the quieting of my mind.

Lewis Thomas, On the Uncertainty of Science

Introduction

In this final chapter, the main concepts and methods introduced in the preceding chapters are summarized and further interpreted. Random field theory is applicable to a wide spectrum of distributed disordered systems (d.d. systems) within an appropriate range on the scale of time or distance, providing methodology for the description and analysis, and in some cases, prediction and control of complex and random systems.

Systems with randomly varying properties abound in nature. The random quantities may be, for example, chemical or physical properties of matter on earth or in space, concentration of nutrients or contaminants in water or soil, transport properties of cell membranes, or metereological quantities such as temperature or rate of precipitation. In some cases the random attribute is a property of the d.d. system while in others it describes the state of excitation or response of the system.

A central thesis of this work is that it is seldom useful or necessary to describe in detail the local point-to-point variation occurring on the microscale in time or space. Even if such information were desired, it may be impossible to obtain. Random field models need only provide information about random variation on the microscale sufficient to represent behavior under (at least some) local averaging. A more detailed description of the variability and the correlation structure on the microscale tends to unnecessarily overload the model.

The approach based on local averages resolves a long-standing problem of how to deal with the nonexistence of mean square derivatives of (many common models for) stationary random functions. Even a small amount of local averaging suffices to smooth the microscale fluctuations, and the resulting local average processes are shown to possess stable mean square derivatives and related level-crossing and extreme-value statistics. These quantities all depend simply and directly on the variance function, which fully captures second-order information about the random process.

The marginal probability distribution characterizing the point-to-point variation may be of little use if (as is often the case) the performance of the d.d. system depends on the behavior of local averages. As mentioned, a similar concept in continuum mechanics – Saint-Venant's principle – holds that erratic patterns of local stresses in a solid body may be replaced by locally averaged stresses for the purpose of overall force-deformation analysis, since the details of the microstructure of the material affect behavior on the macroscale only through their effect on local averages. Based on this concept, continuum mechanics becomes meaningful at the meso- or macro-scale and there is no need to consider details of a microstructure characterized by grains and voids, or cells, or atoms.

The theory based on local averages applies not only to random processes that are originally continuous in state and parameter but also to discrete-state and discrete-parameter random fields and to random impulse processes. Related, many economic and social phenomena can also be examined using random field theory, but the spatial coordinates may need to be replaced by, or linked to, discrete units. For instance, crop yield in a region or the rate of unemployment in a state may be seen as varying randomly with time and among the units or subdivisions comprising the "system". Each subdivision is characterized by a local average (or local integral) of contributions from all the units within it, and for each subdivision, or for the system as a whole, one can then evaluate the statistics of different kinds of "rare events", typically defined in terms of level excursions and extremes values.

Definitions and Notation

Second-Order Description of Random Fields

A real, scalar random field $X(\mathbf{t})$ is a collection of random variables at points with coordinates $\mathbf{t} = (t_1, \ldots, t_n)$ in an n-dimensional "parameter space". Second-order information about point-to-point variation is contained in the covariance function $B(\mathbf{t}, \mathbf{t}')$, the covariance between values of the random field at two locations $\mathbf{t}$ and $\mathbf{t}'$. If the random field is *homogeneous* (or, for $n = 1$, "stationary"), the covariance function will depend only on the differences $\tau_i = t_i - t_i'$, $i = 1, \ldots, n$, the components of the lag vector $\boldsymbol{\tau}$. The generalized Wiener-Khinchine equations state that the n-dimensional Fourier transform of $B(\boldsymbol{\tau})$ equals the spectral density function (s.d.f.) $S(\boldsymbol{\omega})$, where $\boldsymbol{\omega}$ identifies a point in the (generalized) frequency domain. Normalizing these functions with respect to the variance σ^2 yields another Fourier transform pair: the correlation function $\rho(\boldsymbol{\tau})$ and the normalized s.d.f. $s(\boldsymbol{\omega})$.

If the *spatial* character of the random field requires emphasis, the problem is usually formulated in terms of Cartesian spatial coordinates $\mathbf{u} = (u_1, u_2, u_3)$. If the random field is specifically a *space-time process*, it will depend on the vector of spatial coordinates $\mathbf{u}$ and on time t. Spatial lags, denoted by $\nu_i = u_i - u_i'$, are transformed into wave numbers k_i.

Random Fields (Processes) of Lower Dimensionality

If one or more of the parameters of a homogeneous n-dimensional random field $X(\mathbf{t})$ is held constant, one obtains a homogeneous field of dimensionality less than n. The notation $X(t_i)$ is used to refer to a stationary one-dimensional random process on a line parallel to the t_i-axis, while $X(t_i, t_j)$ refers to random variation on a plane parallel to the axes t_i and t_j. Related second-order quantities are also identified by simple subscripts. For example, the two-dimensional (2-D) random field $X(t_i, t_j)$ is characterized by the covariance function $B(\tau_i, \tau_j)$ and the spectral density function $S(\omega_i, \omega_j)$.

Quadrant Symmetry

The correlation structure of a homogeneous random field is *quadrant symmetric* (q.s.) if the covariance function $B(\boldsymbol{\tau})$ is *even* with respect to each component of the lag vector $\boldsymbol{\tau} = (\tau_1, \ldots, \tau_n)$. Both $B(\boldsymbol{\tau})$ and $S(\boldsymbol{\omega})$ then possess quadrant symmetry, and assorted cross-spectral density functions become real (their imaginary part vanishes). The second-order properties of "q.s. processes" are fully characterized by the covariance function $B(\boldsymbol{\tau})$ defined for positive lags only ($\boldsymbol{\tau} \geq \mathbf{0}$), or by the spectral density function $G(\boldsymbol{\omega}) = 2^n S(\boldsymbol{\omega})$ defined for positive frequencies (wave numbers) only.

The correlation structure is quadrant symmetric if it is (1) isotropic, (2) separable, or (3) ellipsoidal (defined in terms of principal coordinates). In an applications context it is often convenient to describe and analyze random fields in the quadrant symmetric mode (Chap. 3).

Moments of the Spectral Density and Correlation Functions

Two sets of moments play important roles in theoretical developments throughout the book. The first set, the *spectral moments*, are all characterized by the symbol λ with appropriate addscripts. For example, $\lambda_k^{(i)}$ is the kth moment of the direction-dependent s.d.f. $S(\omega_i)$, the integral over frequency of $\omega_i^k S(\omega_i)$. The joint spectral moment $\lambda_{kl}^{(ij)}$ is defined as the integral of $\omega_i^k \omega_j^l S(\omega_i, \omega_j)$ over the plane of frequencies (ω_i, ω_j); for a homogeneous n-dimensional random field these joint moments can be assembled into an n by n matrix $\mathbf{\Lambda}_{kl}$.

The second set of *moments*, defined in terms of the *n-dimensional correlation function* $\rho(\boldsymbol{\tau})$, are identified by the symbol θ for $n = 1$, α for $n = 2$, and β for $n = 3$; the symbol η is used to refer to a space-time process with $n = 4$ or when the dimensionality of the random field is unspecified. For example, $\theta_2^{(i)}$ is the second moment of the correlation function $\rho(\tau_i)$, while $\alpha_{11}^{(ij)}$ is the joint moment of second order of the two-dimensional (2-D) correlation function $\rho(\tau_i, \tau_j)$. The moment of order zero is just the integral under the correlation function, and the subscript zero is dropped when referring to the *integral* correlation measures θ, α, β, or η.

Spectral Bandwidth Measures

For a homogeneous n-dimensional random field, two n by n matrices $\mathbf{\Delta}$ and $\mathbf{E}$ are defined in terms of joint spectral moments (Sec. 4.1). For $n = 1$, these matrices reduce to the following dimensionless measures of spectral bandwidth: $\Delta \equiv \delta^2 = 1 - \lambda_1^2/\lambda_0\lambda_2$ and $\mathbf{E} \equiv \varepsilon^2 = 1 - \lambda_2^2/\lambda_0\lambda_4$.

Bounds and Existence Conditions in the One-Dimensional Case

The upper and lower bounds on the spectral bandwidth measures δ and ε are, respectively, *zero* (for a sinusoid with random phase angle) and *one* (for certain wide-band processes whose spectral density decays with frequency). They provide a quantitative measure of the degree of disorder of a random phenomenon and have important interpretations in terms of the behavior of sample functions and their envelopes (Sec. 4.4). Values of δ and ε corresponding to various common spectral density functions are presented in Sec. 4.7.

Note that δ will exist only if the second spectral moment λ_2 is finite, and ε if λ_4 is finite. In Sec. 5.4 it is shown that a small amount of local averaging of a random process guarantees that λ_2 will indeed be bounded and hence that δ will exist. This implies that the theory of level excursions and extremes in Chap. 4 (based on the assumption $\lambda_2 < \infty$) can legitimately be applied to the random process of local averages.

Different Lower Bounds for Isotropic Random Fields

In Sec. 4.3, the *envelope* of a homogeneous Gaussian random field is introduced (as an extension of the one-dimensional Cramér-Leadbetter envelope). It is shown that the matrix $\mathbf{\Delta}$ is closely related to the covariance matrix of the partial derivatives of the envelope field.

The fact that the determinant of the matrix $\mathbf{\Delta}$ must be nonnegative leads to interesting lower bounds on the bandwidth measure δ of one-dimensional random variation associated with *isotropic* n-dimensional random fields (Sec. 4.1). In particular, the constraint $\delta \geq \sqrt{(n-1)/n}$ implies the respective lower bounds $\sqrt{1/2}$ and $\sqrt{2/3}$ for $n = 2$ and 3.

If a distributed random property is subjected to sustained local averaging (as, for example, in diffusion), the bandwidth factor δ gradually increases and tends toward its upper limit ($\delta = 1$). Similar results are obtained in Chap. 4 for the spectral bandwidth measure ε.

Statistics of Level Excursions and Extreme Values

For homogeneous Gaussian random fields that satisfy the necessary continuity conditions (requiring that various spectral moments are bounded), we derive in Chap. 4 estimates for (1) the mean rate of occurrence of isolated regions of excursion (harboring local maxima) above a high level, (2) the mean size of these excursion regions, and (3) a measure of the tendency for high local maxima to occur in clumps or clusters. Similar results are obtained for the envelope of a Gaussian field.

The level excursion statistics depend importantly on the determinant of the matrix $\mathbf{\Lambda}_{11}$ of joint spectral moments of second order. The latter equals the covariance matrix of the partial derivatives of the random field at a point in the parameter space (Sec. 4.2). When a homogeneous random field is locally averaged (over a rectangular box with sides parallel to the coordinate axes,) the off-diagonal terms of the matrix $\mathbf{\Lambda}_{11}$ vanish and its determinant $|\mathbf{\Lambda}_{11}|$ reduces to a product of mean square partial derivatives (Secs. 6.4 and 7.7).

Theory of Local Averages of One-Dimensional Random Processes

Motivation: Lack of Stability of Mean Square Derivatives

Consider a one-dimensional continuous-parameter stationary random process $X(t)$ with mean zero and unit variance. ($X(t)$ may represent random variation on a line in the parameter space of a homogeneous n-dimensional random field.) The correlation structure of $X(t)$ is commonly represented either by the correlation function $\rho(\tau)$, where $\tau = t - t'$ is the "time lag," or by its Fourier transform, the one-sided, unit-area spectral density function $g(\omega)$ (defined for nonnegative frequencies only).

Selection of an analytical model for $\rho(\tau)$ or $g(\omega)$ is usually based on the quality of fit in the range of observed (observable) values of τ or ω, which, due to physical or sampling-related limitations, does not include the "microscale" ($\tau \to 0$ or $\omega \to \infty$). Any model choice does, of course, imply an assumption about the nature of random variation on the microscale. This assumption, although basically unverifiable, greatly affects predictions of quantities related to the *pattern* of fluctuation such as the mean square derivative, mean threshold crossing rates and extreme value statistics.

Many common correlation models predict that the mean square derivative and the mean threshold crossing rates are non-existent (infinite), while in reality, in the sample paths these models should represent, they are obviously finite. If the second spectral moment does exist, its value tends to depend strongly on assumptions about the very high frequency portion of $g(\omega)$. This excessive sensitivity to the choice of correlation model is a major weakness of the classical theory of level crossings and extremes. It is shown (Chaps. 5, 6 and 7) that this problem can be sidestepped, for all practical purposes, by permitting a small amount of local averaging, sufficient to smooth the microscale fluctuations.

One-Dimensional Variance Function

Consider the random process $X_T(t)$ obtained by local averaging of $X(t)$ over the interval T. The variance function $\gamma(T)$ is defined as the ratio of the variance of the process $X_T(t)$ and the "point" variance of the original process $X(t)$. The function $\gamma(T)$ decreases monotonically with T, its upper bound is $\gamma(0) = 1$, and for processes that are ergodic in the mean, $\gamma(\infty) = 0$. The function $\gamma(T)$ fully characterizes the correlation structure of $X(t)$ and contains the same information as $\rho(\tau)$ or $g(\omega)$. Another form of the variance function, $\Delta(T) = T^2\gamma(T)$, is defined in reference to the random process $I_T(t)$ obtained by integrating $X(t)$ over the moving window T. Specific analytical models for the variance function(s) are considered in Secs. 5.1 and 5.3, and also in Secs. 6.3 and 7.2.

Scale of Fluctuation

If a random process is ergodic in the mean, its scale of fluctuation θ can be expressed alternately in terms of the variance function $\gamma(T)$, the correlation function $\rho(\tau)$, or the (unit-area) spectral density function $g(\omega)$. By definition the variance function $\gamma(T)$ converges to the expression θ/T when the averaging window T grows without bound. Also, θ equals the area under the correlation function (from $-\infty$ to $+\infty$) and is proportional to the zero-frequency ordinate of the normalized spectral density function (Sec. 5.1).

The condition for validity of the three interpretations of θ is $\dot{g}(0) = 0$, where $\dot{g}(\omega)$ is the first derivative of $g(\omega)$, or $|\theta_2| < \infty$, where θ_2 is the second moment of the correlation function $\rho(\tau)$.

For broad-band processes, θ captures information not only about low frequency content, since $\theta = \pi g(0)$ and $\dot{g}(0) = 0$, but also about high frequency content, since contributions to the variance associated with frequencies above $2(\pi\theta)^{-1}$ become negligible (owing to the fact that $g(\omega)$ has unit area). Random processes for which $\theta = 0$ have a very different correlation structure; their variance decays in inverse proportion to T^2 (rather than T) under extended local averaging.

When a process with variance σ^2 and scale θ is integrated over a window of finite length T, the variance of the integral converges to $\sigma^2\theta T$ when $T \gg \theta$. The product $\sigma^2\theta$ is itself an important parameter which in some applications (as in diffusion problems where mainly the variance of particle displacement matters) provides sufficient characterization of the random phenomenon under study (σ^2 and θ need not be known separately). The quantity θ is also closely related to the intensity of white noise and to the mean rate of a Poisson arrival process (Sec. 5.6).

Statistics of Level Crossings and Extreme Values

In the one-dimensional case, the variance of $X_T(t)$ is $\text{Var}[X_T] = \sigma^2\gamma(T)$, and the ratio of the variances of $\dot{X}_T(t)$ and $X_T(t)$ is $2[1-\rho(T)]/\Delta(T)$. For $T \to 0$, $\text{Var}[\dot{X}_T]$ converges toward $\sigma^2_{\dot{X}}$, provided $X(t)$ is mean square differentiable. In general, the scale θ controls the behavior of a random process under long-term averaging (when $T \gg \theta$). For wide-band processes, however, even if the averaging window T is only a fraction of θ, the stochastic behavior of the family of local average processes $X_T(t)$ can be described satisfactorily in terms of the mean m, variance σ^2 and scale of fluctuation θ of the original process $X(t)$. Details of the "correlation structure" are of secondary importance, and their importance fades as the averaging interval T grows.

The three-parameter (m, σ^2, θ) description suffices to obtain *exact* asymptotic results (for $T \gg \theta$) of appealing analytical simplicity for the mean square derivative [Eq. (5.4.13)], the mean threshold crossing rate [Eq. (5.5.3)], and the probability distribution of extreme values [Eq. (5.5.4)] for the family of derived "local average" processes. In this context, the role of the Gaussian and Poisson processes as fundamental limiting processes is clearly demonstrated, and it is shown how their parameters are related to those of the original process. The latter need not be Gaussian, and the theory applies also to discrete-parameter and discrete-state processes, as well as to random impulse processes.

Theory of Local Averages of Homogeneous Random Fields

Two-Dimensional Variance Function and Measure of Correlation

The concept of the variance function and the idea of defining a correlation parameter in terms of the asymptotic form of the variance function remain valid (and gain appeal) in multi-dimensional situations. In the two-dimensional case the variance function $\gamma(T_1, T_2)$ is defined as the ratio of the variance of the local average $X_A(t_1, t_2)$ over a rectangular area $A = T_1T_2$ (with sides T_1 and T_2 parallel to the two axes) and the "point" variance of the original process $X(t_1, t_2)$.

The "correlation area" α has basic properties similar to those of θ in the 1-D case; in particular, the 2-D variance function converges to the expression α/A when the dimensions T_1 and T_2 of the averaging window much larger than their respective (direction-dependent) scales of fluctuation $\theta^{(1)}$ and $\theta^{(2)}$ (Secs. 6.1 and 6.2).

Conditional Variance Function and Conditional Scale of Fluctuation

The 2-D variance function $\gamma(T_1, T_2)$ can be expressed as the product of two 1-D variance functions, one "marginal" and the other "conditional" (in the same way as one would express the joint probability density function of two dependent random variables). The conditional variance function $\gamma(T_2|T_1)$ depends on the *conditional scale of fluctuation* $\theta^{(2)}_{T_1}$. These quantities characterize the 1-D random process dependent on t_2 and are obtained by averaging a 2-D homogeneous random field within a band of prescribed width T_1 parallel to the t_1-axis (Sec. 6.3).

When $T_1 = 0$, there is no averaging, and the conditional scale of fluctuation reduces to the scale $\theta^{(2)}$. As the width T_1 of the averaging band grows, $\theta^{(2)}_{T_1}$ tends to increase slowly and approaches the asymptotic value $\alpha/\theta^{(1)} = c_\alpha\theta^{(2)}$, where the dimensionless constant $c_\alpha = \alpha/\theta^{(1)}\theta^{(2)}$ characterizes the correlation between the patterns of fluctuation along the principal axes. If the 2-D correlation structure is "separable," that is, $\rho(\tau_1, \tau_2) = \rho(\tau_1)\rho(\tau_2)$, then $\alpha = \theta^{(1)}\theta^{(2)}$ and $c_\alpha = 1$.

n-Dimensional Variance Function

The properties of homogeneous n-dimensional random field can be described by an n-dimensional variance function that can be expressed as the product of n one-dimensional (marginal and conditional) variance functions. Additional correlation parameters (and conditional scales of fluctuation) are introduced in the same way as in the two-dimensional case. The 3-D correlation measure is denoted by β, while the symbol η is used in the general case (Secs. 7.1 and 7.2).

Statistics of Level Excursions and Extreme Values

In Sec. 7.7 an approximate (and asymptotically exact) solution is derived for the mean rate of local maxima and the mean size of regions of excursion above a high threshold level, for an entire family of local averages of homogeneous (not necessarily Gaussian) n-dimensional random fields. The results depend on the correlation parameters of the original random field, the dimensions of the averaging domain, and the threshold level.

Stochastic Analysis of Space-Time Processes

A two-dimensional homogeneous space-time process $X(u,t)$ is characterized by temporal and spatial scales of fluctuation, θ^t and θ^u, respectively. A decomposition of the process in the frequency domain leads to the introduction of the *frequency-dependent spatial* scale of fluctuation $\theta^u(\omega) \equiv \theta^u_\omega$ and the frequency-dependent variance function, which describes how the frequency content of the space-time process changes under local spatial averaging.

The function $\theta^u(\omega)$ tends to decrease monotonically with frequency; its largest value occurs at the frequency origin where it equals $\theta^u(0) = \alpha/\theta^t = c_\alpha \theta^y$. Unless the 2-D correlation function is separable (in which case $\theta^u(\omega) \equiv \theta^u$), the scale $\theta^u(\omega)$ will vary in proportion to ω^{-1} at high frequencies.

Similar results are obtained for space-time processes that depend on two or three spatial coordinates. In particular the frequency-dependent "correlation area" $\alpha^u(\omega)$ of the space-time process $X(u_1, u_2, t)$ varies in proportion to ω^2 at high frequencies and is largest at $\omega = 0$ where, for a spatially isotropic process, it equals $\alpha^u(0) = \beta/0^t$ (Sec. 7.3).

Important Related Concepts and Methods

"Laws" of Disorder

Two general laws appear to govern the behavior of homogeneous "distributed disordered systems" under local averaging, or more generally under linear transformation. The first law is a statement of "conservation of uncertainty," as measured by the product $\sigma^2\theta$ for a 1-D process and by $\sigma^2\eta$, $\sigma^2\theta^{(i)}$, etc., for general random fields. These quantities remain *invariant* under any linear transformation that preserves the mean. If the mean is nonzero, a more general statement is that the quantities $\sigma^2\theta/m^2$, $\sigma^2\eta/m^2$, $\sigma^2\theta^{(i)}/m^2$, etc., remain invariant under linear transformation.

The second law states that the *degree of disorder* of a homogeneous random field, as measured by the direction-dependent spectral bandwidth factor $\delta^{(i)}$, *tends to increase* when a random field is subjected to local aggregation (averaging or integration). Under repeated local averaging, this measure of disorder is expected to approach its upper bound, $\delta^{(i)} = 1$ (which in some sense characterizes complete chaos). A similar statement can be made with reference to the bandwidth measure $\varepsilon^{(i)}$ (Secs. 5.6, 6.6, and 7.8).

Covariances of Local Averages and Stochastic Finite Elements

To evaluate covariances of local averages of a random field, methodology based on variance functions is proposed. It offers considerable ease and economy of computation as it involves relatively simple *algebraic* operations (sums and products) instead of multiple integration (Secs. 5.2, 6.4, and 7.5).

The methodology is ideally suited for stochastic finite element analysis which requires as input the means and covariances of the random element properties, local averages of the random property within each finite element in a distributed disordered system. The approach can also be used to represent spatial-temporal random excitation and evaluate system response in random vibration and wave propagation problems (Secs. 6.4 and 7.6).

Optimal Linear Prediction

In Secs. 5.9 and 7.6 it is argued that the classical theory of optimum linear prediction (reviewed in Sec. 2.6), combined with the results for local averages (developed in Chaps. 5 through 7), provides powerful new methodology to formulate and solve problems involving prediction, updating, and optimal design of data acquisition networks. For single-scale random fields, the scale of fluctuation plays a key role in determining optimal sampling intervals and assessing residual uncertainty (see Sec. 5.8).

Composite Random Fields Modeling and the Scale Spectrum

For $n = 1$ the three-parameter representation (mean m, variance σ^2, scale θ) provides a flexible tool for practical modeling of wide-band random variation. A random process with a complex correlation structure can be modeled as a sum of two or more uncorrelated single-scale processes, each characterized by the parameters m, σ^2, and θ. This permits efficient description and analysis of patterns of random variation that resemble $1/f$ (or fractional) noise. Such a composite model can also accommodate uncorrelated random measurement errors (for which $\theta = 0$), perfectly correlated systematic errors (for which $0 = \infty$), and "$\theta = 0$" narrow-band components (Sec. 5.9).

Multi-scale random variation can often be modeled as a sum of statistically independent single-scale component random processes. The scale spectrum of the composite process expresses the component variances as a function of their respective scales of fluctuation. Of particular interest is a scale spectrum that is uniform across a wide range of scales, when the latter form either an arithmetic or a geometric array. The density of scales and the limits on the range of scales will control the behavior (specifically, the patterns of fluctuation) of the composite process.

The methods of composite modeling extend effortlessly to n-dimensional random fields; for the general case, see Sec. 7.1.

Decisions and Control

A dynamic, decision-oriented approach to random fields involves interaction between the stochastic environment and the decision maker. The latter may be able to exert a degree of control over the behavior of the random field, its properties (such as m, σ^2, and θ), or the systems it represents or excites. The control may involve a one-time design decision or on-line adjustment of controllable variables. Data acquisition also requires important decisions: where and how often to sample a random field in order to achieve a proper balance between the cost of data acquisition and the amount of residual uncertainty.

The theory of extremes of local averages offers exciting new possibilities for practical application of concepts of feedback and control to distributed disordered systems that are sensitive to local extremes of system properties or response measures. The system attribute may be the level of shear stress in the case of a dam or an earthquake fault, the level of contaminant or carcinogen in an organ of the human body, or the inflation rate in an economic system. The idea is that the system as a whole is affected when a threshold (a tipping point) is reached or exceeded somewhere (anywhere) in the parameter space in terms of which the system is defined. This initiating event (local threshold exceedance) may cause irreversible and perhaps catastrophic changes in the system as a whole. In the examples cited, catastrophe means alternately dam failure, earthquake, cancer, and runaway inflation. Homogeneous random field theory provides the methodology to evaluate the probability of the initiating event. The change in the system caused by the initiating event may then manifest itself as a change in the systemwide mean, variance, correlation measures, tolerable limit, or a combination of these. In non-catastrophic situations (the image summoned is that of pimples or potholes) the theory predicts mean occurrence rates of local excursions above critical levels. Depending on the specific circumstances, it may be possible to monitor the behavior of the system through sampling and estimation, or to exert a degree of control over the random field parameters that affect system behavior.

Quantum-Physics-Based Random Variables and Random Fields

Sec. 2.8 introduces a family of probability distributions derived from Planck's blackbody radiation spectrum (which heralded the dawn of the age of quantum physics). They characterize the random properties, suitably scaled, of single energy quanta associated with local thermal equilibrium. The distributions all have a simple analytical form and tractable statistical properties, and promise to be very useful in applications of random field theory in physics and biology. One such application is described below.

In research started soon after RF1 [137] was published, the writer sought to develop a (multi-scale, isotropic) random field model of the energy density fluctuations in the very early universe. As described in *Quantum Origins of Cosmic Structure* [140], such a model arises as quite naturally as the end product of quantum-physical processes during the inflation phase of the Big Bang. In particular, the theory envisions a cascade of particle proliferation and simultaneous propagation of uncertainty, quantifiable in terms of the kinds of probability distributions mentioned above (and described in Sec. 2.8). The model implies highly informative stochastic initial conditions – extreme local spatial variability, along with overall statistical homogeneity, in energy densities – on how the universe evolves after inflation. The initial random field of energy density gives rise to two coupled random fields, namely the field of local temperatures (becoming gradually smoother, consistent with observed tiny fluctuations in the Cosmic Microwave Background) and the field of matter densities (becoming gradually more clumpy, due to gravity). The theory robustly predicts, in ways that are quantitative and testable, spatially variable primordial synthesis of chemical elements, and a specific scenario of how cosmic structure develops on the scales of clusters of galaxies, galaxies, and solar systems.

Bibliography

[1] Abramowitz, M. and Stegun, I.A. *Handbook of Mathematical Functions.* New York: Dover Publ. (1965).

[2] Adler, R. "On the Envelope of a Gaussian Random Field", *J. Appl. Prob.*, vol. 15, no. 3, pp. 502-513 (1979).

[3] Adler, R. *The Geometry of Random Fields.* Wiley, New York (1981).

[4] Agterberg, F.P. "Autocorrelation Function in Geology", In *Geostatistics.* Ed. by D. F. Merriam. Plenum Press, Toronto (1970).

[5] Aki, K. and Richards, P.G. *Quantitative Seismology: Theory and Methods.* Vols 1 and 2. Freeman, San Francisco (1980).

[6] Andrade, J.E., Baker, J.W. and Ellison, K.C. "Random Porosity Fields and their Influence on the Stability of Granular Media", *Int. J. for Numerical and Analytical Methods in Geomechanics*, 32 (10), pp. 1147-1172 (2008).

[7] Bachelier, L. "Theorie de la Speculation", *Ann. Sci. Norm. Sup.*, vol. 3, pp. 21-86 (1900).

[8] Bachelier, L. *Calcul des Probabilities.* Gauthier-Villars, Paris (1912).

[9] Bartlett, M.S. *An Introduction to Stochastic Process.* Cambridge University press, Cambridge, England (1956).

[10] Bath, M. *Spectral Analysis in Geophysics.* Elsevier, Amsterdam (1974).

[11] Batchelor, G.K. *The Theory of Homogeneous Turbulence.* Cambridge University Press, Cambridge, England (1953).

[12] Belayev, Y.K. "Distribution of the Maximum of a Random Field and Its Application to Reliability Problems", *Eng. Cybernet.* vol. 2, pp. 269-275 (1970).

[13] Belayev, Y.K. "Point Processes and First Passage Problems", *Proc. 6th Berkeley Symposium Math. Stat. Prob.* vol. 2, pp. 1-17 (1972).

[14] Bendat, J.S. and Piersol, A.G. *Engineering Applications of Correlation and Spectral Analysis.* Wiley, New York (1980).

[15] Beran, M.J. *Statistical Continuum Theories.* Wiley Interscience, New York (1968).

[16] Berman, S.M. "Excursions above High Levels for Stationary Gaussian Process", *Pacif. J. Math.* vol. 36, no. 1, pp. 63-79 (1971).

[17] *Bessel Functions.* vol. 6, part 1. Mathematical Tables of the British Association for the Advanceent of Science. Cambridge University Press, Cambridge, England (1937).

[18] Blackman, R.B. and Tuckey, J.W. *The Measurement of Power Spectra from the Point of View of Communication Engineering.* Dover, New York (1959).

[19] Boissières, H.P. and Vanmarcke, E.H. "Estimation of Lags for a Seismograph Array: Wave Propagation and Composite Correlation", *Soil Dynamics and Earthquake Engineering,* vol. 14, No. 1 (1995).

[20] Boissières, H.P. and Vanmarcke, E.H. "Spatial Correlation of Earthquake Ground Motion: Non-parametric Estimation", *Soil Dynamics and Earthquake Engineering,* vol. 14, No. 1 (1995).

[21] Bolotin, V.V. "Statistical Methods in Structural Mechanics", (Transl. from Russian), Holden-Day, Inc., San Francisco (1969).

[22] Brillinger, D.R. *Time Series: Data Analysis and Theory.* Holt, Rinehart and Winston, New York (1975).

[23] Brockett, R. *Finite Dimensional Linear Systems.* Wiley, New York (1970).

[24] Bruining, J., Van Batenburg, D., Lake, L.W. and Yang, A.P. "Flexible Spectral Methods for the Generation of Random Fields with Power-law Semivariograms", *Mathematical Geology,* 29 (6), pp. 823-841 (1997).

[25] Bryson, A. and Ho, Y. *Applied Optimal Control.* Ginn/Blaisdell, Waltham, Mass. (1969).

[26] Clarkson, B.L. (Ed.) *Stochastic Problems in Dynamics.* Pitman Press, London (1977).

[27] Caughey T.K. and Stumpf, H.J. "Transient Response of a Dynamic System under Random Excitation", *J. Appl. Mech.,* vol. 28, pp. 563-566 (1961).

[28] Chakravorty M. "Transient Spectral Analysis of Linear Elastic Structures and Equipment under Random Excitation", Research Report R 72-18. MIT Department of Civil Engineering, Cambridge Mass. (1972).

[29] Chandrasekhar, S. "Stochastic Problems in Physics and Astronomy", *Rev. Mod. Phys.,* vol. 15, pp. 1-89 (1943).

[30] Chaudhuri, A. and Sekhar, M. "Stochastic Finite Element Method for Probabilistic Analysis of Flow and Transport in a Three-dimensional Heterogeneous Porous Formation", *Water Resources Research,* 41 (9), art. no. W09404, pp. 1-10 (2005).

[31] Chaudhuri, A. and Sekhar, M. "Probabilistic Analysis of Pollutant Migration from a Landfill Using Stochastic Finite Element Method", *J. Geotechnical and Geoenvironmental Engineering,* 131 (8), pp. 1042-1049 (2005).

[32] Chernov, L.A. *Wave Propagation in a Random Medium.* Dover, New York (1967).

[33] Choot G. E.G. "Stochastic Underseepage Analysis in Dams", Masters thesis. MIT Department of Civil Engineering, Cambridge Mass. June (1980).

[34] Churchman, C.W. and Ratoosh, P., ED. *Measurement: Definitions and Theories.* Wiley, New York (1959).

[35] Cleary, M.P., Chen, I-W. and Lee, S-M. "Self-Consistent Techniques for Heterogeneous Media", *J. Eng. Mech. Div.,* ASCE, pp. 861-887, October (1980).

[36] Coles, S. *An Introduction to Statistical Modeling of Extreme Values.* Springer-Verlag. Berlin (2001).

[37] Corotis, R.B. and Marshall, T.A. "Oscillator Response to Modulated Random Excitation", *J. Eng. Mech. Div.*, ASCE, EM4, pp. 501-503 (1977).

[38] Corotis, R.B., Vanmarcke, E.H. and Cornell, C.A. "First Passage of Nonstationary Random Processes", *J. Eng. Mech. Div*, ASCE, vol. 98, p. 402 (1972).

[39] Corotis, R.B. and Vanmarcke, E.H. "Time Dependent Spectral Content of System Response", *J. Eng. Mech. Div.*, ASCE, EM5, pp. 623-637 (1975).

[40] Cramér, H.E. "On the Intersections between the Trajectories of a Normal Stationary Stochastic Process and a High Level", *Arkiv. Mat.* , vol. 6, p 337 (1966).

[41] Cramér, H.E. and Leadbetter, M.R. *Stationary and Related Stochastic Processes.* Wiley, New York (1967).

[42] Crandall, S.H. "First-Crossing Probabilities of the Linear Oscillator", *J. Sound Vib.*, vol. 12, no. 3, p. 285 (1970).

[43] Crandall, S.H. "Random Vibration of One- and Two-Dimensional Structures", in *Developments in Statistics*, vol. 2, chap. 1, Ed. by P. R. Krishnaish. Academic Press, New York (1979).

[44] Crandall, S.H., Chandiramani, K.L. and Cook, R.G. "Some First-Passage Problems in Random Vibration", *J. Appl. Mech.*, vol. 33, trans. ASCE, vol. 88, series E, p. 532 (1966).

[45] Crandall, S.H. and Mark. W.D. *Random Vibration in Mechanical Systems.* Academic Press, New York (1963).

[46] Darling, D.A. and Siegert, A.J.F. "The First Passage Problem for a Continuous Markov Process", *Ann. Math. Statist.*, vol. 24, p. 624 (1953).

[47] Davenport, W.B. and Root, W.L. *Introduction to the Theory of Random Signals and Noise.* McGraw-Hill, New York (1958).

[48] Davenport, A.G. "Note on the Random Distribution of the Largest Value of Random Function with Emphasis on Gust Loading", *Proc. Inst. of Civil Engineers*, London, vol. 28 (1964).

[49] de Haan, L. "On Regular Variation and its Application to the Weak Convergence of Sample Extremes", Report. Mathematisch Centrum, Amsterdam (1970).

[50] Ditlevsen, O. "Extremes and First Passage Times with Applications in Civil Engineering", Ph.D. dissertation. Technical University of Denmark, Copenhagen (1971).

[51] Drake, A.W. *Fundamentals of Applied Probability Theory.* McGraw-Hill, New York (1967).

[52] Dwight, H.B. *Tables of Integrals and other Mathematical Data.* 4th Ed., MacMillan, New York (1961).

[53] Dyer, I. "Response of Plates to a Decaying and Convecting Random Pressure Field", *J. Acoust. Soc. AM.*, vol. 31, pp. 922-928 (1959).

[54] Einstein, A. "Investigations on the Theory of the Brownian Movement", Ed. by R. Furth. Dover, New York, 1956; original article in *Ann. d. Physik*, vol. 17, p. 549 (1915).

[55] Embrechts, P.C., Klüppelberg, C. and Mikosch, T. *Modeling Extremal Events for Insurance and Finance.* Springer-Verlag, Berlin (1997).

[56] Feller, W. "Fluctuation Theory of Recurrent Events", *Trans. Am. Math. Soc.*, vol. 67, no. 111, pp. 98-119 (1949).

[57] Feller, W. *An Introduction to Probability Theory and its Application.* vol. 1, 3rd Ed., Wiley, New York, 1967; vol. 2, 2nd ED, Wiley, New York (1966).

[58] Feng, L.-L., Pando, J. and Fang, L.-Z. "Intermittent Features of the Quasar Lyα Transmitted Flux: Results from Cosmological Hydrodynamic Simulations", *Astrophysical Journal*, 587 (2 I), pp. 487-499 (2003).

[59] Fenton, G.A. and Griffiths, D.V. *Risk Assessment in Geotechnical Engineering.* Wiley (2008)

[60] Fenton, G.A. and Vanmarcke, E.H. "Simulation of Random Fields of Earthquake Ground Vibration", *J. Eng. Mech.*, vol. 116, No. 8, pp. 1733-1749 (1990).

[61] Fishman, G.S. *Spectral Methods in Econometries.* Harvard University Press, Cambridge, Mass. (1969).

[62] Frimpong, S. and Achireko, P.K. "Conditional LAS Stochastic Simulation of Regionalized Variables in Random Fields", *Computational Geosciences*, 2 (1), pp. 37-45 (1998).

[63] Gelb, Arthur (Ed). *Applied Optional Estimation.* The MIT Press, Cambridge, Mass (1974).

[64] Geol, N.S. and Richterdyn, N. *Stochastic Models in Biology.* Academic Press, New York (1974).

[65] Gnedenko, V.V. "Sur la Distribution Limite du Terme Maximum d'une Serie Aleatoire", *Ann. Math.*, vol. 44, pp. 423-453 (1943).

[66] Granger, C.W.J. *Spectral Analysis of Economic Time Series.* Princeton University Press, Princeton N.J. (1964).

[67] Gumbel, E.J. "Statistics of Extremes", Columbia University Press, New York (1958).

[68] Hachich, W. and Vanmarcke, E.H. "Probabilistic Updating of Pore Pressure Fields", *J. Geot. Eng. Div.*, ASCE, vol. 109, pp. 373-385 (1983).

[69] Hammond, J.K. "On the Response of Single and Multi Degree of Freedom Systems to Non-Stationary Random Excitation", *J. Sound Vib.*, vol. 7, pp. 393-416 (1968).

[70] Harichandran, R.S. and Vanmarcke, E.H. "Stochastic Variation of Earthquake Ground Motion in Space and Time", *J. Eng. Mech.*, vol. 112, No. 2 (1986).

[71] Hasofer, A.M. "The Mean Number of Maxima above High Levels in Gaussian Random Fields", *J. Appl. Prob.*, vol. 13 (1976).

[72] Heisenberg, W. "Multi-body Problem and Resonance in the Quantum Mechanics", *Zeitschrift für Physik 38 (6/7)*, pp. 411-426 (1926).

[73] Isserlis, L. "On a Formula for the Product-Moment Coefficient in Any Number of Variables", Biometrica, vol. 12, no. 1-2, pp. 134-139 (1918).

[74] Jenkins, G.M. and Watts D.G. *Spectral Analysis and its Applications.* Holden-Day, San Francisco (1968).

[75] Kalman, R. "A New Approach to Linear Filtering and Prediction Problems", *J. Basic Eng. series D*, vol. 82, p. 35, March (1960).

[76] Kalman, R. and Bucy, R. "New Results in Filtering and Prediction Problems", *J. Basic Eng. series D*, vol. 83, p. 95, March (1961).

[77] Khinchine, A.I. "Korrelationstheorie der Stationaren Stochastischen Prozesse", *Math. Ann.* vol. 109, pp. 605-616 (1934).

[78] Khinchine, A.I. *Mathematical Foundations of Statistical Mechanics.* Dover, New York (1949).

[79] Kolmogorov, A.N. "Interpolation and Extrapolation of Stationary Random Sequences", *Izv. Akas. Nauk SSSR*, Ser. mat., vol. 5, no. 3 (1941).

[80] Krige, D.G. "Two-Dimensional Weighted Moving Averaging Trend Surfaces for Ore Evaluation", *Proc. Symposium on Math. Stat. and Computer Applications for Ore Evaluation*, Johannesburg, S.S., pp. 13-18 (1966).

[81] Leadbetter, M.R., Lindgreen, G. and Rootzén, H. *Extremes and Related Properties of Random Sequences and Processes.* Springer-Verlag. Berlin (1983).

[82] Lee. Y.W. *Statistical Theory of Communications.* Wiley, New York (1960).

[83] Lin, Y.K. *Probabilistic Theory of Structural Dynamics.* McGraw-Hill, New York (1967).

[84] Longuet-Higgins, M.S. "On the Statistical Distribution of the Height of Sea Waves", *J. Marine Research*, vol. 11, no. 3, pp. 246-266 (1952).

[85] Lumley, J.L. *Stochastic Tools in Turbulence.* Academic Press, New York (1970).

[86] Lutes, L.D. and Sarkani, S. *Stochastic Analysis of Structural and Mechanical Vibrations*, Prentice-Hall, Inc., Englewood Cliffs, NJ (1997).

[87] Lyon, R.H. "On the Vibration Statistics of Randomly Excited Hard-Spring Oscillator II", *J. Acoust. Soc. Am.*, vol. 33, no. 10, p. 1395 (1961).

[88] Mandelbrot, B.B. "Une Classe de Processus Homothetiques a Soi; Application a la Loi Climatologique de H.E. Hurst", *C.R. Acad. Sci.*, Paris, 260, pp. 3274-3277 (1965).

[89] Mandelbrot, B.B. and Van Ness, R. "Fractional Brownian Motions, Fractional Noise and Applications", *SIAM Review*, vol. 10, no. 4, pp. 422-437 (1968).

[90] Marani, M. "On the Correlation Structure of Continuous and Discrete Point Rainfall", *Water Resources Research*, 39 (5), pp. SWC21-SWC28 (2003).

[91] Marani, M. "Non-power-law Scale Properties of Rainfall in Space and Time", *Water Resources Research*, 41 (8), pp. 1-10 (2005).

[92] Mark, W.D. "Spectral Analysis of the Convolution and Filtering of Non-Stationary Processes", *J. Sound Vib.*, vol. 11, no. 1, pp. 19-63 (1970).

[93] Matern, B. *Spatial Variation: Stochastic Models and Their Application to Some Problems in Forest Surveys and other Sampling Investigations.* Swedish Forestry research Institute, vol. 49, no. 5 (1960).

[94] Matheron, G. *Les Variable Regionalisees et leur Estimation.* Masson, Paris (1965).

[95] Matheron, G. "Kriging or Polynomial Interpolation Procedures", *Canadian Inst. of Mining Bulletin*, vol. 60, p. 1041 (1967).

[96] Middleton, D. *An Introduction to Statistical Communication Theory.* McGraw-Hill, New York (1960).

[97] Moan, T. and Shinozuka, M, (Eds.) *Structural Safety and Reliability.* Elsevier, Amsterdam (1981).

[98] Monin, A.S. and Yaglom, A.M. *Statistical Fluid Mechanics.* Ed. by J. L. Lumley. The MIT Press, Cambridge, Mass, 1971; orig. publ. by Nauka Press, Moscow (1965).

[99] Newland, D.E. *An Introduction to Random Vibrations and Spectral Analysis,* Longman Inc., New York (1984).

[100] Nosko, V.P. "The Characteristics of Excursions of Gaussian Homogeneous Fields above a High Level", *Proc. USSR-Japan Symposium on Probability,* Novosibirsk (1969).

[101] Ornstein, L.S. and Uhlenbeck, G.E. "On the Theory of the Brownian Motion", *Physical Review,* vol. 36, pp. 823-841, Spetember (1930).

[102] Page, C.H. "Instantaneous Power Spectra", *J. Applied Physics,* vol. 23, no. 1, pp. 103-106 (1952).

[103] Parzen, E. *Stochastic Processes.* Holder-Day, San Francisco (1962).

[104] Paulson, K.S. "Spatial-temporal Statistics of Rainrate Random Fields", *Radio Science,* 37 (5), pp. 21/1-21/8 (2002).

[105] Perrin, J. *Atoms.* Constable Press, London (1916).

[106] Petocz, E. "Upcrossing by Oscillatory Processes and their Envelopes", Ph. D. dissertation. The University of New South Wales, Sydney, Australia, June 1981.

[107] Planck, M., Verh. Deutsche Phys. Ges., 2: 237; and in [2] Abramowitz, M. and Stegun, I.A. (1965), *Handbook of Mathematical Functions.* New York: Dover Publ. (1900).

[108] Priestley, M.B. "Power Spectral Analysis of Non-Stationary Random Processes", *J. Sound Vib.,* vol. 6, no. 1, pp. 86-97 (1967).

[109] Pugachev, V.S. *Theory of Random Functions and its Application to Control Problems.*Transl. from Russian, Pergamon Press, Oxford (1965).

[110] Rainal, A.J. "First and Second Passage Times of Sine Wave Plus Noise", *Bell System Technical Journal',* 47, pp. 2239-2258 (1968).

[111] Rainal, A.J. "Origin of Rice's formula", *IEEE Trans. Inf. Theory,* vol. 34, No.6, pp.1383-1387 (1988).

[112] Reinick, S.I. *Extreme Values, Regular Variation and Point Processes.* Springer-Verlag. Berlin (1987).

[113] Rice, S.O. "Mathematical Analysis of Random Noise", *Bell System Technical J.,* vol. 32, p. 282, 1944; vol. 25, p. 46, 1945. Also reprinted in N. Wax, *Selected Publications on Noise and Stochastic Processes.* Dover, New York (1954).

[114] Roberto, V., Andreani, P. and Wamsteker, W. "Numerical Simulation of Non-Gaussian Random Fields with Prescribed Correlation Structure", *Astronomical Society of the Pacific,* 113 (786), pp. 1009-1020 (2001).

[115] Rodriguez-Iturbe, I. and Meija, J.M. "The Design of Rainfall Networks in Time and Space", *Water Resources Research* (1973).

[116] Savage, V.J. *The Foundations of Statistics.* Wiley, New York (1954).

[117] Schweppe, F.C. *Uncertain Dynamics Systems.* Prentice-Hall, Englewood Cliffs, NJ (1973).

[118] Seshadri, V. and Lindenberg, K. "Extreme Statistics of Wiener-Einstein Processes in One, Two and Three Dimensions", *J. Stat. Phys.*, vol. 22, no. 1 (1980).

[119] She, J. and Nakamoto, S. "Spatial Sampling Study for the Tropical Pacific with Observed Sea Surface Temperature Fields", *J. Atmospheric and Oceanic Technology,* 13 (6), pp. 1189-1201 (1996).

[120] Slepian, D. "On Bandwidth", *Proc. of the IEEE,* vol. 64 pp. 292-300 (1976).

[121] Slutzky, E. "The Summation of Random Causes as the Source of Cyclic Processes", In *Problems of Economic Conditions.* Ed. by the Conjecture Institute, Moskov, vol. 3, no. 1, 1937; published in *Econometrica,* vol. 5, pp. 105-146 (137).

[122] Smoluchowski, M.V. "Drei Vortrage uber Diffusion, Brownsche Bewegung und Koagulation von Kollidteilchen", *Physik Zeits,* vol. 17, p. 557 (1916).

[123] Spanos, P.D. and Zeldin, B.A. "Monte Carlo Treatment of Random Fields: A Broad Perspective", *Applied Mechanics Reviews,* 51 (3), pp. 219-237 (1998).

[124] Spitzer, R. *Principles of Random Walk.* Springle-Verlag, Berlin (1976).

[125] Stratonovitch, R.L. *Topics in the Theory of Random Noise.* Gordon and Breach (1967).

[126] Taylor, G.I. "Statistical Theory of Turbulence", *Proc. Royal Soc.* vol. 151, series A, pp. 421-478 (1935).

[127] Taylor, G.I. "Diffusion by Continuous Movements", *Proc. London Math. Soc. (2),* vol. 20, pp. 196-211, 1921. See also G.I. Taylor, *Scientific Papers.* vol. 2, Cambridge University Press (1960).

[128] Tennekes, H. and Lumley, J.L. *A First Course in Turbulence.* The MIT Press, Cambridge University Press (1960).

[129] Torquato, S. *Random Heterogeneous Materials,* Springer-Verlag, New York (2002).

[130] Tribus, M. *Rational Decriptions, Decisions and Designs.* Pergamon Press, New York (1969).

[131] Tsuji, H. "The Transformation Equations between One- and n-Dimensional Spectra in the n-Dimensional Isotropic Vector or Scalar Fluctuation Field", *J. Physical Society of Japan,* vol. 10, no. 4 (1955).

[132] Vanmarcke, E.H. "First-Passage and Other Failure Criteria in Narrow-Band Random Vibration: A Discrete State Approach", Research Report No. R69-68. MIT Department of Civil Engineering, Cambridge, Mass (1969).

[133] Vanmarcke, E.H. "Parameters of the Spectral Density Function: Their Significance in the Time and Frequency Domains", Research Report R70-58. MIT Department of Civil Engineering, Cambridge, Mass (1970).

[134] Vanmarcke, E.H. "Properties of Spectral Moments with Applications to Random Vibration", *J. Eng. Mech. Div.*, ASCE, vol. 98, p. 425 (1972).

[135] Vanmarcke, E.H. "On the Distribution of the First-Passage Time for Normal Stationary Random Processes", *J. Appl. Mech.*, Trans. ASME, vol. 42, pp. 215-220 (1975).

[136] Vanmarcke, E.H. "Structural Response", in *Seismic Risk and Engineering*

Decisions. Ed. by C. Lommitz and E. Rosenblueth. Elsevier, Amsterdam (1976).

[137] Vanmarcke, E.H. "Random Fields: Analysis and Synthesis", the MIT Press, Cambridge, Mass. and London, England (1983).

[138] Vanmarcke, E.H. "Extreme Value Statistics Compatible with Random Field Theory", in Probabilistic Methods in the Mechanics of Solids and Structures, Eds., S. Eggwertz and N.C. Lind. Invited Paper at Weibull Symposium of IUTAM, Stockholm (1984). Published by Springer-Verlag (1985).

[139] Vanmarcke, E.H. "Natural Fractal Variation: Cosmic Origin and Earthbound Applications", Keynote Lecture, Proceedings 7th International Conference on Applications of Statistics and Probability (ICASP-7), Paris, France, 1995; A.A. Balkema Publishers, Rotterdam, vol. 3, pp. 1377-1386 (1996).

[140] Vanmarcke, E.H. *Quantum Origins of Cosmic Structure.* Rotterdam, Holland and Brookfield, VT: Balkema Publ. (1997).

[141] Vanmarcke, E.H. "Quantum-Physics-Based Probability Models with Applications to Reliability Analysis", Chapter 16 in Reliability-Based Design Handbook, Nikolaidis, E., Ghiocel, D. and Singhal, S., Eds., CRC Press, pp. 16.1-13 (2005).

[142] Vanmarcke, E.H. "Random Field Modeling of the Void Phase of Soils", Georisk, vol. 1, no.1, pp. 57-68 (2007).

[143] Vanmarcke, E.H. and Fenton, G.A. "Conditional Simulation of Local Fields of Earthquake Ground Motion", *Structural Safety,* vol.10, pp. 247-264 (1991).

[144] Vanmarcke, E.H. and Fenton, G.A. (Eds) "Probabilistic Site Characterization at the National Geotechnical Experimentation Sites (NGES)", *Geotechnical Special Publication* No. 121, Geo-Institute of ASCE, Published by ASCE (2003).

[145] Vanmarcke, E.H. and Grigoriu, M. "Stochastic Finite Element Analysis of Simple Beams", *J. Eng. Mech. Div.*, ASCE (1983).

[146] Vanmarcke, E.H., Heredia-Zavoni, E. and Fenton, G.A. "Conditional Simulation of Spatially Correlated Earthquake Ground Motion", *Journal of Engineering Mechanics, ASCE*, vol.119, pp. 2333-2352 (1993).

[147] Van Dyck, J.F. "Envelopes of Broad-band Processes", Masters thesis. MIT Department of Civil Engineering, Cambridge, Mass, June (1981).

[148] Vogel, H. and Zuckerkandl, E. "Randomness and Thermodynamics of Modecular Evolution", in *Biochemical Evolution and the Origin of Life.* Ed. by E. Schoffeniels. North-Holland, Amsterdam (1971).

[149] Von Karman, T. *Collected Works.* Von Karman Institute for Fluid Dynamics, Rhodest-Genese, Belguim (1975).

[150] Whittle, P. "On the Variation of Yield Variance with Plot Size", *Biometrika*, vol. 43, pp. 337-343 (1956).

[151] Whittle, P. "Topographic Correlation, Power-law Covariance Functions, and Diffusion", *Biometrika*, vol. 49, pp. 305-314 (1962).

[152] Wiener, N. "Generalized Harmonic Analysis", *Acta Math.*, vol. 55, pp. 117-258 (1930).

[153] Wiener, N. *Extrapolation, Interpolation and Smoothing of Stationary Time Series.* The MIT Press, Cambridge, Mass. (1949).

[154] Wirsching, P.H., Paez, T.L. and Ortiz, K. *Random Vibrations: Theory and Practice*, John Wiley & Sons, New York, NY (1995).

[155] Wunsch, C. "The Interpretation of Short Climate Records, with Comments on the North Atlantic and Southern Oscillations", *Bulletin of the American Meteorological Society*, 80 (2), pp. 245-255 (1999).

[156] Wunsch, C. "Greenland - Antarctic Phase Relations and Millennial Time-scale Climate Fluctuations in the Greenland Ice-cores", *Quaternary Science Reviews*, 22 (15-17), pp. 1631-1646 (2003).

[157] Yaglom, A.M. "On the Problem of Linear Interpolation of Stationary Random Sequences and Processes", *Ups. Mat. Nauk*, vol. 4, no. 4, p. 173 (1949).

[158] Yaglom, A.M. *An Introduction to the Theory of Stationery Random Functions.* Transl. by R.A. Silverman. Dover, New York, 1973; orig. publ. by Prentice-Hall, Englewood Cliffs. N.J. (1962).

[159] Yang, J.N. "First Excursion Probability in Non-Stationary Random Vibration", *J. Sound Vib.*, vol. 27, pp. 165-182 (1973).

[160] Zadeh, L. and Descoer, C. *Linear System Theory, the State Approach.* McGraw-Hill, New York (1963).

[161] Zerva, A. *Spatial Variation of Seismic Ground Motions: Modeling and Engineering Applications.* CRC Press (2009).

[162] Ziman, J.M. *Models of Disorder.* Cambridge University Press, Cambridge, England (1979).

Index